T0291898

CAMBRIDGE PHYSICAL SERIES

EXPERIMENTAL OPTICS

EXPERIMENTAL OPTICS

A MANUAL
FOR THE LABORATORY

BY

G. F. C. SEARLE, Sc. D., F.R.S.

UNIVERSITY LECTURER IN EXPERIMENTAL PHYSICS
AND
DEMONSTRATOR OF EXPERIMENTAL PHYSICS
AT THE CAVENDISH LABORATORY
SOMETIME FELLOW OF PETERHOUSE

CAMBRIDGE
AT THE UNIVERSITY PRESS
1925

CAMBRIDGE
UNIVERSITY PRESS

University Printing House, Cambridge CB2 8BS, United Kingdom

Published in the United States of America by Cambridge University Press, New York

Cambridge University Press is part of the University of Cambridge.

It furthers the University's mission by disseminating knowledge in the pursuit of education, learning and research at the highest international levels of excellence.

www.cambridge.org
Information on this title: www.cambridge.org/9781107683686

© Cambridge University Press 1925

First published 1925
First paperback edition 2014

A catalogue record for this publication is available from the British Library

ISBN 978-1-107-68368-6 Paperback

PREFACE

THIS Manual describes the optical experiments done in my practical class at the Cavendish Laboratory. The students attending the class vary greatly as regards previous experience of optical work, and a wide range of experiments is required to satisfy their needs. One term in each year is now devoted entirely to Optics instead of to a combination of Electricity and Optics, and this change has led to some extension of the course in Optics, for we have now to be prepared to provide experiments for perhaps 70 students working simultaneously.

Much of Optics depends on purely geometrical principles and is independent of any particular theory of light. Where the actual nature of the vibrations is involved, I employ Maxwell's electromagnetic theory and use the simple methods due to Oliver Heaviside. Maxwell's kindness in showing me some experiments at the Cavendish Laboratory, when I visited it with my father as a birthday treat nearly 50 years ago, drew me to the Laboratory, where I have now taught for 37 years. It is therefore natural that I should use Maxwell's theory.

Optical measurements depend so closely upon the mathematical treatment of the subject that a good deal of mathematical work was inevitable. In order to shorten the accounts of the experiments, the theoretical discussions have, for the most part, been placed in chapters or in sections by themselves.

The experiments have been tested by use in the practical class; most of the "Practical examples" are the records of students' work. If I have used some of my own measurements, it is merely because they were at hand and not because any special accuracy is claimed for them.

The apparatus required for most of the experiments is of a simple nature and can be modified without loss of efficiency, provided that the optical parts be of good quality and that the measuring scales be accurate. The greater part of the apparatus used in my class has been made by Mr Lincoln and Mr Roff, instrument makers at the Laboratory, by my laboratory assistants, or by myself.

I have authorised Messrs W. G. Pye & Co., of Cambridge, to supply apparatus made to my designs.

I owe the method of Experiment 46 to the late Prof. R. B. Clifton, of the Clarendon Laboratory, Oxford.

Towards the preparation of the manuscript, much help has been given by my wife and also by Messrs G. M. B. Dobson, W. L. Carter, D. A. Keys, C. F. Sharman and C. Underwood. I am indebted to my colleague Mr G. Stead for his account of Experiment 8. The Cambridge Philosophical Society, Messrs W. G. Pye & Co., and Mr W. Wilson have lent blocks for several figures. Mr C. G. Tilley, my laboratory assistant, has kindly made a large number of the drawings. Many others have helped in other ways; my thanks are due to them all.

Miss J. M. W. Slater, Lecturer in Chemistry and Physics, Newnham College, assistant demonstrator in my practical class since 1919, has revised the proofs. Her intimate knowledge of the experiments, and of the difficulties encountered by students, has made her generous help of much value. The proofs have also been read by Mr O. A. Saunders, of Trinity College, and by Mr C. Underwood, of Peterhouse.

In the preface to the Manual on *Experimental Elasticity* published in 1908, I expressed the hope that an *Experimental Optics* would follow in a few months. But months have grown to years. A decline in strength culminated in 1910 in a severe breakdown which stopped all work for more than a year, and the preparation of the *Optics* for several years. When I recovered in 1915, the War was raging. On my return from the Royal Aircraft Establishment to Cambridge in 1919, the demands of the abnormal number of students prevented me from making a fresh start. Prof. Sir E. Rutherford came to the rescue and released me from the duty of teaching in the Long Vacation of 1920 and of subsequent years; his kindness has given me the opportunity to prepare the manuscript and to see the book through the press.

Restoration to health was essential to the completion of the work. I should be more than ungrateful if I did not confess that a little book, *The Living Touch*, in which Miss Dorothy

Kerin has recorded her own miraculous healing, was used to teach me that God does mighty works of healing to-day. From that time I have been given full health and strength. I could tell of others who have come to know, by happy experience, that "with His stripes we are healed."

May this book be for the glory of the great Creator, Who in the beginning said, "Let there be light."

G. F. C. SEARLE.

CAVENDISH LABORATORY,
    CAMBRIDGE.
    4 *September*, 1925.

CONTENTS

CHAPTER I

ELEMENTARY LAWS OF OPTICS

CHAPTER II

SOME APPLICATIONS OF THE LAWS OF REFLEXION AND REFRACTION

CHAPTER III

EXPERIMENTS WITH PLANE SURFACES

CHAPTER XI

ASTIGMATISM AND FOCAL LINES

CHAPTER XII

INTERFERENCE AND POLARISATION BY REFLEXION

CHAPTER I

ELEMENTARY LAWS OF OPTICS

1. Introduction. Optical calculations are based upon the laws of reflexion and refraction. When we seek reasons for these laws, we must turn to some theory of light and endeavour to deduce the mathematical laws from the assumptions of the theory. We shall adopt Maxwell's electromagnetic theory as being most nearly in agreement with all that is known upon the subject.

2. Maxwell's electromagnetic theory of light. On Maxwell's theory, light is an electromagnetic disturbance involving both electric and magnetic forces. The electric displacement (or the electric polarisation) and the magnetic induction are in the wave front, and thus the vibrations which occur are transverse. The theory accounts for reflexion and refraction and leads to Fresnel's wave surface in crystals. The velocity of electromagnetic waves, calculated in terms of quantities found by electrical experiments, agrees closely with the observed velocity of light.

3. The electromagnetic vectors. Maxwell's theory involves four vectors. (1) The electric force E. (2) The magnetic force H. (3) The electric displacement D. (4) The magnetic induction B. In an isotropic medium, D is in the same direction as E and

$$D = KE/4\pi, \quad \dots\dots\dots\dots\dots\dots(1)$$

where K is the specific inductive capacity. When the electric displacement varies, it acts as an electric current whose strength is equal to the rate of increase of the displacement. In an isotropic medium, B is in the same direction as H and

$$B = \mu H, \quad \dots\dots\dots\dots\dots\dots(2)$$

where μ is the magnetic permeability.

4. The electromagnetic relations. These are:

I. The total flux of displacement outwards from any closed surface is equal to the electric charge within it.

II. The total flux of magnetic induction outwards from any closed surface is zero.

III. The work done upon a unit positive charge by the field, when the charge is taken once round any circuit, is equal to the rate of diminution of the flux of induction through the circuit, the positive direction round the circuit being connected with the positive direction of the induction in the same way as the rotation and translation of a right-handed screw working in a fixed nut.

IV. The work done upon a unit positive pole by the field, when the pole is taken once round any circuit, is 4π times the total current through the circuit, the positive directions being connected as in III.

5. Plane waves. Heaviside's method. We now consider a plane wave of the type devised by Dr Oliver Heaviside *. When the wave reaches any point, the electric and magnetic vectors at that point suddenly jump from zero to definite constant values E, H, D and B. Let any one of the vectors be represented by a line drawn perpendicular to the axis OX (Fig. 1), the direction of propagation of the wave. Then the

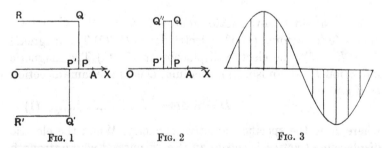

FIG. 1   FIG. 2   FIG. 3

wave is represented by $XPQR$, and the vector jumps from zero to the constant value PQ when the wave front reaches A. If, very soon after the first wave is started, a second wave of the

* "On the electromagnetic wave surface," *Electrical Papers*, Vol. II, p. 1, or *Phil. Mag.* June 1885, p. 397.

same character be started, the vectors in this wave being equal in magnitude but opposite in direction to those in the first wave, the second wave may be represented by $XP'Q'R'$. The second wave will neutralise all but a very small portion of the first wave, and we shall be left with the very thin wave represented by $XPQQ''P'O$ in Fig. 2. This wave produces only a momentary disturbance as it passes over A. Any sort of wave, as for instance the sinusoidal wave represented in Fig. 3, can now be built up of a succession of infinitely thin waves similar to $PQQ''P'$, and hence, at every part of a train of plane waves of any description, the relations between the vectors will be the same as in the Heaviside wave.

6. Directions of D and B. Take a fixed cylinder with its curved surface parallel to the direction of propagation of the wave and its plane ends p, q parallel to the wave front, as shown in section in Fig. 4, and suppose that the wave front lies between the ends of the cylinder and is moving towards p. By relation I, the total outward flux of displacement from the cylinder is zero,

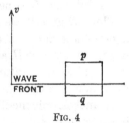

FIG. 4

since it encloses no charge. There is no flux through the end p, for the wave has not yet reached it, and therefore D is zero there. If there be any outward flux through one part of the curved surface, it is exactly compensated by a flux inwards through the remainder of that surface. Thus the outward flux of displacement through the end p and through the curved surface is zero, and hence D can have no component perpendicular to the end q. Hence D is parallel to the wave front.

If we use relation II, a similar argument shows that B is parallel to the wave front.

7. Directions of E and H. Let O be a point on the wave front, let Ov (Fig. 5) be the normal to the wave front through O, and let OB, at right angles to Ov, be the direction of the

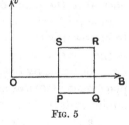

FIG. 5

magnetic induction B in the wave. Let $PQRS$ be a rectangle fixed in space with two sides parallel to Ov and two to OB, the line OB lying between PQ and RS. Since, at any point in the wave, B is parallel to the plane of this rectangle, there is no flux of induction through it, and therefore the flux does not vary with the time as the wave front advances. Hence, by relation III, no work is done in taking a charge round the rectangle. But along RS the electric force is zero, since the wave has not reached RS, and the amounts of work done along QR and SP neutralise each other. Hence the work done along PQ is zero, for the work done in the whole circuit is zero. Thus the electric force parallel to B is zero, or E is perpendicular to B.

By considering a rectangle with two sides parallel to D and two parallel to the normal to the wave front, we find, by relation IV, that H is perpendicular to D.

Thus, D and B are perpendicular to the wave normal and E is perpendicular to B and H to D. But it does not follow that, in general, E and H are perpendicular to the wave normal, for in a crystalline medium D is not necessarily in the same direction as E, nor B in the same direction as H.

In an isotropic medium, D is in the same direction as E, and B in the same direction as H, and then E and H are perpendicular to the wave normal and each is perpendicular to the other.

8. Relations between E and H in an isotropic medium.
Take a rectangle $abcd$ (Fig. 6) with two sides parallel to H and two to the wave normal Ov, and let ab be l cm., this side being ahead of the wave front. Since E is perpendicular to Ov and to OH, it is normal to the plane of the rectangle; we shall suppose the positive direction of E to be towards the reader. If the velocity of the wave be v cm. sec.$^{-1}$, the disturbed

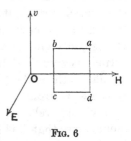

Fig. 6

area increases by lv square cm. per sec. Hence the flux of displacement increases at the rate lvD, or the current through the circuit is lvD. It is only along cd that there is any magnetic force and hence the work done by the field when a unit pole is taken once round the circuit in the direction $abcd$ is lH.

By relation IV, $lH = 4\pi lvD$. Thus, by (1),

$$H = KvE. \quad\quad\quad (3)$$

If we take a rectangle with two sides parallel to E and two to Ov, we find, by relation III, that

$$E = vB = \mu vH. \quad\quad\quad (4)$$

Right-handed rotation through $\tfrac{1}{2}\pi$ about the forward direction of propagation turns E round to H.

9. Velocity of propagation. From (3) and (4), we obtain

$$v = 1/\sqrt{K\mu}. \quad\quad\quad (5)$$

It can be shown that $\{K_{air} \times \mu_{air}\}^{-\frac{1}{2}}$ is numerically equal to the ratio of the electromagnetic unit to the electrostatic unit of electricity, the medium "air" being specified because the units are defined in connexion with that medium. Hence the velocity of an electromagnetic wave through air is numerically equal to the ratio of the units.

10. Periodic wave motion. When a single electron executes regular vibrations about a mean position O, it generates a succession of waves. At any fixed point the disturbance goes through cycles in a regular manner. At any instant the disturbance is not the same at all points on a straight line drawn from O. If we take a point P, where the disturbance is zero at any instant, there will at the same instant be zero disturbance at all those points on OP whose distances from O differ from OP by integral multiples of λ, the distance traversed by the waves during one cycle of the electron.

11. Periodic plane waves. When the distance from the electron is very great compared with the greatest width of the part of the wave under consideration, that part may be treated as a portion of a plane wave. In a train of plane waves there is no change in the amplitude as the waves advance; thus the disturbance will be periodic with respect to the distance measured along the wave normal as well as regards the time.

At every point of a fixed plane parallel to the planes of the waves the electric force is in the plane and has the same magnitude and direction. Let the electric force in this origin plane be

$$E_0 = a \sin pt. \quad\quad\quad (6)$$

At a point at distance x measured from the origin plane in the direction in which the waves advance, the electric force at time t is the same as it was in the origin plane at the time $t - x/v$, and thus

$$E = a \sin p\, (t - x/v). \quad\dots\dots\dots\dots(7)$$

The electric force at time t has the same value at all points for which x is a multiple of $2\pi v/p$, and hence this distance is the wave length λ. Thus, $\lambda = 2\pi v/p$. Hence,

$$E = a \sin \frac{2\pi}{\lambda}\, (vt - x). \quad\dots\dots\dots\dots(8)$$

The quantity $(2\pi/\lambda)\,(vt - x)$ is called the phase of the vibration at the time t at a point defined by x.

The number of complete vibrations per second is called the frequency and is denoted by n.

Thus $n\lambda = v.$ $\dots\dots\dots\dots\dots(9)$

The magnetic force is at right angles both to the electric force and to the wave normal. By (3),

$$H = KvE = Kva \sin \frac{2\pi}{\lambda}\, (vt - x). \dots\dots\dots(10)$$

In the train of plane waves which we have considered, the electric force is everywhere parallel to a fixed straight line which is perpendicular to the wave normal. This character is expressed by saying that the train of waves is polarised; the "plane of polarisation" is perpendicular to the electric force.

12. Rays. A luminous point sends out a succession of waves, which, in their progress, may be reflected or refracted. When the medium is isotropic, a straight line PN drawn normal to the wave front which is passing over a point P at any instant is called the ray through P.

When we consider doubly refracting media, such as some crystals, it is necessary to distinguish between the wave normal and another straight line which is then called a "ray." With isotropic media, the expressions "wave normal" and "ray" are interchangeable.

13. Laws of reflexion. Let v_1 be the velocity of propagation of waves in the medium [1] containing the incident wave, and

v_2 the velocity in the second medium [2]. Let the plane of Fig. 7 cut the surface of separation at right angles in AB, and

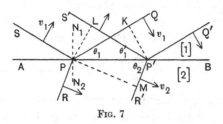

FIG. 7

also be perpendicular to the incident wave fronts. Let PQ be the section of the front of the incident wave at time $t = 0$, and let $P'Q'$ correspond to time t. Then, if $P'K$ be the perpendicular from P' on PQ, $P'K = v_1 t$.

If θ_1 be the angle QPB between the surface and the front of the incident wave, $PP' = P'K/\sin \theta_1 = v_1 t/\sin \theta_1$. If θ_1' be the angle SPA between the surface and the front of the reflected wave, $PP' = v_1 t/\sin \theta_1'$, since the perpendicular PL from P on $P'S'$ is $v_1 t$. Hence

$$\sin \theta_1' = \sin \theta_1. \qquad \qquad \text{......................(11)}$$

Thus the reflected wave is inclined to the surface at the same angle as the incident wave but is on the other side of $N_1 PN_2$, the normal to the surface at P, as indicated by PS and PQ in Fig. 7.

The incident ray, or the normal to the incident wave at P, is perpendicular to PQ, while the reflected ray, or the normal to the reflected wave at P, is perpendicular to PS, and each is in the plane of the paper. The plane containing the incident ray and the normal to the surface is called the plane of incidence. The acute angle between the incident ray and the normal to the surface is called the angle of incidence, and the acute angle between the reflected ray and the normal to the surface is called the angle of reflexion. These angles are equal to those between the surface and the two wave fronts. The result, in terms of the rays, is as follows:

Laws of reflexion. (1) The reflected ray is in the plane containing the incident ray and the normal to the surface at the point of incidence. (2) The incident and reflected rays make equal angles with the normal: or the angles of incidence and reflexion are equal.

14. Laws of refraction; Snell's law. The perpendicular PM from P on the refracted wave front $P'R'$ is $v_2 t$. Hence, if θ_2 be the angle RPA between the surface and the front of the refracted wave,

$$v_2 t/\sin \theta_2 = PP' = v_1 t/\sin \theta_1.$$

Thus, $$\frac{\sin \theta_1}{v_1} = \frac{\sin \theta_2}{v_2}. \quad \ldots\ldots\ldots\ldots\ldots(12)$$

The incident ray at P is perpendicular to PQ while the refracted ray is perpendicular to PR, and each is in the plane of the paper. The acute angle between the refracted ray and the normal $N_1 P N_2$ to the surface is called the angle of refraction. It is clear that the angle of incidence is equal to θ_1, and that the angle of refraction is equal to θ_2. Hence the result is as follows:

Laws of refraction. (1) The refracted ray is in the plane containing the incident ray and the normal to the surface at the point of incidence. (2) The sine of the angle of refraction bears a constant ratio to the sine of the angle of incidence (Snell's law).

15. Remarks on the laws of reflexion and refraction. The method by which we investigated the laws of reflexion and refraction does not give the relative magnitudes of the disturbances in the three waves; it does not even tell us that the reflected and the refracted waves exist. It merely tells us the directions they must have if they do exist. For a given angle of incidence, the relative magnitudes and the directions of the electric forces in the reflected and the refracted waves depend upon the direction of the electric force in the incident wave. By the electromagnetic theory, the directions and magnitudes of the electric forces in the reflected and refracted waves can be calculated when the direction and the magnitude of the electric force in the incident wave are given (see Chapter XII).

16. Refractive index. The ratio of the sine of the angle of incidence to that of refraction is called the refractive index of the second medium relative to the first; the ratio will be denoted by $_1\mu_2$. Thus,

$$\sin \theta_1/\sin \theta_2 = {}_1\mu_2. \quad \ldots\ldots\ldots\ldots\ldots(13)$$

Hence, by (12), $$_1\mu_2 = v_1/v_2. \quad \ldots\ldots\ldots\ldots\ldots(14)$$

If $_2\mu_1$ be the refractive index of the first medium relative to the second, i.e. the ratio of $\sin \theta_2$ to $\sin \theta_1$, we have, by (12)

$$_2\mu_1 = \sin \theta_2/\sin \theta_1 = v_2/v_1,$$

and hence $\qquad _2\mu_1 = 1/_1\mu_2.$ ....(15)

The refractive index of any medium relative to a vacuum is called the absolute refractive index, or simply the refractive index, of the medium. If v be the velocity of a wave in a vacuum and μ_1, μ_2 be the absolute refractive indices of the media [1], [2], then $\mu_1 = v/v_1$, $\mu_2 = v/v_2$. Hence, by (14),

$$_1\mu_2 = v_1/v_2 = \mu_2/\mu_1.$$ ....(16)

Since $v_1/v_2 = \mu_2/\mu_1$, we have, by (12),

$$\mu_1 \sin \theta_1 = \mu_2 \sin \theta_2,$$ ....(17)

the form in which Snell's law is usually employed.

In most experiments, it is the refractive index of the medium relative to air which is determined. By (16), we can find the absolute refractive index of the medium (M), since that of air (A) is known. Thus,

$$\mu_M = {}_A\mu_M \times \mu_A.$$ ....(18)

At 0° C. and 760 mm., the absolute refractive index, μ_A, of air for sodium light is 1·0002918.

17. Dispersion. In the case of glass, water and many other substances and within the range of visible light, the refractive index increases as the frequency increases. Since the frequency in blue is greater than in red light, the refractive indices of these media for blue are greater than for red light, and thus blue is often said to be more refrangible than red light. But it is not universally true that the refractive index increases with the frequency, for with some substances there is a narrow range of frequency within which the refractive index *diminishes* as the frequency increases.

The dependence of the refractive index upon the frequency of the disturbance is called *dispersion*.

CHAPTER II

SOME APPLICATIONS OF THE LAWS OF REFLEXION AND REFRACTION

18. Relations between incident and reflected rays. Let
PA (Fig. 8) be a normal to the plane reflecting surface AHK,
K any point on AHK and
KN the normal at K. Let a
plane HAP containing PA
cut the surface in AH. Since
PA is normal to AHK, HA,
KA cut AP at right angles.
Let the ray PK be reflected
along KL. Since KN is parallel to AP, the plane $PKLN$,

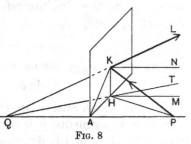

FIG. 8

containing the rays and the normal KN, is identical with the plane
KPA. Hence KL, produced backwards, cuts PA in Q. Draw
KH perpendicular to AH. Since KH is at right angles to AH
and to AP, it is normal to the plane HAP. Since KN is parallel
to PQ and since $LKN = PKN$, we have $KQP = KPQ$. Hence
KA bisects PQ at right angles in A, and thus $QA = PA$. Then,
by symmetry, all corresponding distances and angles are equal.
Hence HP and QHT, the *projections* on the plane HAP of the
rays PK and KL, make equal angles with HM, the normal to
the surface at H. Since KH is normal to HAP, the angles
which KP, KQ make with the plane HAP are KPH, KQH,
and by symmetry these are equal. Thus:

(1) The *projections* of the incident and reflected rays on any
plane normal to the reflecting surface obey the laws of reflexion.
(2) The incident and reflected rays are equally inclined to any
plane normal to the reflecting surface.

19. Image of a point by reflexion at a plane surface.
By § 18, the position of Q is independent of the direction of the
ray PK. Hence, if P be a luminous point, *all* the reflected rays,
if produced, pass *accurately* through Q. The point Q is called
the image of P. Hence the image of a point formed by a plane

reflecting surface lies on the normal through the point, and the point and its image are on opposite sides of the surface and equidistant from it.

In Fig. 8, the rays diverge from a point P, and we have an object which is called "real," because the rays start from it, and an image which is called "virtual," because the reflected rays do not actually pass through it.

In many cases we have a virtual object, i.e. the incident rays converge towards a single point but strike the surface before they reach the point. Fig. 8 will show this case, if the direction of the ray KL be reversed. The image of Q is at P and is real, since the reflected rays actually pass through P.

20. Relations between incident and refracted rays. Let PA (Fig. 9) be a normal to the plane surface AHK separating the media [1], [2] of refractive indices μ_1, μ_2. Let K be any point on AHK and KN be the normal at K. Let a plane HAP containing PA cut the surface in AH. Then HA, KA cut AP at right angles. Let the ray PK in [1] be refracted at K and become KL in [2]. Since KN is parallel to AP, the plane $PKLN$, containing the rays and the normal KN, is identical with the plane KPA. Hence KL, produced backwards, cuts AP in Q. Draw KH perpen-

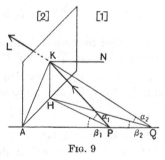

FIG. 9

dicular to AH; then KH is normal to the plane HAP. The lines PH, QH are the projections of PK, QK on the plane HAP, and the angles $KPH = \alpha_1$, $KQH = \alpha_2$, are the angles which PK, QK make with the plane HAP. Let $HPA = \beta_1$, $HQA = \beta_2$. Since KN is parallel to AP,

$$\mu_1 \sin APK = \mu_1 \sin PKN = \mu_2 \sin QKN = \mu_2 \sin AQK.$$

Thus $\mu_1 . AK/PK = \mu_2 . AK/QK$. Hence

$$PK/\mu_1 = QK/\mu_2. \qquad \qquad \text{......................(1)}$$

But $$KH = PK \sin \alpha_1 = QK \sin \alpha_2. \qquad \text{............(2)}$$

Thus, by (1), $$\mu_1 \sin \alpha_1 = \mu_2 \sin \alpha_2. \qquad \text{....................(3)}$$

Since $AH = \sin \beta_1 . PH = \sin \beta_1 . PK \cos \alpha_1 = \sin \beta_2 . QK \cos \alpha_2$, we have, by (1),

$$\mu_1 \cos \alpha_1 \sin \beta_1 = \mu_2 \cos \alpha_2 \sin \beta_2. \quad \ldots\ldots\ldots\ldots(4)$$

Hence, by (3), $\mu_1 \cos \alpha_1 \sin \beta_1 = (\mu_2{}^2 - \mu_1{}^2 \sin^2 \alpha_1)^{\frac{1}{2}} \sin \beta_2$. (5)

Equations (3) and (5) determine α_2 and β_2 when α_1 and β_1 are known.

For a given value of α_1, the sine of the angle β_1 between the normal at H and the *projection* of the incident ray on the plane HAP, which is normal to the refracting surface, bears a constant ratio to the sine of the angle β_2 between the normal and the *projection* of the refracted ray. These projections are in one plane with the normal at H and thus are related in the same way as if they were parts of an actual ray passing from a medium of refractive index $\mu_1 \cos \alpha_1$ to one of index $\mu_2 \cos \alpha_2$ or $(\mu_2{}^2 - \mu_1{}^2 \sin^2 \alpha_1)^{\frac{1}{2}}$.

When both α_1 and β_1 are small, we have

$$\mu_1 \alpha_1 = \mu_2 \alpha_2, \qquad \mu_1 \beta_1 = \mu_2 \beta_2. \quad \ldots\ldots\ldots\ldots(5a)$$

21. Geometrical image of a point by refraction at a plane surface. By (1), $QK = PK . \mu_2/\mu_1$, and thus the distance QA depends upon the angle KPA. When this angle is very small, PK, QK may be regarded as equal to PA, QA, and then

$$QA = PA . \mu_2/\mu_1. \quad \ldots\ldots\ldots\ldots\ldots(6)$$

Hence, if P be a luminous point, all the rays which meet the surface at very small angles of incidence will proceed in the new medium as if they came from a point at Q defined by (6). The point Q is called the geometrical image of P.

22. Effect of a parallel plate.
In many experiments, a medium [1] is separated from a medium [2] by a plate with plane parallel faces. In Fig. 10, let the plane of the paper be perpendicular to the faces of the plate and to the front of the incident wave, and let it cut the faces of the plate in

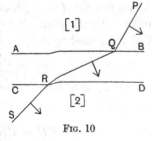

FIG. 10

AB, CD and the wave front in PQ. Let QR, RS be the intersections of the plane of the paper with the wave fronts in the plate and in the second medium respectively. Then the velocity of R along CD equals that of Q along AB, and hence the angle CRS depends upon PQB and upon the velocities of waves in [1] and [2] in the same way as if the plate were absent and the first medium extended right up to the second. In other words, the direction of the ray in the second medium is unaltered by the presence of the plate.

23. Total reflexion. When a wave in a medium [1] is incident upon a surface separating [1] from [2], and the angles of incidence and refraction are θ_1, θ_2,

$$\mu_1 \sin \theta_1 = \mu_2 \sin \theta_2. \quad \ldots\ldots\ldots\ldots\ldots(7)$$

When μ_2 is greater than μ_1, we see, by (7), that $\sin \theta_2$ is less than $\sin \theta_1$, and thus $\sin \theta_2$ is less than unity. Hence θ_2 is less than $\frac{1}{2}\pi$ for all values of θ_1. The maximum value of θ_2 occurs when $\theta_1 = \frac{1}{2}\pi$; then $\sin \theta_2 = \mu_1/\mu_2$.

When μ_2 is less than μ_1, equation (7) shows that $\sin \theta_2$ is greater than $\sin \theta_1$ and that $\sin \theta_2$ becomes unity when $\sin \theta_1 = \mu_2/\mu_1$. If $\sin \theta_1$ be greater than μ_2/μ_1, the quantity $\mu_1 \sin \theta_1/\mu_2$ is greater than unity, and then it is impossible to find any real angle whose sine has this value. We therefore conclude that, for all values of $\sin \theta_1$ greater than μ_2/μ_1, there is no refracted wave. The energy brought up to the surface by the incident wave does not pass across the surface and hence the reflected will be as vigorous as the incident wave. When no refracted wave exists, the light is said to suffer total reflexion.

The angle whose sine is μ_2/μ_1, where $\mu_1 > \mu_2$, is called the critical angle for the two media; we shall denote it by α. Thus

$$\sin \alpha = \mu_2/\mu_1 = {}_1\mu_2, \qquad \operatorname{cosec} \alpha = \mu_1/\mu_2 = {}_2\mu_1. \quad \ldots\ldots(8)$$

When $\sin \theta_1$ is less than μ_2/μ_1, or θ_1 is less than α, the critical angle, there is a refracted as well as a reflected wave, but, when θ_1 is greater than α, the reflected wave alone exists.

When the angle of incidence is equal to the critical angle, the sine of the angle of refraction is unity, the angle of refraction is $\frac{1}{2}\pi$, and the refracted ray grazes the surface of separation.

24. Deviation of a ray by refraction. Let the incident ray
PQ (Fig. 11) in medium [1] meet the
surface AB at Q and be refracted into
medium [2] along QR. Let the angles
of incidence and refraction be θ_1, θ_2.
Then

$$\mu_1 \sin \theta_1 = \mu_2 \sin \theta_2. \quad \ldots\ldots(9)$$

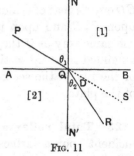

FIG. 11

If the refractive index of [2] relative to
[1] be μ, we have

$$\sin \theta_1 = \mu \sin \theta_2. \quad \ldots\ldots(10)$$

When μ is greater than unity, $\sin \theta_1$ is
greater than $\sin \theta_2$, and, since neither angle exceeds a right
angle, θ_1 is greater than θ_2. Hence, when a ray passes from a
medium of lower into one of higher index, the ray is bent towards the normal. If the straight line PQ be continued along
QS, the angle RQS is called the deviation; it will be denoted
by D. Thus

$$D = \theta_1 - \theta_2. \quad \ldots\ldots\ldots\ldots\ldots(11)$$

Since $\sin D = \sin \theta_1 \cos \theta_2 - \cos \theta_1 \sin \theta_2$, we have, by (10),

$$\sin D = \sin \theta_2 \{\mu \sqrt{1 - \sin^2\theta_2} - \sqrt{1 - \mu^2 \sin^2\theta_2}\}$$
$$= (\mu^2 - 1) \sin \theta_2/\{\mu \sqrt{1 - \sin^2\theta_2} + \sqrt{1 - \mu^2 \sin^2\theta_2}\}.(12)$$

The numerator increases and the denominator diminishes as θ_2
increases. Since θ_1 increases as θ_2 increases, D increases when
either θ_1 or θ_2 increases.

25. Deviation of a nearly normal ray. When θ_1 is small,
θ_2 is small also, and thus in (9) we may replace $\sin \theta_1$ and $\sin \theta_2$
by θ_1 and θ_2. Then $\mu_1 \theta_1 = \mu_2 \theta_2$, or $\theta_2 = \mu_1 \theta_1/\mu_2$. Hence

$$D = \theta_1 - \theta_2 = \left(1 - \frac{\mu_1}{\mu_2}\right)\theta_1 = \frac{\mu_2 - \mu_1}{\mu_2}\theta_1 = \frac{\mu_2 - \mu_1}{\mu_1}\theta_2. \ldots(13)$$

If the refractive index of [2] relative to [1] be μ, we have, by
§ 16, $\mu_2 = \mu . \mu_1$. Hence

$$D = (\mu - 1)\theta_1/\mu = (\mu - 1)\theta_2. \quad \ldots\ldots\ldots(14)$$

26. Deviation by a prism of a ray in a principal plane.
A prism is a body having two plane but not parallel faces. The
straight line in which these faces meet is called the edge of

the prism, and the angle between the faces is called the angle of the prism. A plane perpendicular to the edge, and therefore perpendicular to both faces, is called a principal plane of the prism. The section of a prism by a principal plane is called a principal section.

We consider a ray which passes through a prism of angle i, in a principal plane, and suppose that μ, the index of the prism relative to the surrounding medium, is greater than unity.

Let QOR (Fig. 12) be a section of the prism by a principal plane, and let $PQRS$ be the ray. Let MQL and $NRLH$, the

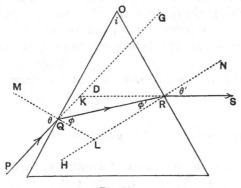

FIG. 12

normals at Q and R, meet in L. Let the angles which the ray makes with the normals at Q and R be as follows:

$$PQM = \theta, \qquad RQL = \phi, \qquad QRL = \phi', \qquad SRN = \theta'.$$

The angles θ, ϕ are positive when P and O are on opposite sides of QM; θ', ϕ' are positive when S and O are on opposite sides of RN. If the straight lines $PQKG$ and KRS intersect at K, the angle GKS is called the deviation; it will be denoted by D.

Since QLH, the angle between the normals QL, RL, is equal to that between QO and RO, $QLH = i$. But $QLH = \phi + \phi'$. Hence

$$i = \phi + \phi'. \qquad \qquad \text{...............(15)}$$

Similarly, $D = KQR + KRQ$. But $KQR = \theta - \phi$ and $KRQ = \theta' - \phi'$, and hence $D = \theta + \theta' - \phi - \phi'$, or, by (15),

$$D = \theta + \theta' - i. \qquad \qquad \text{...............(16)}$$

Since $\sin \theta = \mu \sin \phi$, $\sin \theta' = \mu \sin \phi'$, ...(17)

we obtain, by (15),

$$\sin \theta' = \mu \sin (i - \phi) = \mu \sin i \cos \phi - \mu \cos i \sin \phi$$
$$= \sin i (\mu^2 - \mu^2 \sin^2 \phi)^{\frac{1}{2}} - \mu \cos i \sin \phi$$
$$= \sin i (\mu^2 - \sin^2 \theta)^{\frac{1}{2}} - \cos i \sin \theta. \quad\quad\quad(18)$$

This gives θ' in terms of θ. Then D can be found by (16).

It can be shown, by § 24, that, when the refractive index of the prism relative to the surrounding medium is greater than unity, the deviation of a ray passing through the prism in a principal plane is always from the edge, and D is positive.

If the refractive index of the prism be less than that of the surrounding medium, the deviation is towards the edge, and D is negative.

If the media to the left and right of the prism in Fig. 12 have indices μ_1, μ_2, while the index of the prism is μ, equations (17) become

$$\mu_1 \sin \theta = \mu \sin \phi, \quad\quad \mu_2 \sin \theta' = \mu \sin \phi'.........(18a)$$

27. Deviation by a prism of small angle of a ray in a principal plane. When θ, the angle of incidence, is small, ϕ is small, and thus, by (15), ϕ' is small, since the angle of the prism is small, and θ' is small. Hence, replacing sines by angles,

$$\theta = \mu\phi, \quad\quad \theta' = \mu\phi'. \quad\quad\quad\quad(19)$$

Since $\phi + \phi' = i$, we have $\theta + \theta' = \mu i$, and, by (16),

$$D = \theta + \theta' - i = (\mu - 1) i. \quad\quad\quad\quad(20)$$

Thus, when a ray in a principal plane strikes nearly normally one face of a prism of *small* angle i, the deviation depends only upon μ and i and not on the angle of incidence. This result is of great importance in the theory of lenses.

28. Deviation of a symmetrical ray in a principal plane. In this case $\theta = \theta'$ and $\phi = \phi'$. Hence, by (15), $\phi = \frac{1}{2}i$. If Δ be the deviation, we find, by (16), $\frac{1}{2}(\Delta + i) = \theta$. But, by (17), $\sin \theta = \mu \sin \phi$. Hence

$$\sin \tfrac{1}{2}(\Delta + i) = \mu \sin \tfrac{1}{2}i. \quad\quad\quad\quad(21)$$

Equation (21) is used for finding μ from the observed values of Δ and i. Thus we have

$$\mu = \sin \tfrac{1}{2} \, (\Delta + i)/\sin \tfrac{1}{2} i. \qquad \ldots\ldots\ldots\ldots(22)$$

29. Minimum deviation of a ray in a principal plane.
The deviation, D, of a ray passing in a principal plane is positive when the deviation is away from the edge. It can be shown, by aid of § 24, that, of the rays passing in this plane, the symmetrical ray affords the minimum value of $|D|$.* When μ, the refractive index of the prism relative to the surrounding medium, is greater than unity, D is positive, but when $\mu < 1$, D is negative. In either case, the stationary value of the deviation is given by (21).

30. Effect of a prism on a ray not in a principal plane.
Let KLM (Fig. 13) be a section of the prism by a principal plane, the angle KLM being i. Let $EFGH$ be the *projection* on the principal plane of a ray $E'F'G'H'$ which passes through the prism, but not in a principal plane. Let the rays $E'F'$, $F'G'$, $G'H'$ be inclined at the angles α, α_1, α_2 to the principal plane and let

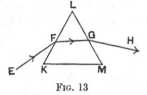

FIG. 13

EF, FG, the *projections* of $E'F'$, $F'G'$ upon that plane, make angles θ, ϕ with the normal at F, while the *projections* FG, GH make angles ϕ', θ' with the normal at G.

By equation (3) of § 20, we have, for the refractions at F', G',

$$\sin \alpha = \mu \sin \alpha_1, \qquad \mu \sin \alpha_1 = \sin \alpha_2. \qquad \ldots\ldots\ldots(23)$$

Hence $\qquad\qquad\qquad \alpha_2 = \alpha. \qquad \ldots\ldots\ldots\ldots\ldots\ldots(24)$

Thus the rays $E'F'$, $G'H'$ are equally inclined to the principal plane.

By equation (4) of § 20, we have, for the refractions at F', G',

$$\cos \alpha \sin \theta = \mu \cos \alpha_1 \sin \phi, \qquad \mu \cos \alpha_1 \sin \phi' = \cos \alpha_2 \sin \theta'. \quad (25)$$

Since ϕ and ϕ' are the angles between FG, the *projection* of $F'G'$, and the normals at F and G,

$$\phi + \phi' = i. \qquad \ldots\ldots\ldots\ldots\ldots\ldots(26)$$

* The symbol $|D|$ denotes the magnitude of D without regard to sign. Thus $|5| = 5$, and also $|-5| = 5$.

From (25) and (26), since $\alpha_2 = \alpha$,

$$\sin \theta' = \frac{\mu \cos \alpha_1 \sin \phi'}{\cos \alpha} = \frac{\mu \cos \alpha_1}{\cos \alpha} (\sin i \cos \phi - \cos i \sin \phi).$$

Using the value of $\sin \phi$ given by (25), we obtain

$$\sin \theta' = \sin i \left\{ \frac{\mu^2 \cos^2 \alpha_1}{\cos^2 \alpha} - \sin^2 \theta \right\}^{\frac{1}{2}} - \cos i \sin \theta. \quad\dots\dots(27)$$

If we eliminate α_1 by aid of (23), we have

$$\sin \theta' = \sin i \left\{ \mu^2 - \sin^2 \theta + (\mu^2 - 1) \tan^2 \alpha \right\}^{\frac{1}{2}} - \cos i \sin \theta. \quad (28)$$

This equation gives the direction of the *projection* of the emergent ray on the principal plane. The inclination (α_2) of the emergent ray itself to that plane is equal, by (24), to the inclination (α) of the incident ray.

If D_0 denote the deviation of $EFGH$, the *projection* of the ray, we have, as in (16), § 26,

$$D_0 = \theta + \theta' - i. \quad\dots\dots\dots\dots\dots(29)$$

The deviation is the angle, less than π, between two straight lines drawn *from* a point parallel to the forward directions of the incident and emergent rays. Let the plane $ABCO$ (Fig. 14) be parallel to a principal plane of the prism, let OP, OR be parallel to the incident and emergent rays and let $OR = OP$. The perpendiculars PA, RC to the plane AOC are in one plane. Then POR equals D, the deviation of the ray, while AOC equals D_0, the deviation of the *projection* of the ray. Since POA, ROC are each equal to α, $PA = RC$ and $OA = OC$. Thus, if K be the middle

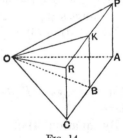

Fig. 14

point of PR, the perpendicular KB on the plane AOC bisects AC at B. Since the triangles POR, AOC are isosceles, KO, BO bisect POR, AOC. Now $PK = OP \sin \frac{1}{2} D$, and $AB = OA \sin \frac{1}{2} D_0$, while $OA = OP \cos \alpha$. Thus

$$AB = OP \sin \tfrac{1}{2} D_0 \cos \alpha.$$

But $PK = AB$, and hence*

$$\sin \tfrac{1}{2}D = \sin \tfrac{1}{2}D_0 \cos \alpha. \qquad\qquad(30)$$

When $\alpha = 0$, the ray is its own projection. If θ_0' be the value of θ' in this case, we have, by (18), § 26,

$$\sin \theta_0' = \sin i\,(\mu^2 - \sin^2\theta)^{\frac{1}{2}} - \cos i \sin \theta. \qquad\ldots(31)$$

When α is small, we write α for $\tan \alpha$ in (28); then, expanding the square root as far as α^2 and comparing the result with (31), we have

$$\sin \theta' = \sin \theta_0' + \frac{\alpha^2\,(\mu^2 - 1)\sin i}{2\,(\mu^2 - \sin^2\theta)^{\frac{1}{2}}}. \qquad\ldots(32)$$

If $\theta' = \theta_0' + \epsilon$, $\sin \theta' = \sin \theta_0' + \epsilon \cos \theta_0'$, since ϵ is small.

Hence

$$\epsilon = \frac{\alpha^2\,(\mu^2 - 1)\sin i}{2 \cos \theta_0'\,(\mu^2 - \sin^2\theta)^{\frac{1}{2}}}. \qquad\ldots(33)$$

If D' be the deviation of a ray which on incidence coincides with EF, the *projection* of $E'F'$ (Fig. 13), $D' = \theta + \theta_0' - i$, where θ_0' is given by (31).

If D_0 be the deviation of $EFGH$, the *projection* of $E'F'G'H'$, we have

$$D_0 = \theta + \theta' - i = D' + \epsilon. \qquad\ldots(34)$$

---

* R. S. Heath, *Geometrical Optics*, Second Edition, 1895, obtains $\cos \tfrac{1}{2}D = \cos \tfrac{1}{2}D_0 \cos \alpha$. This erroneous result is quoted in the "Optics" of Winkelmann's *Handbuch der Physik*, 1906, in Vol. I of Kayser's *Handbuch der Spectroscopie*, 1900, and in other books. The result (30) was given by Larmor (*Proc. Cambridge Phil. Soc.* Vol. IX, p. 108, 1898).

CHAPTER III

EXPERIMENTS WITH PLANE SURFACES

31. Testing of glass plates by telescope. In many experiments, it is important that the surfaces of the plates employed should be at least very nearly plane. The quality of a plate may be readily tested by aid of a telescope fitted with cross-wires. The telescope should have a magnifying power of at least 10 and should give a perfectly distinct image of a distant object, such as a lightning conductor. If absence of parallax cannot be secured both for horizontal and for vertical motions of the eye, the telescope is of poor quality.

Reflexion test. The telescope is first focused on an object at least 100 metres distant. It is then turned to view the same object by reflexion at one surface of the plate, the other surface being coated with vaseline or soap to stop regular reflexion. If the object be still in focus, as shown by the absence of parallax, for both vertical and horizontal motions of the eye, the reflecting surface is satisfactory. The second surface is tested in a similar manner.

If each surface be sufficiently nearly plane, the plate is cleaned and the object is again viewed by reflexion. Two images, at least, will now be seen, unless the surfaces be parallel, and the images will move relative to each other if the plate be turned round in its own plane. The "plate" is thus shown to be in reality a prism of very small angle. For most purposes, flatness of surfaces is of more importance than entire absence of prism error; if the surfaces have been found to be good, the plate may be considered satisfactory. If one image be good and the other or others cannot be satisfactorily focused, the glass is lacking in homogeneity.

Transmission test. When the plate is required for transmitting and not for reflecting light, a simpler test may suffice. The telescope is turned to view the distant object, and the plate is then interposed between telescope and object. It may happen that this interposition causes a displacement of the image but no change of focus. If the displacement remain unchanged when the plate is given a motion of translation in any direction in its own plane, the "plate" is a prism of small angle. If, by a change of focusing, a perfect image can be brought to the cross-wires, the "plate" is a very weak lens, and a motion of the "plate" in its own plane will cause the image to move.

EXPERIMENT 1. **Verification of Snell's law.**

32. Method. In Fig. 15, $ABCD$ is a tank with vertical sides, containing liquid. The side AB is of plane parallel glass. A plane

mirror M, with its plane vertical, is held immersed in the liquid by a support which turns about a vertical axis. An arm E, also turning about a vertical axis, carries a converging lens L and a needle point P, at the focus of the lens. The rays which fall upon the lens L from the tip P are reduced by L to a parallel beam. The beam falls upon AB and is refracted, and is reflected by M.

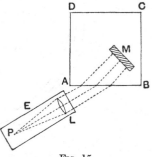

FIG. 15

If the rays strike M normally, they are reflected back along their own paths and form an image of the tip P coincident with P itself. The image of the needle is inverted. Since AB is of plane parallel glass, the direction of the rays in the liquid is, by § 22, the same as if the glass were absent.

The angle of incidence, θ, is the angle through which the arm E has been turned from the zero position, in which an image of P coincident with P is obtained by reflexion at AB. The angle of refraction, ϕ, is the angle through which M has been turned from its zero position, in which M is parallel to AB.

By Snell's law, (17), § 16,

$$\sin \theta = \mu \sin \phi, \qquad \ldots\ldots\ldots\ldots\ldots\ldots(1)$$

where μ is the refractive index of the liquid relative to air.

A special form of spectrometer shown diagrammatically in Fig. 16 is convenient. The spectrometer table is a metal disk having a central raised platform G. The edge of the table is divided into degrees, and a circular V-groove is cut near the scale. Each of the carriages E and F has an index, two spherical feet R, S resting in the V-groove and a levelling screw

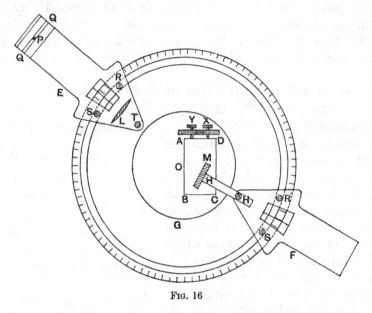

FIG. 16

resting on the table near the platform. A geometrical slide with five points of contact is thus obtained. The carriage has only one degree of freedom and this allows it to move round the axis of the table. The piece QQ sliding on E carries a vertical plate pierced by a slit which expands into a rectangular opening (Fig. 17). The slit is covered by a right-angled prism, as in

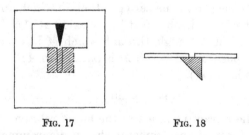

FIG. 17   FIG. 18

Fig. 18, and a needle is attached to the plate so that its point is close to the upper end of the slit. When a flame is placed in the plane of the plate and to the right of the prism in Fig. 18, the slit is brightly illuminated. The image of one edge of the slit formed by rays reflected at M is made to coincide with the needle. The carriage F bears an arm H supporting M. This arm projects from an adjustable vertical column and is capable of adjustment about a horizontal axis. The plane of M is set perpendicular to the plane of the spectrometer table by aid of a set-square.

The tank is adjusted in azimuth by the screws X, Y. The face AB is approximately over the centre O, and the tank is levelled so that AB is perpendicular to the plane of the table. The complete apparatus is shown in Fig. 19*.

33. Practical details. To focus the collimator, the slit is illuminated by aid of the prism and the light is allowed to fall on the mirror M; then Q is adjusted until there is no parallax between the tip of the needle and one edge of the image of the slit. A magnifying lens is used.

To level the lens carriage, the plane of M is set perpendicular to the platform by a square, and the levelling screw T of the

FIG. 19

carriage is then adjusted so that the lower end of the *image* of the slit is on a level with the upper end of the slit.

The lens carriage is at its zero when the image of the slit, due

* In the later form of spectrometer shown in Fig. 19 the piece Q takes the form of a tube sliding in V's and fitted with a slit and a reflecting prism. The carriage which holds the mirror in the present experiment has two V's in which slides a tube having a slit at one end. Each tube can be clamped to its carriage. Each carriage bears a removable lens. The tank is secured by a clamp which allows proper adjustment.

to reflexion at AB, coincides with the needle. If the index do not point exactly to a degree mark, one of the screws X, Y is adjusted. To prevent reflexion from CD, an opaque plate is placed in the tank. The zero reading of the mirror carriage is obtained by adjusting that carriage so that, when the lens carriage is in *its* zero position, the image of the slit due to reflexion at M coincides with the needle. During these operations, the slit may be illuminated by *white* light.

Care must be taken not to throw the mirror out of adjustment by contact with the tank.

In testing Snell's law, a sodium flame must be used.

The lens carriage is turned through $10°$ from its zero position and the mirror carriage is then adjusted so that the image of the slit coincides with the needle, and the index of the mirror carriage is read. This operation is repeated for angles of incidence of $20°$, $30°$ ..., on each side of the zero. The mean of the two angles of refraction found for a given angle of incidence is used in the calculation.

For a given value of θ, there is a second position of M for which the image coincides with the needle. For, if the angle between M and AB be $\frac{1}{2}\phi$, the rays reflected at M will fall normally upon AB and will be reflected along their own paths, suffering a second reflexion at M. Accurate measurements can be made, but the image is not very bright.

34. Practical example. The table gives a few readings obtained by G. F. C. Searle. Temperature of water 17° C. By (1), sin θ/sin $\phi=\mu$.

Lens Reading	Mirror Reading	Lens Reading	Mirror Reading	θ	Mean	$\dfrac{\sin\theta}{\sin\phi}$
180°	359°·7	180°	359°·7			
200	14 ·6	160	344 ·8	20°	14° 54′	1·330
220	28 ·6	140	330 ·9	40	28 51	1·332
240	40 ·2	120	319 ·2	60	40 30	1·334
260	47 ·3	100	312 ·1	80	47 36	1·334

Thus Snell's law is closely verified. The mean value of μ is 1·332.

EXPERIMENT 2. **Determination of refractive index of a liquid by total reflexion.**

35. Introduction. Let [1] and [2] be two media of indices

μ_1 and μ_2. If $\mu_1 > \mu_2$ and if α be the critical angle, we have, by § 23,

$$\operatorname{cosec} \alpha = \mu_1/\mu_2. \qquad \ldots\ldots\ldots\ldots\ldots\ldots(1)$$

Hence we can find μ_1/μ_2 by measuring the critical angle.

Let $ABCD$ (Fig. 20) be a tank with ends AB, CD formed of vertical plane glass plates. This tank contains the liquid of refractive index μ_1. Within the tank is placed a cell $XYVU$ formed of two parallel plates of glass, cemented together at the edges and enclosing a thin film of the liquid or gas of refractive index μ_2. The cell is supported by an axle turning about a vertical axis and fitted with a divided circle. The plates XY, UV are vertical.

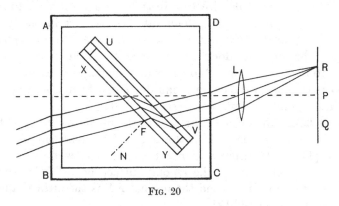

Fig. 20

A ray which enters the tank through AB and falls upon the face XY of the cell will pass through XY into the film if the angle between the ray and the normal FN be not too great. But, if this angle be increased beyond a certain value, the ray suffers total reflexion at the second surface of XY and does not emerge through CD.

A sodium flame is placed about 30 cm. from one end of the tank. At the other end is placed a lens L with a pin P at its focus; the axis of the lens points towards the flame. As the cell is turned round from its position when XY is parallel to AB, an eye placed beyond P will, after a time, see the bright field gradually invaded by darkness; the boundary between brightness and darkness is slightly curved, the tangent to the curve being vertical at the point corresponding to the focus for rays which are horizontal in the tank. The radius of curvature of the curve formed in the focal plane of L is $\mu_1 f \tan \alpha$, where f is the focal length of L. The boundary can be brought to coincide with P. There are two positions of the cell which secure this result; for one position the right part of the field is bright and the left dark, while for the other position the reverse is the case.

Let P, Q, R be three points in the focal plane of the lens L, the line QPR being horizontal. If R were a luminous point, the rays from R, after passing through L, would form a parallel beam. The direction of this beam would be changed by refraction at the plane faces of CD, UV, XY, but its quality of parallelism would remain. Reversing the rays, we see that all rays having a suitable direction in the liquid in the tank to the left of XY will come to a focus at the point R in the focal plane of L; the *direction* of these rays may be said to correspond to R. The method does not require that the rays which ultimately pass through R come from a single point at infinity. They may, and in the experiment do, come from a number of luminous points in different parts of the flame.

By § 22, the direction of the rays in the medium contained by the cell is the same as if the plates XY, UV were absent and the media [1] and [2] were in contact, and hence, when the rays suffer total reflexion at the second surface of XY, the angle between the rays in [1] and the normal FN is the critical angle for the media [1] and [2].

If the cell be in such a position that the angle between FN and those rays in the liquid in the tank which come to a focus at P, or correspond to P, is the critical angle α, these rays will then be totally reflected on reaching the medium contained by the cell. The angle of incidence of the rays whose directions correspond to R is less than α; these rays therefore pass through the cell and come to a focus at R. But the angle of incidence of the rays corresponding to Q is greater than α, and hence these rays are totally reflected on reaching the medium in the cell.

Thus, all the rays which came to foci between P and Q, when FN was parallel to BC, now suffer total reflexion, while those which then came to foci between P and R still come to the same foci. Hence, half of the field of view will be bright and half dark, the boundary between the bright and dark parts corresponding to the direction of those rays which strike XY at the critical angle.

If, in Fig. 20, the rays corresponding to P be just totally reflected, these rays graze the second surface of the plate XY and do not meet the plate UV. The rays reaching UV are those which, in the medium in the cell, are sufficiently inclined to XY to cross the film within the distance between the edges of the cell. If the film be of appreciable thickness, the light crossing the whole film in any direction will diminish with the angle which the corresponding rays in the film make with XY, for a larger and larger proportion of the rays will fall upon the cement between the plates. The result, as seen at P, will be that the illumination will fall off *gradually*, reaching zero a little before the ideal boundary. The remedy is to use a very thin film.

Usually the cell contains air so that $\mu_2 = 1$, and then

$$\operatorname{cosec} \alpha = \mu, \quad \ldots\ldots\ldots\ldots\ldots\ldots(2)$$

where μ is the refractive index of the liquid in the tank.

By filling the cell with water and the tank with a liquid of higher refractive index, such as turpentine, the index of the liquid can be compared with that of water. To allow the introduction of the liquid into the cell, the top edge must be left open.

36. Experimental details. The apparatus is shown in Fig. 21. Two index points are provided for determining the angles turned through by the divided head. (§ 57.)

The pin is adjusted so that, when the cell is set to bring the boundary between brightness and darkness to the pin, there is no parallax between the pin and the boundary. The pin is then in the focal plane of the lens. No plane mirror, collimator or other focusing device is needed.

The cell may be made of two pieces of "patent" plate glass, kept apart by a narrow band of tinfoil near their edges and held together by marine glue or shellac varnish applied to the edges. The boundary will not be sharp unless the plates be of good quality. The plate forming the end of the tank nearer the lens should also be of good quality.

Dark bands may be seen crossing the bright part of the field near the boundary. They are known as Herschel's fringes and are due to interference.

Since the index of refraction of water for red rays is less than for blue rays, the critical angle for red rays is greater than for

Fig. 21

blue rays. We suppose that the cell is adjusted so that when blue light is used P lies on the boundary between brightness and darkness. Then, if red light be substituted for blue, and the cell be in the position shown in Fig. 20, the boundary will no longer be at P but at Q. Thus, if an ordinary white flame be used instead of a sodium flame, there will be a transition band PQ. At P there will be full illumination and at Q there will be darkness; as we pass from P to Q the light becomes more and more nearly pure red.

One face of the cell is marked. This face is turned towards the flame, the cell is adjusted to each of the critical positions and the scale readings of each pointer are taken. The mean of the differences of readings is 2α. The marked face is then turned towards the lens; the readings furnish a second value for 2α.

Since the refractive index depends somewhat on the temperature of the liquid, this should be noted.

37. Practical example. Temperature of water = 18° C. approx. The following readings of the two index points were obtained, the first reading in each case corresponding to darkness on the left of the pin.

Marked face of cell facing flame		Marked face of cell facing lens	
Point A	Point B	Point A	Point B
261°·1	80°·8	81°·2	261°·1
164 ·0	344 ·0	344 ·2	163 ·8
97 ·1	96 ·8	97 ·0	97 ·3

Mean value of $a = \frac{1}{2}(97°\cdot05) = 48° 31'\cdot5$.
Refractive index of water $= \mu = \operatorname{cosec} 48° 31'\cdot5 = 1\cdot335$.

EXPERIMENT 3. **Determination of refractive index of glass block by microscope.**

38. Method. The experiment is made by aid of a microscope carried on a sliding carriage (Fig. 22). The base board carries a steel rod and a steel millimetre scale, the rod and the scale being parallel to each other. The carriage, of cast iron, has

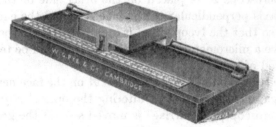

FIG. 22

three feet on its lower side. In two of these a V-groove is cut and these two V's rest upon the steel rod, the remaining foot resting upon the scale. The carriage has a pointer by which the scale readings are taken.

By a bolt passing through a hole in the carriage, any object can be secured to the carriage. For most purposes it is convenient to fit a vertical steel rod to the carriage and to fix any object to the rod by a suitable clamp.

The microscope (Fig. 23) with an objective of 5 to 15 cm.

FIG. 23

focal length, has its axis parallel to the direction of motion of
the carriage.

Let AP (Fig. 24) represent a glass block with parallel faces. A
few grains of lycopodium are placed on one face of the block, one
grain being at P. Let PA be normal to the face AR. A ray PR,
after refraction at R, proceeds along RT as if it had come

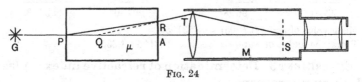

FIG. 24

from Q. If μ be the index of the block, $\sin RQA = \mu \sin RPA$,
or $RA/RQ = \mu \cdot RA/RP$. When RA/PA is very small, we may
replace RP, RQ by AP, AQ, and then

$$\mu = AP/AQ. \qquad \qquad (1)$$

The microscope M is placed on the other side of the block,
with its axis perpendicular to the face AR, and the carriage is
adjusted so that the lycopodium is seen in focus. If the microscope have a micrometer scale S, the focusing can be tested by
the method of parallax.

A little lycopodium is now placed at A on the face nearer the
microscope and then, without altering the optical adjustment
of the microscope, the carriage is moved so that the grains are
again focused. The distance through which the carriage has
been moved is equal to AQ. The thickness AP is measured by
calipers. The refractive index of the glass relative to air is
given by (1).

The lycopodium may be illuminated by white light, or, better,
by a sodium flame placed at G. Since, however, the variation of
the refractive index over the brightest part of the spectrum is
not more than two per cent., it will be difficult to detect any
difference between the settings of the microscope required for
sodium light and for white light.

39. Practical example.   In an experiment by Mr W. Burton upon
a block of glass, the readings of the microscope gave for AQ, 3·03, 3·02, 3·05,
3·03, 3·07 cm. Mean value of $AQ = 3 \cdot 040$ cm.

Thickness of glass measured by sliding calipers $= AP = 4 \cdot 56$ cm. By (1),
refractive index $= \mu = AP/AQ = 1 \cdot 500$.

EXPERIMENT 4. **Determination of refractive index of a liquid by microscope.**

40. Geometrical image of a point by refraction at any number of parallel plane surfaces. We consider four media separated by the three surfaces AK, BL, CM (Fig. 24 A).

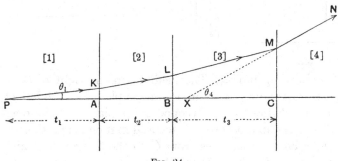

FIG. 24 A

Let $PKLMN$ be the ray, and let MN in the medium [4] cut the axis at X, the distance CX being counted positive when X lies on the same side of C as P. Let θ_1, θ_2, θ_3, θ_4 be the small angles between the common normal PC and successive parts of the ray, and let $PA = t_1$, $AB = t_2$, $BC = t_3$. Then, using the angles instead of their sines, $\mu_1\theta_1 = \mu_2\theta_2 = \mu_3\theta_3 = \mu_4\theta_4$. Let $AK = h_1$, $BL = h_2$, $CM = h_3$. Then

$$h_1 = t_1\theta_1, \qquad h_2 = h_1 + t_2\theta_2, \qquad h_3 = h_2 + t_3\theta_3,$$

and thus

$$h_3 = t_1\theta_1 + t_2\theta_2 + t_3\theta_3 = \mu_1\theta_1\{t_1/\mu_1 + t_2/\mu_2 + t_3/\mu_3\}.$$

But $CX = h_3/\theta_4 = h_3\mu_4/\mu_1\theta_1$, and thus

$$CX = \mu_4\{t_1/\mu_1 + t_2/\mu_2 + t_3/\mu_3\}. \qquad \dots\dots\dots\dots(1)$$

This equation gives the position of the geometrical image of P.

41. Method. A microscope moving along a *horizontal* scale (§ 38) is used. The liquid is contained in a tank with ends formed of parallel plates of glass of good quality (Fig. 25).

The distance, D, between the objective of the microscope and an object, when the instrument is focused on it, must be greater than the length of the tank. If necessary, D can be increased by attaching a diverging lens to the objective.

The thicknesses PA, BC (Fig. 25) of the plates are t_1 and t_3, and the distance AB between them is t_2. The plates PA, BC

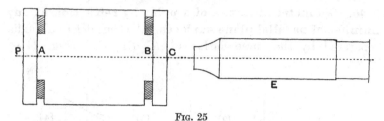

FIG. 25

have indices μ_1 and μ_3. The medium to the right of C is air and has unit index. A sodium flame is placed to the left of P.

The tank being empty, lycopodium is placed on the faces A and B and the microscope E is focused first on A and then on B. Let w, x be the distances from C to the images of A and B. Then, putting $t_1 = 0$ in (1) for the case of A, and $t_1 = 0$, $t_2 = 0$ for the case of B, and noting that $\mu_4 = 1$ and $\mu_2 = 1$, we find that

$$w = t_2 + t_3/\mu_3, \qquad x = t_3/\mu_3.$$

Hence $\qquad\qquad\qquad t_2 = w - x.$ ....(2)

Thus the difference between the scale readings of the microscope when focused on A and on B is equal to t_2.

The lycopodium is now removed from the faces A, B and some is placed on the faces P, C. The microscope is focused first on the image of P and then on C; the tank remains empty. If y be the distance from C to the image of P, we have by (1), since $\mu_2 = 1$,

$$y = t_1/\mu_1 + t_2 + t_3/\mu_3. \qquad ....(3)$$

The tank is now filled with liquid, and the microscope is again focused first on P and then on C. If z be the distance from C to the image of P, we have by (1), since μ_2 is now equal to μ, the refractive index of the liquid,

$$z = t_1/\mu_1 + t_2/\mu + t_3/\mu_3. \qquad ....(4)$$

Subtracting (4) from (3), we have $t_2 (1 - 1/\mu) = y - z$, and hence

$$\mu = \frac{t_2}{t_2 - (y - z)}. \qquad ....(5)$$

42. Practical example. When the tank was empty, the mean bench readings of the microscope, when focused (1) on A, (2) on B, were 17·390, 12·504 cm. Hence $t_2 = w - x = 4·886$ cm. The readings for P and C gave $y = 5·108$ cm.

When the tank contained water, the readings for P and C gave $z = 3·890$ cm.

Hence
$$\mu = \frac{t_2}{t_2 - (y - z)} = \frac{4·886}{4·886 - 1·218} = \frac{4·886}{3·668} = 1·332.$$

EXPERIMENT 5. **Determination of refractive index of liquid by concave mirror.**

43. Method. Let O (Fig. 26) be the centre of curvature of the surface AQ of a concave mirror; let A be vertically below O and let the radius $AO = r$. Liquid of refractive index μ is placed on the mirror so that the surface cuts the vertical OA in C. Let $AC = t$. The pool of liquid must be large enough to ensure that, over a considerable central area, the surface is not distorted by the capillary action at the edge. If the radius of curvature of

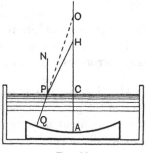

FIG. 26

the mirror be large, it is convenient to place the mirror in a shallow tank, so that' it is *completely* immersed in the liquid. Let H be a point on OC such that a ray HP on striking the surface at P is refracted so as to meet the mirror normally at Q. Then this ray will retrace its path after reflexion at Q. The angles which the rays HP, PQ make with PN, the normal at P, are equal to PHC, POC respectively. Hence, $\sin PHC = \mu \sin POC$, or $PC/PH = \mu . PC/PO$, or $\mu = PO/PH$. If we use only those rays which make small angles with OA, we may put CO, CH for PO, PH, and then

$$\mu = CO/CH. \qquad \qquad \dots\dots\dots\dots\dots\dots(1)$$

The mirror should rest on a *firm* support. A needle, held in a suitable clip, is adjusted so that the tip coincides with its own image, i.e. is self-conjugate. The tip is now at O, the centre of curvature of AQ. The distance OA is measured; the device shown in Fig. 129 may be used. The liquid is then placed on the mirror and the needle is adjusted so that its tip is again self-conjugate.

The tip is now at H. In order that the surface may be un-ruffled, the air should be free from draughts. The distance $HA = h$ is found by direct measurement. The distance CA or t is the depth of the liquid; this may be found by a sphero-meter, the tip of the screw being adjusted to touch first C and then A. An alternative method is to place a cycle ball in the mirror and to adjust the amount of liquid so that the plane of the surface touches the ball at its highest point. If the mirror be immersed in the liquid in a tank, the ball may be run to one side of the tank, clear of the mirror though still *immersed*, when the optical observations are made. Then

$$\mu = CO/CH = (r - t)/(h - t). \quad\quad\dots\dots\dots(2)$$

A mirror of speculum metal or of glass silvered on the front is suitable for the experiment. Care should be taken not to scratch the mirror with the measuring rod.

44. Glass mirror silvered at the back. If the upper sur-face of the glass and the silvered surface have a common centre, the experiment may be carried out just as in § 43, since the ray which falls normally on one surface falls normally on the other also.

If the surfaces be not concentric, it will be easy to find at least three self-conjugate points when the mirror contains no liquid. One image (the brightest) is due to reflexion at the silvered surface, one to reflexion at the front surface, and one to two reflexions at the silvered surface and one internal reflexion at the front surface. Other images may be found, if the needle be brightly illuminated. Only the brightest image is considered below.

Let O (Fig. 27) be the centre of the front surface AQ, and

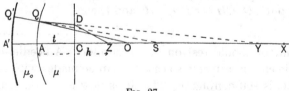

FIG. 27

S the centre of the silvered surface $A'Q'$, and let $OA = r$. Let the index of the glass be μ_0. First suppose that the whole space to the right of AQ is filled with a medium of index μ. Let a ray $Q'Q$, directed to S, be refracted at Q along QDY, where $AY = y$. Then $\mu_0 \sin OQS = \mu \sin OQY$, or, for small angles, $\mu_0 (QOA - QSA) = \mu (QOA - QYA)$. Since $QOA = QA/r$, ... we have

$$\mu_0 (1/r - 1/SA) = \mu (1/r - 1/y). \qquad (3)$$

If the medium μ extend only as far as the plane CD, where $AC = t$, the ray QD, which is directed to Y while in this medium, will be refracted at D into air and will cut the axis in Z, where $AZ = h$. Then $CY = \mu CZ$, or $y - t = \mu (h - t)$. Hence (3) becomes

$$\mu_0 \left(\frac{1}{r} - \frac{1}{SA} \right) = \mu \left(\frac{1}{r} - \frac{1}{\mu h - (\mu - 1) t} \right). \qquad (4)$$

The point Z is that point which is self-conjugate when the mirror contains liquid.

If the whole medium to the right of AQ be air, and if X be now self-conjugate, where $AX = x$, then, putting $\mu = 1$, $h = x$, in (4), we have

$$\mu_0 (1/r - 1/SA) = 1/r - 1/x. \qquad (5)$$

By (4) and (5),

$$\mu = 1 + \frac{r}{h - (\mu - 1) t/\mu} - \frac{r}{x}, \qquad (6)$$

a quadratic equation for μ. When t is small, a method of approximation is convenient. As a first approximation, $(\mu - 1) t/\mu$ is neglected in comparison with h, giving $\mu = 1 + r (1/h - 1/x)$. This value of μ is used in the small term $(\mu - 1) t/\mu$, and then μ is found by (6).

Since O is not, in general, self-conjugate, r differs from x and must be found by a spherometer.

45. Practical example. The refractive index of water was measured by G. F. C. Searle.

Mirror of speculum metal. The temperature of water 21·7° C. The depth of the water was adjusted by aid of a ball $\frac{1}{4}$ inch or ·64 cm. in diameter. Hence $t = ·64$ cm.

Radius $= r$, 107·16, 107·06, 107·16. Mean 107·13 cm.

Distance $HA = h$, 80·56, 80·62, 80·52. Mean 80·57 cm.

Hence, by (2), $\mu = (r - t)/(h - t) = 106·49/79·93 = 1·332$.

Mirror of glass silvered at the back. The temperature of water 25° C.

Distance $XA = x$, 19·50, 19·51, 19·52. Mean 19·510 cm.

Radius $OA = r = 20·491$ cm., measured by spherometer.

Depth of water $= t = ·255$ cm.

Distance $ZA = h$, 14·83, 14·84, 14·83. Mean 14·833 cm.

The first approximation, $\mu = 1 + r (1/h - 1/x)$, gives $\mu = 1·331$ and $(\mu - 1) t/\mu = ·063$ cm. Then, by (6),

$$\mu = 1 + 20·491 \left(\frac{1}{14·833 - ·063} - \frac{1}{19·510} \right) = 1·337.$$

CHAPTER IV

THE SPECTROMETER

46. Introduction. The optical elements of a spectrometer are shown in Fig. 28. Here AB represents a collimator. A slit S,

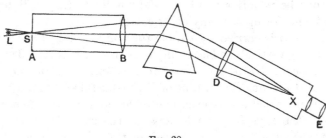

FIG. 28

perpendicular to the plane of the diagram, is fixed at one end of a tube and a lens B is fixed at the other end. The slit S is illuminated by a monochromatic source of light L. The rays which proceed from any given *point* of the slit form a parallel beam after passing through B, if the slit be in the focal plane of B. When this is the case, the collimator is said to be "set for infinity." The prism C is so placed that its edge is approximately perpendicular to the plane of the diagram, which is therefore nearly a principal plane of the prism (§ 26).

Since the collimator is set for infinity, the rays due to the given *point* of the slit strike the prism as a parallel beam and form a parallel beam both in the prism and on emergence from it. The emergent rays are received by the object glass D of a telescope DE and are brought to a focus at X in the focal plane of D. Thus X is the image of the *point* in S from which the rays proceeded. To serve as an index mark, a pair of cross-wires or a pointer is provided. When the cross-wires are in the focal plane of D, the telescope is said to be set for infinity. The cross-wires, as well as the image X, are viewed by aid of the eye-piece E, and the two are seen in focus at the same time when the telescope is set for infinity and the eye-piece is adjusted to the observer's eyesight.

The straight line from the point of intersection of the cross-wires to the corresponding nodal point (§ 191) of the objective, when the telescope is set for infinity, is called the *line of collimation* of the telescope. All the rays which come to a focus at the point of intersection are parallel to the line of collimation before striking the lens D.

The lenses B and D are achromatic; if the collimator and telescope be in adjustment for sodium light, they will be in adjustment for light of any other colour.

The prism is carried by a table turning about an axis, which is generally vertical, and is usually provided with a divided circle. The telescope DE is carried by an arm turning about an axis intended to coincide with that of the prism table, and a divided circle is provided for determining its angular displacements.

Since the rays from each *point* of the slit form a parallel beam when they strike the lens D of the telescope, they come to a focus in the focal plane of D. Thus the whole slit will be seen in focus and not merely a single point of it, and this is true whatever the direction of the edge of the prism relative to the plane of the diagram.

We suppose that the prism is adjusted so that the axis about which the telescope turns is perpendicular to a principal plane of the prism. There is one point of the slit (supposed infinitely narrow) such that rays from it, after passing through the lens B, are parallel to the principal plane. The rays from any other point of the slit are *not* parallel to the principal plane after passing through B, and their deviation cannot be calculated by § 26; for these rays we employ § 30.

47. Curvature of image of slit. In § 30 we determined the deviation, D_0, of the *projection* of a ray upon a principal plane of a prism, when the incident ray is inclined at a *small* angle α to that plane. If D' be the deviation of a second ray such that before incidence it coincides with the corresponding part of the *projection* of the first ray and makes the angles θ and θ_0' with the normals on incidence and on emergence,

$$D_0 = D' + \epsilon = D' + \frac{\alpha^2(\mu^2-1)\sin i}{2\cos\theta_0'(\mu^2-\sin^2\theta)^{\frac{1}{2}}}. \quad \ldots\ldots(1)$$

Let Fig. 29 represent the focal plane of the telescope and let O be the point corresponding to rays parallel to a principal plane. Let these rays be due to sodium light.

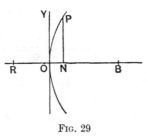

FIG. 29

If we use blue light, the deviation of those rays which are parallel to a principal plane will be greater and the corresponding point will be at B. If red light were used, the point would be at R. Let P correspond to those rays of sodium light, which before and after (§ 30) passing through the prism are inclined at a small angle α to a principal plane. Then, if f be the focal length of the objective, $PN = f.\alpha$.

The distance ON corresponds to the change of deviation of the *projection* of the emergent rays due to the inclination, α, of the incident rays to the principal plane and thus $ON = f.\epsilon$.

Since ϵ is positive, N lies on the side of O towards B, or the image is convex towards the red part of the spectrum. If ρ be the radius of curvature of OP, $2\rho.ON = PN^2$, and thus

$$\rho = \frac{PN^2}{2ON} = \frac{f\alpha^2}{2\epsilon} = \frac{f \cos \theta_0' (\mu^2 - \sin^2\theta)^{\frac{1}{2}}}{(\mu^2 - 1) \sin i} \quad \ldots\ldots\ldots(2)$$

For the position of minimum deviation, $\sin \theta_0' = \sin \theta = \mu \sin \frac{1}{2}i$.

Then
$$\rho = \frac{f\mu (1 - \mu^2 \sin^2 \frac{1}{2}i)^{\frac{1}{2}}}{2 (\mu^2 - 1) \sin \frac{1}{2}i}. \quad \ldots\ldots\ldots\ldots(3)$$

When the prism has $\mu = \frac{3}{2}$ and a refracting angle of $60°$,

$$\sin \theta_0' = \sin \theta = \tfrac{3}{2} \sin 30° = \tfrac{3}{4}, \quad \theta_0' = \theta = 48° \; 35', \quad \rho = 0.794 f.$$

Thus, the radius of curvature of the image of the slit is equal to about four-fifths of the focal length of the telescope objective.

When the edge of the prism and the slit are parallel to the axis of the spectrometer (§ 58), the curvature of the image leads to no error, if the measurements be made on the point O where the tangent is parallel to the axis, for O corresponds to rays which pass through the prism parallel to a principal plane.

48. Focal lines formed by a prism*. Let KLM (Fig. 30) be the section of a prism by a principal plane, supposed hori-

---

* The general theory of focal lines is given in Chapter XI.

zontal, and let O, in this plane, be a real or virtual source of monochromatic rays. Let $OABC$, a ray in the principal plane,

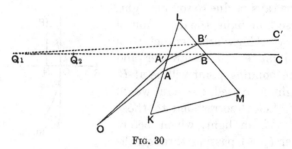

FIG. 30

be the axis of a narrow pencil of rays, and let $OA = u$. Let A be so close to the edge L that AB/u is very small. Let θ, ϕ be the angles between the normal at A and the rays OA, AB, while ϕ', θ' are those between the normal at B and the rays AB, BC. The angle θ is positive when OA lies between KA and the outward normal at A, and θ' is positive when BC lies between MB and the outward normal at B. We call θ the angle of incidence and θ' the angle of emergence.

In the plane of the figure take a ray $OA'B'C'$, with angle of incidence $\theta + d\theta$. Then

$$\mu \cos \phi \, d\phi = \cos \theta \, d\theta, \quad d\phi + d\phi' = 0, \quad \mu \cos \phi' \, d\phi' = \cos \theta' \, d\theta',$$

and $\qquad d\theta/d\theta' = - \cos \theta' \cos \phi/(\cos \theta \cos \phi').$ ....(4)

Hence θ' diminishes as θ increases, and thus $B'C'$, BC meet in a point Q_1, which is on the same side of the prism as O. The primary focal line, which is perpendicular to the principal plane, passes through Q_1. Let the distance of Q_1 from the prism be v_1. Since AB is very close to L, the perpendiculars from B', A' on AB may be taken as equal. Hence $BB' \cos \phi' = AA' \cos \phi$. But $BB' \cos \theta' = - v_1 d\theta'$, $AA' \cos \theta = u d\theta$, and thus

$$- v_1 d\theta' = u d\theta . \cos \theta' \cos \phi/(\cos \theta \cos \phi').$$ ....(5)

By (4), $\qquad v_1 = u \cos^2 \theta' \cos^2 \phi/(\cos^2 \theta \cos^2 \phi') = nu,$ ....(6)

where $\qquad n = \cos^2 \theta' \cos^2 \phi/(\cos^2 \theta \cos^2 \phi').$ ....(7)

If we take a ray OA'', where $A''A$ is perpendicular to the principal plane and $A''OA = d\psi$, the emergent ray $B''C''$ will, by § 30, make an angle $d\psi$ with the principal plane. Since $d\psi$ is infinitesimal, $A''B''$, $B''C''$ have AB, BC for their projections. Hence $B''C''$, BC meet in Q_2, on the same side of the prism as O. The secondary focal line, which lies in the principal plane, passes through Q_2. Let the distance of Q_2 from the prism be v_2. Then $AA'' = u d\psi$, $BB'' = v_2 d\psi$. Since AB is very near L, we put $AA'' = BB''$, and then

$$v_2 = u. \dots\dots\dots\dots\dots\dots\dots\dots\dots\dots(8)$$

Thus, the distance of the secondary line Q_2 from the prism is equal to that of the source O for *all* values of θ. The distance of the primary line Q_1 depends, however, upon θ. When the source O is real, the focal lines are virtual and *vice versa*.

For a given deviation, there are two positions of the prism. In position I the angle of incidence (θ) at A is greater than the angle θ_0 corresponding to minimum deviation, while in position II θ is less than θ_0. Since ϕ increases with θ, it follows that in I $\phi > \phi'$ and that in II $\phi < \phi'$. Substituting for θ and θ' their values in terms of ϕ and ϕ', we have, by (7),

$$n = \frac{v_1}{u} = \frac{(1 - \mu^2 \sin^2\phi')(1 - \sin^2\phi)}{(1 - \mu^2 \sin^2\phi)(1 - \sin^2\phi')} = \frac{C - (\mu^2 - 1)\sin^2\phi'}{C - (\mu^2 - 1)\sin^2\phi},$$

where $C = 1 - \sin^2\phi - \sin^2\phi' + \mu^2 \sin^2\phi \sin^2\phi'$.

When the position is changed from I to II, ϕ and ϕ' interchange values and thus C has the same value in both positions. Hence in I, when $\phi > \phi'$, we have $v_1 > u$, while in II, when $\phi < \phi'$, we have $v_1 < u$. Further, the value of v_1/u or of n in II is the reciprocal of the value of v_1/u in I.

In the position of minimum deviation, $\phi' = \phi$, and $v_1 = u$. In this case, provided AL be infinitesimal, the two focal lines are at the same distance from the prism, and hence the rays of the emergent pencil pass actually or virtually through a single point at a distance u from the prism.

When u is infinite, v_1 and v_2 are both infinite, and thus, for all values of θ, an incident pencil of parallel rays gives rise to an emergent pencil of parallel rays.

The table below gives the values of n, or v_1/u for various values of θ; the deviation D is also given. The angle of the prism is $60°$ and the index is $1\cdot5$. The angle of incidence, θ_0, corresponding to minimum deviation, is $48° 35'$.

θ	θ'	D	n
48° 35′	48° 35′	37° 10′	1·00
58 35	39 56	38 31	1·79
68 35	33 34	42 9	3·70
78 35	29 34	48 9	12·4
88 35	27 56	56 31	786
90 0	27 54	57 54	∞

49. Application to spectrometer. When the collimator is not "set for infinity," the rays from a *point* of the slit do not emerge from the collimator as a parallel beam but diverge from or converge to a point. When the prism is in the position of minimum deviation, the rays still pass through a single point after their passage through the prism and so are brought to a focus at a *point* in the telescope. When the collimator is not set for infinity and the prism is not in the position of minimum deviation, the rays from a point of the slit, after passing through the prism, pass through two focal lines in the positions given by (6), (8), and either of these can be focused on the cross-wires of the telescope by adjusting the distance between the object glass and the wires.

If the telescope be focused on the vertical focal line, the focal lines of successive points of the slit will overlap, and a bright and narrow image will result. If, however, the telescope be focused on the horizontal line, the focal lines of successive points of the slit will not overlap but will be parallel, and a broad illuminated patch will be seen. Thus, when the image of the slit is seen as a narrow vertical line, the telescope is focused on the vertical focal lines.

50. Focusing of telescope and collimator. It is important to put both telescope and collimator into correct adjustment, i.e. to bring (1) the cross-wires and (2) the slit into the focal plane of the corresponding object glass.

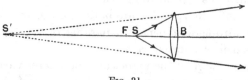

FIG. 31

Let FB or f (Fig. 31) be the focal length of the lens B, let the slit be at S, and let $BS = f - x$. Then if S', the image of S, be at a distance v from B, we have $v = f(f-x)/x$. (See § 125.) When x is small, we may put $v = f^2/x$.

If the diameter of B be h cm. and the angle between the extreme rays be θ radians, $\theta = h/v = hx/f^2$. The uncertainties in the angular measurements will be of the same order of magnitude as θ. If $f = 20$ cm., $h = 2$ cm., then $\theta = 17\cdot2x$ minutes. If θ is to be less than one minute, x must be less than 0·058 cm. When $x = 0\cdot058$ cm., $v = f^2/x = 6896$ cm.

To adjust the eye-piece so that the cross-wires are seen without eye-strain, the observer selects an object at such a distance that he can view it comfortably without strain. He then views the cross-wires through the eye-piece with one eye and the object with the other and adjusts the eye-piece till the cross-wires are in focus.

The simplest method of focusing the telescope is to turn it towards a *very* distant well-defined object, and to adjust the distance between the cross-wires and the objective so that, on testing for parallax by moving the eye from side to side, the wires appear stationary relative to the object. The window must be open, since window glass is generally irregular. The numerical example just given shows that, if the available distance be less than 100 metres, the method is useless except for rough measurements. We then employ other methods as in §§ 51, 52.

When the telescope has been adjusted, the slit of the collimator is illuminated and the telescope is turned to view the slit through the objective of the collimator. The distance of the slit from the lens is then adjusted so that the image of the slit is focused on the cross-wires of the telescope without parallax. The slit should be vertical or, more precisely, should be in a plane passing through the axis of the spectrometer. To test this adjustment, the slit is viewed (1) directly and (2) by reflexion at a face of the prism after it has been levelled (§ 54). If in each case the image of the slit have the same direction as the nominally vertical wire, both the wire and the slit are correctly placed. The plane glass plate of § 53 may be used in place of the prism.

In some spectrometers, an "infinity mark" is engraved on the sliding tube of the collimator, and, if the tube be adjusted to this mark, the collimator is set for infinity. The telescope is then set for infinity by focusing on the slit through the lens of the collimator.

When the telescope and collimator have been put into adjustment, the adjustments should not be disturbed. If a second observer, having "sight" differing from that of the first, wish to see the image of the slit distinctly, he can do so by re-adjusting the *eye-piece*. He will then also see the cross-wires clearly.

51. Auto-collimation method. If the eye-piece allow the cross-wires to be illuminated, or if the sloped plate of § 53 be used, the telescope can be focused "for infinity" by aid of a *plane* reflecting surface. A plate of glass silvered at the back is attached by wax to the spectrometer table, which is adjusted by its levelling screws so that the observer can see the image of the cross-wires formed by reflexion; the distance of the cross-wires from the objective is varied till the absence of parallax shows that the image of the wires is in the same plane as the wires themselves.

The telescope is now directed to the collimator, and the latter is adjusted so that the image of the slit is focused without parallax on the cross-wires of the telescope.

A plate of unsilvered glass, or one face of the prism may be used; the image will, of course, be much less brilliant.

The surfaces of the plate should be *accurately* plane. One way of testing the plate is to focus the telescope by means of the plate and then to turn the telescope to view a *very* distant object. If any change of focus be required, the plate is un-satisfactory.

52. Schuster's method. The prism is placed on the table of the spectrometer with its edge approximately perpendicular to the plane of the divided circle, and the slit, illuminated by monochromatic light, is viewed by the telescope through the prism.

If the collimator be not in correct adjustment, the image which its lens forms of a point of the slit will be at a finite distance u from the prism. For a given position of the telescope, within a certain range, there are *two* positions of the prism for which the slit can be viewed. In position I, $\theta > \theta_0$; in position II, $\theta < \theta_0$ (§ 48). In each position, the rays on emergence from the prism pass through two focal lines, but, by § 49, only the vertical focal line can be focused in the telescope. This line is at a distance v_1 from the prism. In each case, by § 48, $v_1 = nu$, where n depends upon θ. If n_1 be the value of n for position I, n_1 is greater than unity, and the value of n for position II is $1/n_1$. This result leads to Schuster's method. The prism is turned to position I and the telescope is focused on the slit. The vertical focal line viewed by the telescope is at the distance $n_1 u$ from the prism. The prism is now turned to position II; the slit will then be out of focus, since the rays have a vertical focal line at a distance u/n_1 from the prism, while the telescope is in adjustment for seeing a line at a distance $n_1 u$. The collimator is therefore adjusted so that, without changing the adjustment of the telescope, the slit is focused on the cross-wires. If the new distance from the prism to the image of the slit formed by the lens of the collimator be w, the distance from the prism of the vertical focal line is w/n_1. But this is equal to $n_1 u$, since the telescope was set to view a focal line at the distance $n_1 u$. Hence $w = n_1^2 u$.

If the prism be turned back to I and the telescope be refocused, it will be adjusted for seeing a focal line at distance $n_1^3 u$.

One cycle of operations is now complete. The distance of the image of the slit formed by the lens of the collimator has been increased from u to $n_1^2 u$, with a corresponding improvement in the adjustment of the telescope. For a prism of 60° and index 1·5, the Table of § 48 shows that, when $\theta_1 = \theta_0 + 30°$, $n_1 = 12·4$ and $n_1^2 = 154$. Thus, if we start with $u = 1$ metre, a single cycle leaves us with $u = 154$ metres. A second cycle leaves us with $u = 154^2 = 23716$ metres. The progress towards perfect adjustment of both collimator and telescope is very rapid.

53. Levelling the telescope. This operation is not a "student's adjustment" and should be done only by a responsible person. Those who blindly *assume* that the line of collimation is parallel to the axis of the tube of the telescope will use a spirit level. The level is placed on the divided circle, which can be made horizontal if there be levelling screws at the base of the instrument. The level is then placed on the telescope and the tube is levelled. *If* the assumption be correct, the line of collimation is parallel to the plane of the circle.

In accurate work, a glass plate with parallel faces is attached to the table of the spectrometer so that the faces are approximately vertical. The telescope is directed towards the plate, and the image of X, the intersection of the cross-wires, is made to coincide with X by adjusting the levelling screws of the table. The telescope is now turned about the axis of the spectrometer through 180°. If the image of X be now higher or lower than X, the difference is halved, as well as can be judged by eye, by adjusting the telescope. The table is then adjusted to make the image of X again coincide with X. If, on turning the telescope back to the first position, the image of X coincide with X, the line of collimation is perpendicular to the axis of the spectrometer. Any residual error is corrected by repeating the operations.

The necessary illumination may be obtained by using a sloped glass plate, as in EXPERIMENT 19. The eye-piece is removed, and the plate is placed between the cross-wires and a suitable converging lens. The observer views the wires and their images by this lens. If the sloped plate be adjusted so that the lens of the telescope is filled with light, the wires and their images will stand out dark against a bright field.

54. Levelling the prism. Time will be saved if the prism be properly placed relative to K, L, M (Fig. 32), the levelling screws of the table. One of the faces, say FG, should be at right angles to the line joining the screws K, L. When both faces of the prism are to be set at right angles to the plane of the divided circle, the face FG is first set perpendicular to that plane. This setting will not be disturbed if the face EF be adjusted by means of the screw M, for the axis KL, about which the prism turns, is perpendicular to the face FG; in fact FG merely turns in its own plane. It may not be possible to set the line FG quite perpendicular to the axis KL and that axis may not be quite parallel to the plane of the divided circle, and thus the motion of M may slightly disturb

FIG. 32

the setting of the face FG. Any residual error may be corrected by one or two repetitions of the operations.

For the purpose of levelling the prism, a fine wire is fixed, but
not by wax, across the slit at Y, and the image of the slit is viewed
by turning the telescope towards the collimator. The image of Y
is made to coincide with X, the point of intersection of the
cross-wires, by adjusting the fine wire. The prism is fixed to
the table by a little wax, and the telescope is turned to view
the slit by reflexion at the face FG which has been set perpendicular to the line joining the two levelling screws K, L.
The rays from Y, on emerging from the collimator, form a
parallel beam which is reflected at the face of the prism; this
reflected beam will be inclined at the same angle to the plane
of the circle as the incident beam provided the face be perpendicular to that plane, but not otherwise. Hence we can
make the face to be perpendicular to the circle by adjusting
the table so that the image of Y coincides with X, when viewed
by reflexion at the face. The second face EF must now be
adjusted. Either the table or the telescope is turned so that
the slit may be viewed by reflexion at the face EF, and this
face is made perpendicular to the circle by turning the third
levelling screw M until the image of Y coincides with X. The
slit is then again viewed by reflexion at the face FG and any
residual error is corrected by repeating the operations.

It will be seen that, for the purpose of levelling the prism,
it is not *necessary* that the line of collimation of the telescope
be parallel to the plane of the divided circle.

55. Position of prism on table. In determining the refracting angle, the slit is viewed with the telescope by reflexion
at the two faces of the prism in turn, but, unless the prism be
properly placed on the table, it may be impossible to see any
image, although the line of collimation of the telescope be parallel
to the rays reflected at the face of the prism. The difficulty arises
from the finite diameter of the lenses of the collimator and
telescope. In Fig. 33, EFG is the prism with a refracting angle F
of $60°$. A parallel beam from the lens C falls on FG and is
reflected, but no reflected rays enter the telescope, although the
line of collimation of the latter be parallel to the reflected rays.

But, if the prism be moved away from the collimator in the direction FO, where O is the centre of the table, the rays will enter the telescope and an image of the slit will be seen. The prism should be placed so that FO is approximately parallel to the incident rays. Only part of each lens is used, and the rays which fall on FG and those that fall on FE pass through different halves of the lens C.

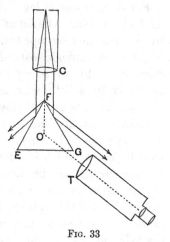

FIG. 33

If F coincide with O, all the light reflected at FG will enter the telescope if the diameter of the lens T be not less than that of the lens C. But only half the light issuing from C enters T.

When the minimum deviation is found, the rays pass through the prism. If F coincide with O, only half the light will be utilised. The amount can be increased by making O lie within the triangle EFG, but it will be at the expense of the light available when the slit is viewed by reflexion, if OF be made too great.

56. Illumination of slit. When a flame is used, it should not be placed so close to the slit as to damage the slit either by heat or by the fumes of salts. Two straight lines drawn from the ends of the horizontal diameter of the collimator lens through the centre of the slit enclose an angle. If the flame be placed within this angle and the sides of the flame touch the two lines, the collimator lens will be filled with light. For smaller distances, no more light will pass through the lens and the additional light which passes through the slit will fall on the blackened walls of the collimator tube. The flame may be placed at a considerable distance from the slit, if a converging lens be used to form an image of the flame upon the slit.

The illumination may be tested by viewing the slit through the lens of the collimator. The flame is adjusted so that, when the eye is moved from side to side over the range determined by the aperture of the lens, the slit appears bright over the whole, or at least the greater part, of that range.

If a narrow vacuum tube be used, it must be placed close to the slit if the lens is to be filled with light. It may be more convenient to use a converging lens to form an image of the source upon the slit.

57. Use of two verniers. In instruments in which an index moves over a divided circle, one of the chief mechanical difficulties is to make the axis about which the index turns coincide with the axis to which the graduation lines on the circle are radial. On a circle of 10 cm. radius, an angle of 15 seconds has an arc of ·000727 cm., and, if the distance between the two axes be as great as $\frac{1}{100}$ mm., the readings will be liable to errors of the order of 15 seconds. The difficulty is overcome by the use of two verniers.

In Fig. 34, let O be the centre from which the graduation lines radiate, C the centre about which the verniers turn, and let $OC = h$. The reading for each vernier is taken from a point on the *edge* of the vernier. Let A, B be the index points of the verniers, and let $CA = a$, $CB = b$. In practice, b is nearly equal to a. For simplicity, we

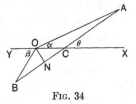

Fig. 34

suppose that B lies on the straight line AC. Draw $YOCX$, and let $ACX = \theta$, $AOX = \alpha$, $BOY = \beta$. If ON be perpendicular to BCA, then $CN = h \cos \theta$, $ON = h \sin \theta$.

Hence $\tan (\theta - \alpha) = \tan NAO = h \sin \theta/(a + h \cos \theta)$,

$\tan (\beta - \theta) = h \sin \theta/(b - h \cos \theta)$.

When h^2/a^2, h^2/b^2 are negligible compared with unity, we write $\theta - \alpha$ for $\tan (\theta - \alpha)$ and $\beta - \theta$ for $\tan (\beta - \theta)$. Then

$$\alpha + \beta - 2\theta = \frac{h \sin \theta \, (a - b + 2h \cos \theta)}{ab - (a - b) \, h \cos \theta - h^2 \cos^2\theta},$$

or, approximately,

$$\alpha + \beta - 2\theta = h \sin \theta \, (a - b + 2h \cos \theta)/ab.$$

If the quantity on the right be negligible for all values of θ, the mean $\frac{1}{2}(\alpha + \beta)$ of the observed angles may be taken as identical with θ, the angle through which the double vernier is turned. In this case, the mean change of reading found by the two verniers for any given displacement may be taken as identical with the angular displacement of the verniers about their centre C.

If $h = \cdot 1$ cm., $a = 10\cdot 1$ cm., $b = 10$ cm., then

$$| \alpha + \beta - 2\theta | < 3 \times 10^{-4} \text{ radian or } < 1' \; 2''.$$

Thus the double vernier secures great accuracy in spite of a large error of centering.

EXPERIMENT 6. **Determination of angle and refractive index of a prism by spectrometer.**

58. Apparatus. Spectrometers are made in many forms, and it must be left to the student to make himself acquainted with the details of the instrument he uses. The instrument shown in Fig. 35 stands on three feet furnished with levelling screws. Attached to the column is a divided circle; a line through the centre of this circle at right angles to its plane is called the axis of the spectrometer.

The telescope, shown on the right, is carried by an arm capable of motion about an axis which, in a *perfect* instrument, coincides with the axis of the spectrometer. The telescope is attached to the arm by a joint which allows the adjustment of the inclina-

FIG. 35

tion of the telescope to the plane of the divided circle. A fitting which can be clamped to the disk of the divided circle, or to the column supporting the disk, is furnished with a "tangent screw" for making a fine adjustment of the direction of the telescope arm in azimuth. This arm is furnished with a vernier moving over the divided circle, or, better, with *two* verniers (§ 57).

The collimator is carried by an arm *fixed* relative to the divided circle; a joint allows the inclination of the collimator to the plane of the circle to be adjusted.

Above the divided circle is a small platform known as the "table" of the spectrometer and this table is capable of motion about an axis, which, in a *perfect* instrument, coincides with the axis of the spectrometer. The platform is provided with three levelling screws by which its inclination to the plane of the divided circle can be varied. The table is furnished with a vernier moving over the divided circle. For accurate work a pair of verniers is used.

59. Determination of angle of prism. The prism, which has been levelled (§ 54), is set with its refracting angle turned towards the collimator, and the illuminated slit is viewed by reflexion at the two faces in turn. In each case X, the intersection of the cross-wires, is made to coincide with a point in the image of the slit. It is not necessary to use a narrow slit if X be made to coincide with a point on *one* edge of the image of the slit; we must, however, keep to the *same* edge of the image throughout. Let P_1Q_1 (Fig. 36) be a ray, parallel to the plane of the divided circle, which falls on the face EF of the prism and is reflected along Q_1R_1. Then, by the law of reflexion, the angles EQ_1R_1, FQ_1P_1 are equal.

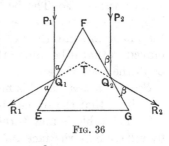

FIG. 36

The collimator forms at infinity an image I of the slit, and J, the image of I by reflexion at the face of the prism, is again at infinity. If both the slit and the face be perpendicular to the plane of the circle, so is also the image J.

Hence the image of the slit formed in the focal plane of the telescope is *straight* and is perpendicular to the plane of the circle. Thus *any* point of the image of the slit may be used as a point of reference for the purposes of measurement.

The angle of the prism may be determined by two methods.

For the second method it is necessary that the table supporting the prism have a vernier moving over the divided circle.

First method. The prism is turned into such a position that an image of the slit can be observed with the telescope by reflexion at the two surfaces EF, FG (Fig. 36), and the prism is then kept *fixed*. Let the projections, on the plane of the circle, of the rays which fall on EF make an angle α with that face, and let β be the corresponding angle for the face FG. Then $EQ_1R_1 = \alpha$ and $GQ_2R_2 = \beta$. Hence, if R_1Q_1, R_2Q_2 intersect in T, and if i be the angle of the prism, $R_1TR_2 = \alpha + \beta + i$. But, since P_2Q_2 is parallel to P_1Q_1, $\alpha + \beta = i$, and thus $R_1TR_2 = 2i$. The telescope is directed to each face in turn, and the intersection of the cross-wires is made to coincide with a point in the image of the slit. If θ be the angle turned through by the telescope, $\theta = R_1TR_2 = 2i$ and thus

$$i = \tfrac{1}{2}\theta. \quad\dots\dots\dots\dots\dots\dots\dots(1)$$

If the magnifying power of the telescope be considerable, a small motion of the telescope causes a large motion of the image relative to the cross-wires, and thus, unless the telescope be moved cautiously, the image may flash across the field of view and escape notice. The best plan is first to view the reflected image by eye, the eye being close to the prism; then, keeping the image in sight, the eye is drawn back, and the telescope is moved so that it lies along the line of sight. The image will then be in the field of view.

Second method. In this method, the telescope is fixed in such a position that the slit can be viewed by reflexion from the face EF. The table is now turned round so that the slit is viewed by reflexion at the face FG. The angle ϕ (*less* than π), through which the prism has been turned, is equal to that between FE and the continuation of GF and thus $\phi = \pi - i$. Hence

$$i = \pi - \phi. \quad\dots\dots\dots\dots\dots\dots\dots(2)$$

60. Determination of refractive index of prism. We suppose that the prism has been levelled and that the collimator and telescope are "set for infinity." The slit, illuminated with the light for which the index of the prism is to be found, is viewed by eye through the prism, and the prism table is turned so that the prism is nearly in the position of minimum deviation.

Keeping the image in view, the eye is drawn back, and the telescope is set along the line of sight. The image will now be in the field of view. The telescope is then adjusted so that on moving the prism the image just reaches X, the intersection of the cross-wires, and then turns back. If the prism be now set so that the image passes through X, the prism is in the position of minimum deviation*. After the readings of the telescope verniers have been taken, the table is turned so that the face of the prism which was towards the telescope is now towards the collimator, and the second position of the prism for minimum deviation is found as well as the corresponding position of the telescope. The telescope verniers are again read. Two or three settings of the prism and of the telescope should be made; the mean of the readings in each of the two positions is taken.

Let the difference between the two readings of the vernier be $2D$. If the edge of the prism be very nearly parallel to the axis of the spectrometer, D will not differ appreciably from Δ, the minimum deviation (§§ 28, 29), and hence we may calculate the index of the prism by the formula

$$\mu = \sin \tfrac{1}{2}\,(D + i)/\sin \tfrac{1}{2} i. \quad\dots\dots\dots\dots\dots(3)$$

To determine D directly, we read the vernier for the zero position of the telescope. If the edge of the prism approximately coincide with the axis of the spectrometer, the prism can be so placed that about half the rays from the collimator pass clear of the prism. These are received by the telescope and the image of the slit is made to coincide with X. The differences between this reading and those for the positions of minimum deviation give two values for D. The mean is identical with half the difference between the readings for minimum deviation.

* For accurate work, either the line of collimation of the telescope must be parallel to the plane of the divided circle, or one of the cross-wires must be perpendicular to that plane. In the first case, the image of the slit is made to pass through the point of intersection of the cross-wires; in the second case, the vertical cross-wire is made to be a tangent to the (curved) image. In either case, the deviation observed corresponds to rays which have passed through the prism parallel to a principal plane, if the edge of the prism be perpendicular to the plane of the circle.

61. Practical example. Observations made by Mr W. R. Harper.

Prism fixed. Readings of telescope verniers A, B:

Reflexion at FE	Reflexion at FG
A, 56° 55′; B, 236° 55′	A, 296° 35′; B, 116° 34′.

$\theta_A = (360° + 56° 55′) - 296° 35′ = 120° 20′$; $\theta_B = 236° 55′ - 116° 34′ = 120° 21′$.

Mean $\theta = 120° 20′ 30″$. Hence $i = \frac{1}{2}\theta = 60° 10′ 15″$.

Telescope fixed. Readings of prism table verniers C, D:

Reflexion at FE	Reflexion at FG
C, 288° 50′; D, 108° 49′	C, 168° 58′; D, 348° 59′.

$\phi_C = 288° 50′ - 168° 58′ = 119° 52′$; $\phi_D = (360° + 108° 49′) - 348° 59′ = 119° 50′$.

Mean $\phi = 119° 51′$. Hence $i = \pi - \phi = 60° 9′$.

Mean angle of prism by the two methods, $i = 60° 9′ 37″·5$.

Readings by verniers A, B of telescope, when focused directly on slit, were A, 0° 0′; B, 180° 0′.

Vernier readings for minimum deviation for sodium light:

Incidence on FE	Incidence on FG
A, 48° 40′; B, 228° 42′	A, 311° 18′; B, 131° 19′
Deviations: 48° 40′, 48° 42′	48° 42′, 48° 41′.

The mean value $\Delta = 48° 41′ 15″$ is, of course, independent of the central readings A, 0° 0′; B, 180° 0′.

Hence
$$\mu = \frac{\sin \frac{1}{2}(\Delta + i)}{\sin \frac{1}{2}i} = \frac{\sin 54° 25′ 26″}{\sin 30° 4′ 49″} = 1·6227.$$

EXPERIMENT 7. **Determination of angle and refractive index of a prism by auto-collimating spectrometer.**

62. Apparatus. In an auto-collimating spectrometer, the telescope and collimator of the ordinary spectrometer are combined in a single optical arrangement.

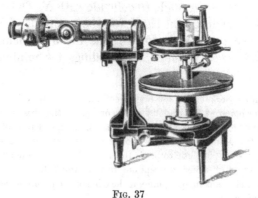

FIG. 37

The eye end of the telescope, Fig. 37, is furnished with an adjustable slit S, which can be illuminated by aid of a reflecting prism R, as in Fig. 38. The slit is observed through the eye-piece E. The plate in which the slit is formed occupies only half the section of the tube, as shown in Fig. 39. Above the slit and in the same plane is fixed a sharp pointer P; in place of P a vertical cross-wire may be used.

When a flame F is placed in the proper position by the side of the spectrometer, rays enter the prism R and are reflected into the telescope T. (See § 56.) It is convenient to use a lens L (Fig. 38) to form an image of the flame upon the prism; the

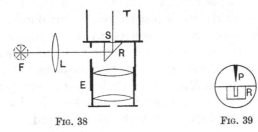

FIG. 38   FIG. 39

distance FR should be about four times the focal length of L. The observer will then not be scorched by the flame. If the slit be in the focal plane of the objective, the rays from any point Q of the slit, on emerging from T, form a parallel beam. If this beam fall *normally* upon a plane reflecting surface, it is reflected along its path and comes to a focus at Q. Those points of the slit which lie below Q have their images above Q, and thus the image of the slit is inverted. Hence, if the point at the top of the slit be self-conjugate, the image of the slit will lie above that point and will be properly placed for comparison with the pointer.

The prism, or grating (EXPERIMENT 65), is carried on a table turning about the axis of the spectrometer. The table has levelling screws; its angular position is read by two verniers. A "tangent screw" is fitted for giving the table a slow motion.

The axis of the telescope is adjusted to be perpendicular to the axis about which the table turns. The adjustment may be tested by the method described in § 53, except that the table is turned round and not the telescope.

63. Determination of angle of prism. The eye-piece is focused on the pointer or cross-wire. This adjustment depends upon the observer's vision and will be different for different observers.

The prism is secured to the prism table by wax. The face LK (Fig. 40) is set perpendicular to the line joining the two levelling screws A, B. A series of lines is ruled on the prism table parallel to AB. If the images of these lines formed by reflexion at the face LK have the same direction as the lines themselves, that face is perpendicular to AB. This face is turned to be approximately normal to the axis of the telescope and is then adjusted by either of the screws A, B so that the lower end of the image of the slit is just above the horizontal edge of the plate in the eye-piece. As soon as any sort of image of the slit is seen, the telescope is focused by the rack and pinion so that a sharp image of the slit is seen in the same plane as the pointer. Then the other face MK is turned to be approximately normal to the axis of the telescope and is adjusted by the third screw C. This adjustment does not disturb the one already made.

If the end of the prism be not perpendicular to the refracting faces, or if much wax be used, the observer may easily find the image due to reflexion at the face LK and yet fail to find the image due to reflexion at MK. The failure arises because the available range of levelling is inadequate to allow the edge of the prism to be made sufficiently nearly parallel to the axis of the prism table. The difficulty can be met by a preliminary adjustment. The prism table is made approximately parallel to the divided circle, and the prism is then adjusted on the table with wax so that, as tested by a set-square, the planes of the faces LK, MK are perpendicular to the table. If this be carefully done, the image of the slit will be easily found.

The face LK is now set normal to the axis OX of the telescope. One edge—say the left—of the image of the slit is selected and is brought up to the pointer by turning the table. Fig. 40 shows

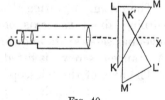

FIG. 40

the prism in the two positions KLM and $K'L'M'$. In each case the face nearer the telescope is normal to OX. Hence, if LK and MK be the intersections of the reflecting faces of the prism with a principal plane, LK and $M'K'$ are parallel. To change the position of the prism from LKM to $L'K'M'$, it must be turned through $\pi - i$ in one direction, or through $\pi + i$ in the opposite direction, where i is the angle of the prism. In each position, the reading of each of the two verniers is recorded. Care must be taken to distinguish vernier I from vernier II.

64. Determination of index of prism. The prism is turned so that the rays from the telescope after refraction at MK fall normally upon LK, as in Fig. 41. The rays are reflected back along their paths, and form an image of the slit coincident with the pointer. The angle between the ray HQ in the glass and HN, the normal to MK, is i. The angle θ between this normal and the ray in air is also the angle through which the

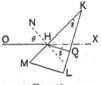

FIG. 41

prism has to be turned from the position just found in order to make HN parallel to XO and so to bring the image by reflexion at MK to the pointer.

The observations are repeated with the face LK towards the telescope. The mean value of θ is used in calculating the refractive index by the equation

$$\mu = \sin \theta / \sin i. \quad\quad\quad\quad\quad (1)$$

Since $\sin \theta$ cannot exceed unity, $\sin i$ must be less than $1/\mu$. If i exceed this critical value, rays refracted at MK in Fig. 41 cannot fall normally on LK and the method fails. For $\mu = 1\cdot6$ the critical value of i is $38° 41'$.

65. Practical example. Vernier readings, by Mr A. Ll. Hughes, for i, the angle of the prism:

	Vernier I	Vernier II
Reflexion at face LK	90° 6′	270° 6′
Reflexion at face MK	236 58	56 56
Differences ...	146 52	213 10

The differences between these angles and 180° give values for i. We thus have $33° 8'$ for I and $33° 10'$ for II. The mean value of i is $33° 9'$.

Vernier readings for θ, the angle of incidence on first face of prism in measurement of μ:

	Vernier I	Vernier II
Reflexion at face LK†	90° 6′	270° 6′
Internal reflexion at MK	26 57	206 56
	63 9	63 10
Reflexion at face MK†	236° 58′	56° 56′
Internal reflexion at LK	300 7	120 6
	63 9	63 10

Readings marked † are the same as those used in finding i. The mean value of θ is 63° 9′ 30″. By (1),

$$\mu = \frac{\sin \theta}{\sin i} = \frac{\sin 63° 9' 30''}{\sin 33° 9' 0''} = 1\cdot6317.$$

EXPERIMENT 8. Spectroscopic study of light from vacuum tube.

66. Introduction*. The relation between the refractive index of a given material and the frequency of the light is important in technical optics, for the design of achromatic lenses depends upon a knowledge of this relation for various kinds of glass.

It is essential to use a source emitting light of a comparatively small number of very definite frequencies. For the present experiment, a vacuum tube containing hydrogen and mercury vapour is suitable. The pressure of the hydrogen is about 025 cm. of mercury, and the pressure of the mercury vapour is its saturation pressure. If the light produced when an electric discharge passes through the tube be observed by aid of a spectrometer, several sharp images of the slit, known as "lines," are seen. Each line corresponds to a definite frequency, and light of intermediate frequencies is absent.

When the index, μ, of a prism has been found for each line and λ, the corresponding wave length in air, has been measured by a grating, we can study the relation between μ and λ.

The frequency corresponding to a given line is more fundamental than the wave length, for the latter depends upon the medium through which the light is passing. Since, however,

* Mr G. Stead allows me to use his account of this experiment.

the quantity directly determined by a grating is the wave
length in air, we use wave lengths rather than frequencies.

67. The vacuum tube. A capillary tube C (Fig. 42) joins
two wide tubes B, B. The electrode fitted to each wide tube is
of stout aluminium wire joined to a platinum wire P. The

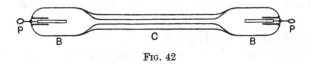

FIG. 42

aluminium wire is covered for about half its length by a narrow
glass guard tube. The platinum wire is sealed into this tube,
which in its turn is sealed into B. The guard tube gives rigidity
to the electrode and prevents discharge from the platinum wire,
which would blacken the tube. If the tube contain 2 or 3 cubic
mm. of mercury, continued use of the tube will not cause the
mercury lines to become faint. The tube is supported with its
axis vertical.

The brightness of the mercury lines can be much increased
by allowing a drop of mercury to rest at the upper end of the
capillary tube, where it is warmed by the discharge. This is
particularly useful when the yellow lines Hg_1 and Hg_2 are being
observed, as it enables them to be distinguished from the
numerous faint hydrogen lines which occur in the orange and
yellow region.

The secondary terminals of an induction coil giving a spark
of 2 cm. in air are joined to the terminals P, P of the tube. The
capillary is placed parallel to the slit of the spectrometer and
about one cm. from it. If the tube *touch* the spectrometer, small
sparks (not dangerous) may pass to the observer's eye.

68. Optical measurements. An ordinary spectrometer
may be used. The telescope and collimator are adjusted (§ 50),
the prism is levelled (§ 54) and its angle i is measured (§ 59).
A sodium flame may be used.

The light from the tube is then examined by the spectrometer.
The principal lines seen are as follows:

	Colour	Element	Symbol	λ	$1/\lambda^2$
				cm.	cm.$^{-2}$
1	Red	Hydrogen	H_a or C	$6\cdot563 \times 10^{-5}$	$2\cdot322 \times 10^8$
2	Yellow	Mercury	Hg_1	$5\cdot791$	$2\cdot982$
3	Yellow	Mercury	Hg_2	$5\cdot770$	$3\cdot003$
4	Green	Mercury	Hg_3	$5\cdot461$	$3\cdot352$
5	Blue	Hydrogen	H_β or F	$4\cdot861$	$4\cdot232$
6	Violet	Mercury	Hg_4	$4\cdot358$	$5\cdot265$
7	Violet, faint	Hydrogen	H_γ	$4\cdot340$	$5\cdot309$

The yellow lines Hg_1, Hg_2 are close together, as are also the violet lines Hg_4 and H_γ, and a narrow slit is necessary for distinct separation.

Theoretically, the prism should be set in the position of minimum deviation for each of the lines in turn, but these positions are so nearly identical that it suffices to use the position of minimum deviation for the green line Hg_3. The cross-wire or pointer of the telescope is now set on each line in turn and the telescope vernier is read. Let these deviations be in the positive direction. The prism table is then turned so that the prism gives minimum deviation in the negative direction for Hg_3, and readings for the lines are again taken. The difference between the readings of the telescope vernier for each line is 2Δ, where Δ is the minimum deviation for that line. Then

$$\mu = \sin \tfrac{1}{2}(\Delta + i)/\sin \tfrac{1}{2}i. \quad\ldots\ldots\ldots\ldots\ldots\ldots(1)$$

The wave lengths may be taken from the table, or may be measured by a diffraction grating.

69. Discussion of results. If Δ be plotted against λ, a calibration curve for the prism is obtained. If Δ for sodium light be measured, the wave length of sodium light can be read on the curve. It is, however, more convenient to plot a curve with λ^{-2} for abscissa and μ for ordinate, since the points will lie very nearly on a straight line, for the range of λ used in the experiment. A straight line drawn (Fig. 43) so as to lie as evenly as possible among the points may be represented by

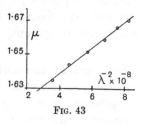

FIG. 43

$$\mu = A + B\lambda^{-2}. \quad\ldots\ldots\ldots\ldots\ldots\ldots\ldots(2)$$

The value of B is found from two points *on this straight line*. If μ_1, μ_2 be the values of μ *given by the line* for λ_1^{-2}, λ_2^{-2},

$$B = (\mu_2 - \mu_1)/(\lambda_2^{-2} - \lambda_1^{-2}). \qquad \ldots\ldots\ldots\ldots\ldots(3)$$

Using this value of B, A is found from

$$A = \mu_1 - B\lambda_1^{-2}. \qquad \ldots\ldots\ldots\ldots\ldots\ldots\ldots(4)$$

If, within any range of λ, a medium have no absorption bands, the curve giving μ in terms of λ^{-2} will be free from sudden changes of curvature, and the constants in Cauchy's formula, $\mu = A + B\lambda^{-2} + C\lambda^{-4} + \ldots$, can be chosen so as to give good agreement between the calculated and observed values of μ over a considerable range. When the range is small, formula (2) is sufficient.

In visual observations, the important part of the spectrum is that between the red hydrogen line H_a or C and the blue hydrogen line H_β or F. The "dispersive power" of the glass is denoted by ω where

$$\omega = (\mu_F - \mu_C)/(\mu_D - 1). \qquad \ldots\ldots\ldots\ldots\ldots(5)$$

Here μ_C, μ_F are the refractive indices for the C and F lines and μ_D is the index for the sodium D line.

70. Practical example. Mr Stead used a flint glass prism of approximately 60° and obtained the following values of μ for the seven lines tabulated in § 68.

H_a or C,	Hg_1,	Hg_2,	Hg_3,	H_β or F,	Hg_4,	H_γ.
1·6306,	1·6366,	1·6371,	1·6405,	1·6489,	1·6598,	1·6604.

When μ was plotted against λ^{-2}, the straight line lying most evenly among the points passed through

$$\lambda_1^{-2} = 2\cdot300 \times 10^8 \text{ cm.}^{-2}, \quad \mu_1 = 1\cdot6300,$$

and through $\qquad \lambda_2^{-2} = 5\cdot350 \times 10^8 \text{ cm.}^{-2}, \quad \mu_2 = 1\cdot6600.$

Hence, by (3), $B = \cdot0300/(3\cdot050 \times 10^8) = 9\cdot836 \times 10^{-11} \text{ cm.}^2$.

By (4), $A = \mu_1 - B\lambda_1^{-2} = 1\cdot6300 - (9\cdot836 \times 10^{-11})(2\cdot300 \times 10^8) = 1\cdot6074$.

Hence the relation between μ and λ^{-2} is

$$\mu = 1\cdot6074 + 9\cdot836 \times 10^{-11} \times \lambda^{-2}. \qquad \ldots\ldots\ldots\ldots\ldots(6)$$

The values of μ, as calculated by (6), for the seven lines, viz.

| 1·6292 | 1·6367 | 1·6369 | 1·6404 | 1·6490 | 1·6592 | 1·6596 |

agree closely with the values found by experiment.

Dispersive power. Since $\lambda_D = 5\cdot893 \times 10^{-5}$ cm., $\lambda_D^{-2} = 2\cdot880 \times 10^8$ cm.$^{-2}$, we find, by (6), $\mu_D = 1\cdot6357$. The F and C lines have $\mu_F = 1\cdot6489$, $\mu_C = 1\cdot6306$, as found by experiment. Hence, by (5),

$$\omega = \cdot0183/\cdot6357 = \cdot0288.$$

CHAPTER V

EXPERIMENTS WITH PRISMS

71. The goniometer. The base of the goniometer (Fig. 44) is formed of a strip of wood furnished at one end with a spherical pivot and at the other with a cross-bar carrying a scale. Angles are measured by means of an arm which turns at one end about the pivot; the other end of the arm moves over the scale on the cross-bar. The optical system consists of an achromatic lens fixed to the arm above the pivot and of a fine vertical wire attached to the other end of the arm and adjusted to be in the focal plane of the lens.

FIG. 44

The spherical pivot is a phosphor-bronze ball attached to the base by a fitting which allows the distance between the centre of the ball and the edge of the scale to be adjusted to a definite value. The ball enters a conical hole turned out of a block of brass attached to the arm. This arrangement destroys three out of the six degrees of freedom of the arm relative to the base. The other end of the arm carries two brass feet which rest upon the cross-bar and thus destroy two degrees of freedom. The remaining degree of freedom allows the arm to turn about an axis through the centre of the ball and perpendicular to the plane of the surface of the cross-bar.

The scale on the cross-bar is divided to millimetres, and the ball is adjusted so that its centre is 40 cm. from the edge of the scale. The readings are taken along the *edge* of the scale by means of a fine wire passing across an opening in the arm and stretched by a spring. The scale is engine-divided on white metal and is provided with an anti-parallax mirror. For small angles, one centimetre along the scale corresponds to $\frac{1}{40}$ radian.

The lens has a focal length of about 35 cm. The vertical wire is held in an adjustable frame, is kept tight by a spring, and is in the focal plane of the lens. The image of a very distant point will then fall upon the wire, if the arm be properly directed. To adjust the frame, a plane mirror is placed so that the lens lies between it and the wire, and the image of the wire formed by the lens and the mirror is made to coincide with the wire itself.

When the instrument is to be used to measure the angle between two beams of parallel light, the arm is moved so as to bring each beam in turn to a focus on the wire; the angle turned through by the arm is then equal to the angle between the beams.

The goniometer does not measure angles but their tangents. If the wire cross the edge of the scale at x cm. from the centre and if the angle between the displaced and the central positions of the arm be θ, then $\tan \theta = x/40$, if the distance from the centre of the pivot to the edge of the scale be 40 cm. When $\tan \theta$ is known, θ can be found in degrees by tables or in radians by the series $\theta = \tan \theta - \frac{1}{3} \tan^3 \theta + \frac{1}{5} \tan^5 \theta - \ldots$, provided $|\tan \theta| \leqq 1$.

The Table gives c, the quantity to be subtracted from $\tan \theta$ to obtain θ in radians. For intermediate values, interpolation may be used. Thus, when $\tan \theta = 0\cdot205$, $c = 0\cdot00280$ and $\theta = 0\cdot205 - 0\cdot00280 = 0\cdot20220$ radian.

$\tan \theta$	c	$\tan \theta$	c	$\tan \theta$	c	$\tan \theta$	c
·02	·00000	·08	·00017	·14	·00090	·20	·00260
·03	·00001	·09	·00024	·15	·00111	·21	·00301
·04	·00002	·10	·00033	·16	·00134	·22	·00345
·05	·00004	·11	·00044	·17	·00161	·23	·00393
·06	·00007	·12	·00057	·18	·00191	·24	·00446
·07	·00011	·13	·00072	·19	·00224	·25	·00502

The goniometer (Fig. 44) is made auto-collimating by the addition of the fitting shown in Fig. 45. The bar EF is clamped to the frame holding the vertical wire by the screws G, H. A vertical slit cut in the upper edge of the bar is covered by a reflecting prism T. When light from a source S falls upon the prism, it is reflected and passes through the slit past the wire WW. The light then passes through the lens of the goniometer; if it suffer reflexion at a plane mirror, those rays which fall normally upon the mirror retrace their paths, and come to a

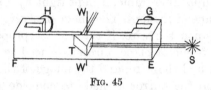

Fig. 45

focus at the point—say X—in the focal plane from which they started. A point in the focal plane below X has its image above X, and thus an inverted image of the slit and wire will be seen (§ 139). By adjusting the mirror about a horizontal axis, the lower edge of the image of the slit may be made to lie on the upper edge of the bar EF, and then, by adjusting the arm of the goniometer or by turning the mirror about a vertical axis, the image of the wire may be made to coincide with the wire itself.

Some light from the slit is reflected at each surface of the lens of the goniometer, and two images of the slit are formed by these reflexions. These images appear as two small *bright* patches when the slit is properly illuminated. In the goniometers at the Cavendish Laboratory, one image is real and one virtual, and both are further from the eye than the wire W.

EXPERIMENT 9. **Measurement of angle and index of prism of small angle.**

72. Method. The prisms of small angle used in spectacles are convenient in many optical experiments. When the refracting angle of a prism is so small that the circular measure, the tangent and the sine of the angle may be treated as identical, the calculations become very simple.

A prism of small angle is used for convenience; the theory does not depend upon the angle being "small."

Let ABC (Fig. 46) be a section of the prism by a principal plane and let the refracting angle BAC be i radians. Let PK

be perpendicular to the face AB. Let the ray KP on passing out of glass into air be refracted along PQ, making an angle θ

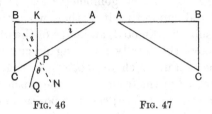

FIG. 46   FIG. 47

with the normal PN. Then, since PK makes an angle i with the normal, we have, if μ be the index,

$$\mu = \sin\theta/\sin i. \quad\quad\dots\dots\dots\dots\dots(1)$$

The two directions PN and PQ are readily identified optically, for PN is the direction of a parallel beam which returns along its own path after reflexion at AC, and PQ is the direction of a beam which returns along its path after two refractions at AC and one reflexion at AB.

The angle i of the prism is easily found. If the prism be turned through two right angles about an axis perpendicular to the plane of the face AB, so that the section of the prism comes into the position shown in Fig. 47, the normal to AC will then make an angle $2i$ with its former direction. Hence this motion of the prism entails a change of direction of $2i$ in the beam which falls normally upon AC in each position.

The prism is held in a simple clamp (Fig. 48). Three cycle balls are soldered to a plate and one face of the prism (AB in Fig. 46) is kept in contact with the spheres by two springs. This clamp is attached to a horizontal rod carried by a heavy stand. The screw at the centre of the plate allows it to be adjusted so that the ledge at the bottom of the plate is horizontal. If the ends of the prism have been made per-

FIG. 48

pendicular to the refracting edge (which is not always the case), the ledge will locate the prism correctly.

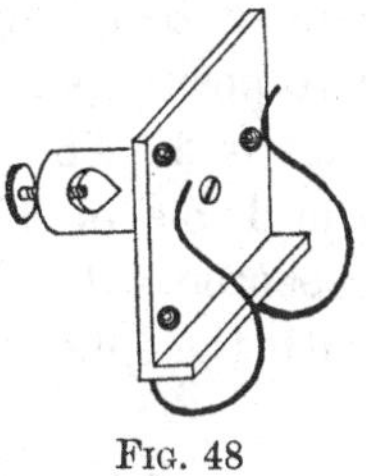

The goniometer (§ 71), fitted with the auto-collimating device, is fixed at such a height that light from a sodium flame passes properly through its reflecting prism. The clamp holding the prism which is to be tested is then set to the corresponding height near the lens of the goniometer and is turned about the horizontal rod supporting it so that, when the arm of the goniometer is suitably directed, an image of the illuminated slit is seen just above the totally reflecting prism. There are two positions of the arm in which an image is seen. In one position the light is reflected at the front face of the prism, in the second it is reflected at the back face; the first position corresponds to PN in Fig. 46, and the second to PQ.

If the setting of the goniometer wire be perfect, as judged by the aid of a *good* plane mirror, and if there be slight parallax between the wire and its image formed by reflexion at the front face of the prism ABC, that face is not quite plane.

If the two images be not at the same height above the bar carrying the totally reflecting prism, the prism to be tested is not accurately adjusted. The error may be corrected by turning this prism through a small angle about an axis perpendicular to the plane containing the points in which the prism touches the spheres.

One face of the prism is marked; this face is placed against the spheres, and the two goniometer readings corresponding to reflexions at the front and back faces are taken. The angle between the positions of the arm is θ in Fig. 46. The prism is turned through 180° about an axis perpendicular to the marked face, which is kept in contact with the spheres, and the observations are repeated; these give a second value of θ. Care must, of course, be taken not to move the clamp. The angle between the two positions of the arm corresponding to reflexion at the front face is $2i$. The goniometer readings are reduced to radians by the Table of § 71 where necessary. The observations may be repeated with the unmarked face, AC, of the prism in contact with the spheres. The index is then found by (1).

73. Practical example. Distance of scale from pivot = 40 cm. Central reading of goniometer is 10·00 cm. Letters F and B indicate readings for reflexion at the front and back faces of prism.

(i) Marked face in contact with spheres:

Mark to right. $F_1 = 7\cdot30$, $B_1 = 11\cdot59$ cm. *Mark to left.* $F_2 = 12\cdot92$, $B_2 = 8\cdot62$ cm.

(ii) Unmarked face in contact with spheres:

Mark to right. $F_3 = 7\cdot08$, $B_3 = 11\cdot38$ cm. Mark to left. $F_4 = 12\cdot70$, $B_4 = 8\cdot41$ cm.

The deviation of arm from central position for F_1 is $-2\cdot70$ cm. and angle is $-\tan^{-1}(2\cdot70/40)$ or $-\tan^{-1}0\cdot06750$. By the Table (§ 71), the angle is $-(0\cdot06750 - 0\cdot00010)$ or $-0\cdot06740$ radian. In the same way, F_2 corresponds to $0\cdot07287$ radian. Hence

$$2i = 0\cdot06740 + 0\cdot07287 = 0\cdot14027 \text{ radian.}$$

From F_1, B_1, $\theta = \tan^{-1}(1\cdot59/40) + \tan^{-1}(2\cdot70/40) = \cdot03973 + \cdot06740 = 0\cdot10713$.
From F_2, B_2, $\theta = 0\cdot10736$; mean $0\cdot10725$ radian.
From F_3, B_3 and F_4, B_4, $2i = 0\cdot14027$; mean $\theta = 0\cdot10725$ radian.
The means are $i = 0\cdot07014$ radian $= 4°\ 1'\cdot1$, $\theta = 0\cdot10725$ radian $= 6°\ 8'\cdot7$.
Hence, by (1), $\mu = \sin\theta/\sin i = 1\cdot528$.

EXPERIMENT 10. **Determination of angle and index of prism of small angle by aid of liquid of known index.**

74. Method. In EXPERIMENT 9 the angle of the prism was not assumed to be "small." In the following experiment, however, to simplify the mathematics, it is necessary that the refracting angle i should be "small."

If the refractive index of the prism be μ and if the deviation of a nearly symmetrical ray be D, we have, by § 27,

$$D = (\mu - 1)\,i. \quad\quad\quad\quad\quad\quad\quad\text{(1)}$$

The prism is placed with its edge vertical in a tank (Fig. 49) with parallel glass sides containing a liquid of refractive index μ_1. If the sides of the tank be plates of plane-parallel glass, they will not (§ 22) affect the deviation, and we may treat the system as a glass prism in a block of water. We then have a glass prism of angle i and two liquid prisms of angles β and γ (Fig. 49), the sum of β and γ being i; β and γ are positive when the refracting angles of these prisms point in the opposite direction to that of the glass prism. By § 22, a film of air interposed between each liquid prism and the glass prism would not affect the devia-tion, and thus each prism may be treated as

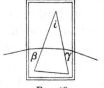

FIG. 49

if it were entirely in air. If a ray fall nearly normally upon the surface of the liquid, it will pass nearly symmetrically through each prism, and formula (20), § 27, may be applied to each prism. If D_1 be the resultant deviation, counted positive when in the same direction as that due to the glass prism alone, we have, since $\beta + \gamma = i$,

$$D_1 = (\mu - 1)\,i - (\mu_1 - 1)\,\beta - (\mu_1 - 1)\,\gamma = (\mu - 1)\,i - (\mu_1 - 1)\,i.$$

Hence $$D_1 = (\mu - \mu_1)\,i. \dots\dots\dots\dots\dots\dots\dots(2)$$

We can find i and μ, if we know μ_1 and observe D and D_1. Since, by (1) and (2),

$$(\mu - 1)/(\mu - \mu_1) = D/D_1, \dots\dots\dots\dots\dots(3)$$

we find $$\mu = (\mu_1 D - D_1)/(D - D_1). \quad\dots\dots\dots\dots(4)$$

Subtracting (2) from (1), we find $D - D_1 = (\mu_1 - 1)\,i$, and thus

$$i = (D - D_1)/(\mu_1 - 1). \quad\dots\dots\dots\dots(5)$$

75. Experimental details. The measurements are made by aid of the goniometer. The collimator AB (Fig. 50) has cross-wires at A. These wires are fixed, for their protection, to the

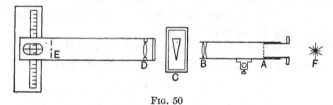

Fig. 50

inner end of a short tube sliding in the main tube of the colli-mator, and are adjusted to lie in the focal plane of the achromatic lens B (of 15 to 20 cm. focal length). The tank C has sides of *plane* glass, but it is not necessary that the two sides should be accurately *parallel*. The collimator is illuminated by a sodium flame F. If the goniometer DE be properly placed, an image of the vertical wire at A will coincide with the vertical wire of the goniometer at E. The tank contains the liquid of index μ_1.

The prism is first placed in the position of minimum deviation between the collimator and the goniometer, the tank being re-moved, and a reading of the goniometer is taken. The prism is then turned about a vertical axis through an angle which is roughly 180°, and a second reading for minimum deviation is taken. The difference between the two readings corresponds to $2D$. The actual angle through which the prism has been turned is $180° \pm D$.

The tank is then put into place and is filled with water, and the prism is placed in the tank. The goniometer reading for minimum deviation is taken and then, without disturbing the tank, the prism is turned through an angle which is roughly $180°$, and a second reading is taken, the difference corresponding to $2D_1$. This reversal of the prism eliminates any error due to want of parallelism between the sides of the tank.

If x and x_1 be the changes of goniometer reading for $2D$ and $2D_1$, the angles will be so small with a suitable prism that we may write $D = x/2l$, $D_1 = x_1/2l$, where l is the length from the centre of the pivot to the edge of the goniometer scale. Equations (4) and (5) then give μ and i. For water and for sodium light,

$$\mu_1 = 1 \cdot 3330 - 0 \cdot 00008\,(t - 20),$$

where $t°$ is the Centigrade temperature. For temperatures near $15°$ C. we may take $\mu_1 = \frac{4}{3}$ and $\mu_1 - 1 = \frac{1}{3}$.

The method may be used to find the index μ_2 of a liquid. If $2D_2$ be the change of deviation when the prism is turned round in the liquid, then, by (5), $D - D_2 = (\mu_2 - 1)\,i$,

and thus $\qquad\qquad \mu_2 = 1 + (D - D_2)/i.$ ....(6)

The auto-collimating device may be used in place of the collimator. A plane mirror is placed to the right of the tank in Fig. 50 and perpendicular to the axis of the goniometer arm when the latter is in its central position. The goniometer wire is made to coincide with its own image in each case.

76. Practical example. The prism of § 73 was tested by auto-collimating method. For goniometer, $l = 40$ cm. Mean readings are given.

Prism in air. Edge to left $11 \cdot 370$; edge to right $8 \cdot 403$ cm. Hence

$$D = \tfrac{1}{2}\,(11 \cdot 370 - 8 \cdot 403)/40 = \cdot 03709 \text{ radian.}$$

Prism in water at $20°$ C.; $\mu_1 = \tfrac{4}{3}$. Readings gave $D_1 = \cdot 01354$ radian. By (5),

$$i = (D - D_1)/(\mu_1 - 1) = \cdot 07065 = 4°\ 2' \cdot 9.$$

By (4), for prism, $\qquad \mu = (\mu_1 D - D_1)/(D - D_1) = 1 \cdot 525.$

Prism in saturated solution of sodium chloride. Readings gave $D_2 = \cdot 01038$ radian. By (6), for solution,

$$\mu_2 = 1 + (D - D_2)/i = 1 \cdot 378.$$

EXPERIMENT 11. **Determination of angle and index of prism of small angle by primary and secondary images.**

77. Method. If a ray, which enters a prism ABC by the face AB, meet the face AC, some of the light passes out through AC, but some is reflected back into the prism. If this reflected light strike AB, some of it is reflected. If this twice reflected light fall on AC, some of the light is refracted and passes out of the prism. If the angle of the prism be small enough, some of the light may suffer 4, 6, ... internal reflexions, but the corresponding emergent rays are exceedingly weak.

The minimum deviation of a ray suffering two reflexions is easily found. Let $PQRSTU$ (Fig. 51) be the ray. Since the path is symmetrical, $AQ = AT$ and $AS = AR$, and SR makes an angle $\frac{1}{2}i$ with the normal to AC at R. But the normal at R bisects SRQ, and hence $QRT = \frac{1}{2}\pi - \frac{1}{2}i$. Thus

$$AQR = QRT - QAR = \frac{1}{2}\pi - \frac{3}{2}i;$$

hence the angle between RQ and the normal NQN' at Q is $\frac{3}{2}i$.

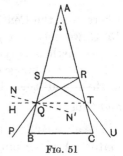

Fig. 51

If TQH be a straight line, then $PQH = \frac{1}{2}D'$, where D' is the deviation. Then the angle between PQ and the normal NQ is $\frac{1}{2}D' + \frac{1}{2}i$. Hence, by Snell's law, $\sin\frac{1}{2}(D' + i) = \mu \sin\frac{3}{2}i$. When i is so small that we can use angles instead of sines,

$$D' = (3\mu - 1)\,i. \dots\dots\dots\dots\dots\dots(1)$$

By § 27, the deviation (D) for a ray which has suffered two refractions but no reflexions is

$$D = (\mu - 1)\,i. \dots\dots\dots\dots\dots\dots(2)$$

From (1) and (2), $\mu = (D' - D)/(D' - 3D),$ $\dots\dots\dots\dots(3)$

and $i = \frac{1}{2}(D' - 3D).$ $\dots\dots\dots\dots\dots(4)$

The measurements of deviation are made as in § 75; a collimator is used. With $\mu = \frac{3}{2}$, $D' = 7D$. A *bright* source of light will be required. The prism is placed as close to the lens of the goniometer as is practicable, and the collimator is placed at a distance (50 cm.) from the prism. If the collimator be close to the prism, the light reflected from the sides of the tube and passing through the prism without suffering reflexion will make it difficult to see the secondary image of the wire of the collimator. The auto-collimating device and a plane mirror *may* be

used, but the image is not easily found. The prism is adjusted
in each case to give minimum deviation.

78. Practical example. The prism of § 73 was used. Mean values
were $D = \cdot 03735$, $D' = 0 \cdot 25165$ radian. By (3) and (4),

$$\mu = (D' - D)/(D' - 3D) = 1 \cdot 535; \qquad i = \tfrac{1}{2}(D' - 3D) = \cdot 0698 \text{ radian} = 4° \, 0' \cdot 0.$$

EXPERIMENT 12. **Determination of angle between two
nearly perpendicular mirrors.**

79. Method. If the planes of two plane mirrors intersect at
any angle, a plane perpendicular to each mirror, and therefore
perpendicular to the line of intersection of their planes, is called
a principal plane of the mirrors. If any ray in a principal plane
suffer two reflexions, one at each mirror, the angle between the
initial and final directions of the ray depends only upon the
angle between the mirrors and not upon the angle of incidence
upon the first mirror.

Let AB, AC (Fig. 52) be a section of the mirrors by a principal plane, and let $BAC = \phi = \tfrac{1}{2}\pi + \theta$. Let the ray be $PQRS$
and let $BQP = \beta$. Then, if PQ, SR meet in F, $FQR = 2\beta$.
Further

$$ARQ = \pi - AQR - QAR = \pi - \beta - (\tfrac{1}{2}\pi + \theta) = \tfrac{1}{2}\pi - \beta - \theta,$$

and thus $\qquad FRQ = 2ARQ = \pi - 2\beta - 2\theta.$

Hence $\qquad PFS = \pi - FQR - FRQ = 2\theta. \qquad \dots\dots\dots\dots(1)$

Hence the angle between PQ and RS does not depend upon
the angle of incidence of the initial ray at Q.

If $\theta = 0$, i.e. if BAC be a right angle, the ray RS is parallel to
PQ.

If the initial ray P_aQ_a be inclined to
the principal plane at angle α, and if
PQ (Fig. 52) be its projection on that
plane, then, by § 18, QR, RS will be
the projections of the parts Q_aR_a,
R_aS_a, and both Q_aR_a and R_aS_a will
be inclined to the principal plane at
angle α.

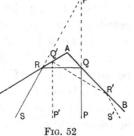

FIG. 52

In Fig. 52, the ray PQ falls first on
AB. If, however, a ray $P'Q'$, parallel to PQ, fall first on AC,

it gives rise to a ray $R'S'$ inclined to $P'Q'$ at the angle 2θ, but the deviation of $R'S'$ from $P'Q'$ is in the *opposite* direction to the deviation of RS from PQ. Hence a *beam* of rays, parallel to PQ and wide enough to fall upon *both* mirrors, gives rise to *two* beams parallel to RS and $R'S'$, these beams deviating by equal angles but in opposite directions from PQ. In this case, we could find by how much BAC differs from $90°$, but we could not decide whether BAC is greater or less than $90°$. To do this, we must ensure that the incident light falls on only *one* of the two mirrors—say on AB.

We can find θ if we measure the angle PFS between the initial and final rays. This can be done by the auto-collimating goniometer (§ 71).

The apparatus is arranged as indicated diagrammatically in Fig. 53. Here AB, AC are the two mirrors, LM is the lens of the goniometer, T the prism of the auto-collimating device, and S a source of light, preferably a small incandescent gas burner. The mirror system should stand on a levelling stand, the line BC being parallel to the line joining two of the screws. By adjusting these screws, the images of the wire W (Fig. 45) can be made parallel to the wire. By the third screw, the height of the images above the prism T can be varied. When the burner is so close to T that the rays which pass through T fill the lens LM with light, *two* images of the wire W will be seen, the wire being midway between the images. The wire cannot be made to coincide with either of these images by moving the goniometer arm, since they move with the wire. If the burner be placed at some distance from T in such a position that the rays transmitted by T fall only on the mirror AB (as indicated in Fig. 53), only *one* image will be seen. If, as in Fig. 53, this single image U be to the left of W, then $BAC > \frac{1}{2}\pi$; if U be to the right of W, $BAC < \frac{1}{2}\pi$. This criterion is due to the late S. D. Chalmers.

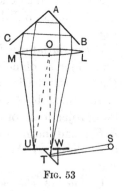

For measuring the small angle 2θ, a piece of glass millimetre scale is fixed to the frame supporting the vertical wire W of

FIG. 53

the goniometer; the divided face of the scale is turned towards the lens and is in contact with W. The flame should be placed so that *both* images are visible.

If O (Fig. 53) be the nodal point of the lens corresponding to the focal plane WU, then $OW = f$, the focal length. By the property of nodal points (§ 191), OW and OU are parallel to the two beams of parallel rays on the other side of the lens. Thus OW and OU correspond to PQ and RS in Fig. 52, and hence, in the case of Fig. 53, the angle between the mirrors is $\phi = \tfrac{1}{2}\pi + \tfrac{1}{2} UOW$. If $WU = y$, then

$$\phi = \tfrac{1}{2}\pi + y/2f. \qquad \dots\dots\dots\dots\dots\dots(2)$$

The focal length may be measured by a modification of EXPERIMENT 43. A single plane mirror is substituted for the double mirror. As the arm of the goniometer is moved through any angle γ, the image of the wire moves across the focal plane, the displacement of the image relative to the wire being the same as if the arm had been at rest and the mirror had been turned through γ. In this case the reflected beam is turned through 2γ. Thus, if a change of goniometer reading of a cm. correspond to a change of reading of the image on the glass scale of b cm. and if l cm. be the distance from the centre of the pivot to the scale, we have $b/f = 2\gamma = 2a/l$ or

$$f = bl/2a. \qquad \dots\dots\dots\dots\dots\dots\dots\dots(3)$$

A mirror system may be constructed of two mirrors fixed to a block of wood. Glass silvered at the back is unsuitable, as there will be a number of images, which will not coincide unless the mirrors be absolutely "plane-parallel." Multiple reflexions may be avoided by using two pieces of unsilvered plate glass, covered at the back with black varnish.

80. Practical example. Distance of centre of pivot from goniometer scale $= l = 40$ cm.

Glass scale	Goniometer scale	Change of reading for 1·5 cm. on glass scale
6·5, 8·0 cm.	9·35, 10·21 cm.	0·86 cm.
7·0, 8·5	9·64, 10·50	0·86
7·5, 9·0	9·93, 10·78	0·85
8·0, 9·5	10·21, 11·06	0·85 Mean 0·855 cm.

By (3), $f = bl/2a = 1\cdot5 \times 40/1\cdot710 = 35\cdot09$ cm.

Readings on glass scale of images due to double mirror were 7·98, 7·65 cm.; half difference $= y = 0\cdot165$ cm. When burner was moved towards mirror system, as in Fig. 53, so as to illuminate AB only, the surviving image U was to left of W, i.e. on opposite side to burner. The angle ϕ between mirrors is therefore greater than a right angle. By (2),

$$\phi = \tfrac{1}{2}\pi + y/2f = \tfrac{1}{2}\pi + 0\cdot165/70\cdot18 = \tfrac{1}{2}\pi + 0\cdot002351 \text{ radian} = 90° 8'\cdot1.$$

EXPERIMENT 13. **Determination of angle of a prism of nearly** $90°$.

81. Method. The method of § 79 is easily adapted to the measurement of such a prism, provided the face opposite the reputed right angle be polished. Let ABC (Fig. 54) be a principal section (§ 26) of the prism, the angle A being nearly $\tfrac{1}{2}\pi$. Let $PQRSXY$ be the path of a ray, the incidence at Q being nearly normal. By equation (1), § 79, if $BAC = \tfrac{1}{2}\pi + \theta$, the angle between QR and XS is 2θ, and, if θ be positive, the point of intersection of QR and XS is on the same side of RS as A, and this is true whatever the angle of incidence of QR at R. If BAC be exactly $\tfrac{1}{2}\pi$, QR and SX are parallel, and then PQ and XY are parallel also. When, however, QR and

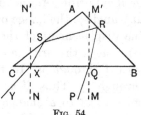

FIG. 54

SX are not parallel, we must take account of the refractions at Q and X. Let MQM', NXN' be the normals at Q and X. Then

$$\sin PQM = \mu \sin RQM' \text{ and } \sin YXN = \mu \sin SXN',$$

where μ is the index of the prism. Now it is easy to place the prism so that the ray PQ which falls upon it from the auto-collimating goniometer is nearly normal to BC, and, if BAC be nearly $\tfrac{1}{2}\pi$, the ray XY will also be nearly normal to BC. We may then use the angles instead of their sines. Thus $PQM = \mu RQM'$, and $YXN = \mu SXN'$, and hence the angle between PQ and YX is μ times the angle between QR and SX, i.e. it is $2\mu\theta$. The angle between PQ and YX is measured by the method of § 79 in the same way as for the double mirror. If this angle be ψ radians, then $\psi = 2\mu\theta$, and the error in the angle BAC is $\psi/2\mu$ radians. The method indicated in Fig. 53 is used to decide whether BAC is greater or less than a right angle. The index of the prism is found by a spectrometer, or, with sufficient accuracy, on a drawing board. (See § 83.)

82. Practical example. The angles of prism were nearly 90°, 45°, 45°. The same goniometer was used as in § 80. The index of prism was measured by a spectrometer. One angle was 44° 59′; with this angle the minimum deviation was 25° 32′. Hence $\mu = 1\cdot509$.

The readings on glass scale of images due to prism were 7·70 and 7·88 cm.; half the difference is 0·09 cm. Then

$$\psi = 0\cdot09/f = 0\cdot09/35\cdot09 = 0\cdot00256 \text{ radian.}$$

When burner was placed so as to illuminate face AB and not AC, the surviving image U (Fig. 53) was to the right of W, i.e. on same side as burner. The angle of prism is therefore *less* than a right angle. Hence,

$$A = BAC = \tfrac{1}{2}\pi - \psi/2\mu = \tfrac{1}{2}\pi - 0\cdot00256/3\cdot018 = \tfrac{1}{2}\pi - 0\cdot00085 \text{ radian} = 89° 57'\cdot1.$$

EXPERIMENT 14. **Measurement of angles of nominally right-angled prism.**

83. Method*. Let ABC (Fig. 55) be a section of the prism by a principal plane; the angles A, B, C are nearly equal to 45°, 45°, 90° respectively. Then the sum $A + B$ and the difference $A - B$ can be accurately measured by an auto-collimating goniometer.

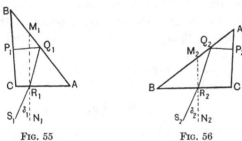

FIG. 55   FIG. 56

Consider a ray $P_1 Q_1$ (Fig. 55) normal to the face BC. After reflexion at Q_1 on AB, it meets AC in R_1 and emerges along $R_1 S_1$. Then $AQ_1R_1 = P_1Q_1B = \tfrac{1}{2}\pi - B$, and $Q_1R_1C = A + (\tfrac{1}{2}\pi - B)$. Let $M_1 R_1 N_1$ be the normal at R_1. Then

$$Q_1 R_1 M_1 = Q_1 R_1 C - \tfrac{1}{2}\pi = A - B.$$

If $S_1 R_1 N_1 = \delta_1$, and if δ_1 be counted positive when $CR_1 S_1$ is *acute*, $\sin\delta_1 = \mu \sin(A - B)$, where μ is the index of the prism. Since $A - B$ is small, we may write $\delta_1 = \mu(A - B)$.

The prism is placed on a levelling table with its edges vertical,

---

* See "Note on the use of the auto-collimating telescope in the measurement of angles." J. Guild, *Proc. Physical Soc. of London*, Vol. XXVIII, p. 242.

and the goniometer is placed so that its arm turns in a horizontal plane and is approximately in its central position when the line of collimation is normal to AC. The reflecting prism of the goniometer (§ 71) sends light from a bright sodium flame through the lens of the goniometer on to the face AC. If the line of collimation be parallel to the normal $N_1 R_1 M_1$, the vertical wire will coincide with its own image. If the line of collimation be now made parallel to $R_1 S_1$, the rays, after normal reflexion at BC, will return along their paths and the wire will again coincide with its image. To distinguish one image from the other, a piece of *wet* (not *damp*) blotting paper is placed on AB; this weakens the image corresponding to $R_1 S_1$.

These two readings should be so recorded as to show whether δ_1 is positive or negative. The difference between the readings divided by the length of the arm gives δ_1 in radians.

The prism is now turned so that BC faces the goniometer. If A be *greater* than B, a ray $P_2 Q_2$ (Fig. 56), which is normal to AC, will, after reflexion at Q_2 and refraction at R_2, emerge along $R_2 S_2$, and the angle $CR_2 S_2$ will be *obtuse*. The angle $S_2 R_2 N_2$ or δ_2 is measured by the goniometer, and is positive when $CR_2 S_2$ is *obtuse*. Since $Q_2 R_2 M_2 = A - B$, we have, since $A - B$ is small, $\delta_2 = \mu (A - B)$.

Since δ_1, δ_2 are theoretically equal, we take the mean $\delta = \frac{1}{2} (\delta_1 + \delta_2)$, and write

$$A - B = \delta/\mu. \quad ....(1)$$

We next find the small angle $C - \frac{1}{2}\pi$. A ray $F_1 G_1 H_1 K_1 L_1$ (Fig. 57), starting from F_1 normal to AB, will, after reflexion

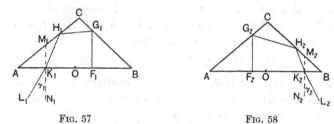

Fig. 57   Fig. 58

first at BC and then at AC, emerge from AB along $K_1 L_1$. Then $M_1 K_1 N_1$ being the normal at K_1, we find in succession

$$CG_1H_1 \quad = F_1G_1B = \tfrac{1}{2}\pi - B,$$
$$AH_1K_1 \quad = CH_1G_1 = \pi - C - (\tfrac{1}{2}\pi - B) = \tfrac{1}{2}\pi + B - C,$$
$$M_1K_1H_1 = (\tfrac{1}{2}\pi - A) - (\tfrac{1}{2}\pi + B - C) = C - (A + B).$$

But $C = \pi - A - B$, and hence, if $C = \tfrac{1}{2}\pi + z$, we have $A + B = \tfrac{1}{2}\pi - z$, and then $M_1K_1H_1 = 2z$. Thus, if $L_1K_1N_1$ be denoted by γ_1, counted positive when AK_1L_1 is *acute*, $\sin\gamma_1 = \mu\sin 2z$. Since z is small, we may write $\gamma_1 = 2\mu z$.

A ray $F_2G_2H_2K_2L_2$, starting from F_2 (Fig. 58) normal to AB and reflected first at AC and then at BC, will emerge from AB along K_2L_2. If $L_2K_2N_2$ be denoted by γ_2, counted positive when BK_2L_2 is *acute*, then $\gamma_2 = 2\mu z$.

Taking the mean $\gamma = \tfrac{1}{2}(\gamma_1 + \gamma_2)$, we have $z = \tfrac{1}{2}\gamma/\mu$. Since $A + B = \tfrac{1}{2}\pi - z$, we have

$$A + B = \tfrac{1}{2}\pi - \tfrac{1}{2}\gamma/\mu. \quad \dots\dots\dots\dots\dots(2)$$

The prism is placed so that the light from the prism of the goniometer falls on the *half AO* of AB. If the prism ABC be small, an opaque screen may be used to prevent the light from falling on the half OB. The line of collimation is made parallel first to the normal K_1N_1 and then to K_1L_1 (Fig. 57), the wire in each case coinciding with its own image. To distinguish one image from the other, wet blotting paper is put on BC to weaken the image corresponding to K_1L_1. The angle between the two positions of the goniometer arm is γ_1. The prism is then placed so that the light falls on the *half OB* of AB, and γ_2 is found.

Since δ and γ are small, an approximate value of μ will suffice. The angle A and the minimum deviation D of a ray passing through AB and AC may be found approximately by aid of a protractor from lines drawn on a drawing board; then $\mu = \sin\tfrac{1}{2}(D + A)/\sin\tfrac{1}{2}A$. More accurate values of A and B are then found from (1) and (2). Finally $C = \pi - (A + B)$.

In the method described above, it is assumed that the prism has *a* principal plane, or, in other words, that the planes of the three faces are parallel to a single straight line. This assumption is unjustified, for, in general, the planes of the three faces will meet in a (distant) point. The prism is then said to have pyramidal error. See EXPERIMENT 16.

84. Practical example. Mr W. E. Aylwin obtained the following results. Goniometer arm, pivot to scale = 40 cm.; number of minutes for one cm. on scale (small angles) = 85·94.

Goniometer readings for (1) $A - B$, (2) $A + B$, each mean of three.

(1) Normal to AC, 9·89; other image, 9·86 cm. Hence $\delta_1 = \cdot 03 \times 85'\cdot 9 = 2'\cdot 6$.

Normal to BC, 9·94; other image, 9·91 cm. Hence $\delta_2 = \cdot 03 \times 85'\cdot 9 = 2'\cdot 6$.

(2) Normal to half face AO, 8·21; other image 8·11 cm. Hence

$$\gamma_1 = \cdot 10 \times 85'\cdot 9 = 8'\cdot 6.$$

Normal to half face BO, 11·70; other image 11·80 cm. Hence

$$\gamma_2 = \cdot 10 \times 85'\cdot 9 = 8'\cdot 6.$$

Hence $\delta = \tfrac{1}{2}(\delta_1 + \delta_2) = 2'\cdot 6$, $\gamma = \tfrac{1}{2}(\gamma_1 + \gamma_2) = 8'\cdot 6$; both δ and γ were *positive*.

Measurements on drawing board gave $D = 27°$, using A, which is nearly $45°$. Hence $\mu = \sin\tfrac{1}{2}(D+A)/\sin\tfrac{1}{2}A = 1\cdot 54$. By (1), (2),

$$A - B = \delta/\mu = 2'\cdot 6/1\cdot 54 = 1'\cdot 7, \qquad A + B = \tfrac{1}{2}\pi - \tfrac{1}{2}\gamma/\mu = \tfrac{1}{2}\pi - 2'\cdot 8 = 89° \ 57'\cdot 2.$$

Hence, $A = 44° \ 59'\cdot 4$, $B = 44° \ 57'\cdot 8$, $C = 90° \ 2'\cdot 8$.

EXPERIMENT 15. The prism refractometer.

85. Theory of refractometer. A prism stands on a table turning about a vertical axis, the edges of the prism being parallel to that axis. Let ABC (Fig. 59) be a principal plane of the prism; then ABC is horizontal. In the experiment, C is nearly $\tfrac{1}{2}\pi$ and A and B are nearly equal. A sodium flame F is so placed that some rays graze the face AB, other rays making small finite angles with that face. Let $BPQR$ be the path of that grazing ray which lies in the principal plane ABC. If any grazing ray make an angle ψ with the principal plane, the rays emerging from AC will also make an angle ψ with that plane. If ψ be small, the projection on ABC

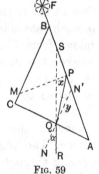

FIG. 59

of the part corresponding to QR will make with QR an angle proportional to ψ^2 (see § 30). Rays will graze every part of the *face AB* and thus, for any given value of ψ, the corresponding grazing rays will give rise to a wide beam of rays emerging from AC and all parallel to the direction determined by ψ. If the emergent rays be received by a telescope of sufficient aperture, all the incident rays which (1) graze the *face AB* and (2) make an angle ψ with the principal plane will be brought to a single point in the focal plane of the objective of the telescope. Those rays which reach the telescope having made finite angles with the face AB will cause the observer to see a not very wide band of illumination. One edge of this band will be slightly curved and will form a *sharp* boundary between light and darkness. The other edge will be irregular, as it depends upon the form of the flame, and will flicker if the flame flicker. The sharp boundary corresponds to those rays which have grazed the face AB. The telescope may thus be "set for infinity" by adjusting it so that there is no parallax between the cross-wire and the sharp boundary. If the *vertical* cross-wire be set to *touch* the boundary, the point of contact will correspond to rays parallel to the principal plane.

The method does not require that the rays which ultimately pass through a single point of the sharp boundary come from a single point at infinity. The rays may, and in the experiment do, come from a number of luminous points in the flame, these points lying in the plane of the face AB.

Let PM, NQN' be the normals at P and Q, N' lying on AB. Let $NQR = \alpha$, and let α be positive, when, as in Fig. 59, RQC is *obtuse*. Let $MPQ = x$, $PQN' = y$. Since

$$QPN' + N'QP = QN'A,$$

we have $\frac{1}{2}\pi - x + y = \frac{1}{2}\pi - A$, or

$$x = A + y. \qquad \dots\dots\dots\dots\dots\dots(1)$$

Since MPQ or x is the critical angle for air and glass, $\sin x = 1/\mu_g$, where μ_g is the index of glass relative to air. Also $\sin y = \sin \alpha/\mu_g$.

By (1), $\sin x = \sin A \cos y + \cos A \sin y$,

and hence

$$1/\mu_g = \sin x = \sin A \{1 - \sin^2\alpha/\mu_g{}^2\}^{\frac{1}{2}} + \cos A \sin \alpha/\mu_g,$$

or $1 = \sin A \{\mu_g{}^2 - \sin^2\alpha\}^{\frac{1}{2}} + \cos A \sin \alpha.$

Hence $$\mu_g{}^2 = \left\{\frac{1 - \cos A \sin \alpha}{\sin A}\right\}^2 + \sin^2\alpha, \qquad \dots\dots\dots(2)$$

from which μ_g can be found when A and α are known.

Let RQ produced meet AB in S, and let $RSA = S$. Since $CQS = \frac{1}{2}\pi - \alpha$, we have $S = CQS - A = \frac{1}{2}\pi - \alpha - A$.

In the experiment, the rays emerging from the face CA enter a *fixed* telescope adjusted for infinity, and the boundary between light and darkness is brought to the vertical wire by adjusting the table carrying the prism. Then QR is parallel to the line of collimation of the telescope.

Now turn the table round and move the flame so that the light which falls on AB emerges from BC, and let β be the angle between the normal to BC and the emergent ray corresponding to the grazing incident ray lying in the principal plane. Then

$$\mu_g{}^2 = \left\{ \frac{1 - \cos B \sin \beta}{\sin B} \right\}^2 + \sin^2\beta. \quad \ldots\ldots\ldots\ldots(3)$$

If the forward direction of the emergent ray make an acute angle T with the direction A to B, we have $T = \frac{1}{2}\pi - \beta - B$.

Since in each case the emergent ray is parallel to the line of collimation, this ray is *fixed* in direction. If D be the angle, less than π, between the first and second positions of the prism, then, in Fig. 60, $D = AOA'$, where $AOBC$, $A'OB'C'$ represent the two positions. If OJ represent the line of collimation, $AOJ = S$, $B'OJ = T$, and hence $D = \pi - S - T$. Thus

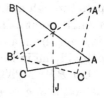

Fig. 60

$$D = A + B + \alpha + \beta. \quad \ldots\ldots\ldots\ldots\ldots(4)$$

The prism employed is nearly isosceles; thus B is nearly equal to A, and β is nearly equal to α. If we write

$$A_0 = \tfrac{1}{2}(A + B), \qquad \alpha_0 = \tfrac{1}{2}(\alpha + \beta), \quad \ldots\ldots\ldots(5)$$

we have

$$\alpha_0 = \tfrac{1}{2}D - A_0. \quad \ldots\ldots\ldots\ldots\ldots\ldots(6)$$

When $A - B$, $\alpha - \beta$ are very small, we use the mean values in finding μ_g. Thus

$$\mu_g{}^2 = \left\{ \frac{1 - \cos A_0 \sin \alpha_0}{\sin A_0} \right\}^2 + \sin^2\alpha_0. \quad \ldots\ldots\ldots\ldots(7)$$

By (1), if $A_0 = \tfrac{1}{4}\pi$, $\sin y = \{1 - \sqrt{(\mu_g{}^2 - 1)}\}/(\mu_g \sqrt{2})$, since $y = x - \tfrac{1}{4}\pi$, and $\sin x = 1/\mu_g$. Hence, if $\mu_g{}^2 > 2$ or $\mu_g > 1\cdot414$, y will be *negative*. Then α will be *negative*, i.e. the angle CQR (Fig. 59) will be *acute* and not *obtuse*.

86. Measurement of index of a liquid.

Suppose that the air to the right of AB in Fig. 59 is replaced by a liquid of index μ

relative to air. Let $BPQR$ now represent the path of the ray which, in the *liquid*, grazes the face AB and lies in the principal plane, the ray emerging into *air* along the fixed direction QR. Let, now, $NQR = \theta$, and let θ be positive when RQC is *obtuse*. If, now, $MPQ = \xi$, $PQN' = \eta$, we have

$$\xi = A + \eta, \qquad \sin \xi = \mu/\mu_g, \qquad \sin \eta = \sin \theta/\mu_g.$$

Since $\sin \xi = \sin A \cos \eta + \cos A \sin \eta$, we have

$$\mu/\mu_g = \sin A \,\{1 - \sin^2\theta/\mu_g{}^2\}^{\frac{1}{2}} + \cos A \sin \theta/\mu_g,$$

or $\qquad \mu = \sin A \,\{\mu_g{}^2 - \sin^2\theta\}^{\frac{1}{2}} + \cos A \sin \theta. \quad \dots\dots\dots(8)$

Now turn the table round and move the flame so that the light which falls on AB emerges from BC, and let ϕ be the angle between the normal to BC and the emergent ray corresponding to a grazing incident ray. If Δ be the angle, less than π, between the two positions of the prism for a fixed direction of the emergent ray,

$$\Delta = A + B + \theta + \phi. \quad \dots\dots\dots\dots\dots(9)$$

If we put $\theta_0 = \frac{1}{2}\,(\theta + \phi)$, so that $\theta_0 = \frac{1}{2}\,\Delta - A_0$, we may use the mean values θ_0 and A_0, and thus

$$\mu = \sin A_0 \,\{\mu_g{}^2 - \sin^2\theta_0\}^{\frac{1}{2}} + \cos A_0 \sin \theta_0. \quad \dots\dots(10)$$

Since the liquid will not remain in contact with the face AB, the device of § 87 is used.

87. The auxiliary prism. An auxiliary prism UVW, similar to ABC, is placed nearly in contact with ABC, as in Fig. 61, the narrow intervening space being occupied by a film of air or of liquid. The emergent ray HK will be parallel to the ray from F incident on VW if VW be parallel to CA and the indices of the two prisms be equal.

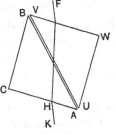

Any ray which enters UVW by the face VW and falls on UV at the critical angle, grazes the face UV in the air or liquid. It, therefore, never reaches the prism ABC. The only rays reaching the face AB are those which are sufficiently inclined to UV to be able to cross the film within the

FIG. 61

distance AB. Hence, the film must be *very thin*, if we are to be able to treat *any* rays in the film as meeting AB at grazing incidence. If the film be of appreciable thickness, the illumination will not fall to zero *abruptly*, but will *gradually* reach zero a little before the ideal boundary (see § 35).

88. Experimental details. The measurements may be made on a spectrometer. We may also use a goniometer (§ 71), kept in a fixed position, in place of the telescope of the spectrometer, and may mount the prism upon any table turning about a vertical axis and provided with a divided circle*. When a spectrometer is used, the angles of the prism ABC can be found by the spectrometer. If an auto-collimating goniometer be used, the angles can be found as in EXPERIMENT 14.

The index μ_g for the prism is found by either § 85 or § 87. In either case, the mid-point O of AB (Fig. 60) is placed over the axis of the table, and the telescope or goniometer is directed towards O. If we use § 85, the flame must be moved so that the plane of AB always passes *through* the flame and not merely to one side of the flame. The band of light seen in the telescope will be narrow, since (1) none of the rays from a flame at F. make large angles with AB, and (2) the emergent rays are more nearly parallel than the incident rays. Thus, if $\mu_g = 1\cdot53$, $A = B = 45°$, all the incident rays in the principal plane, which are inclined at less than $10°$ to AB, emerge within $1° 9'$ of the critical ray. The edge of the band to be used in the measurements does not flicker when the flame flickers.

In the method of § 87, the flame is nearly in the line of collimation of the telescope or goniometer. If the flame be fairly large, the adjustments can be made without moving it.

Two or three pairs of readings for D may be made, the prism being placed in a fresh position on the table for each pair, so as to obtain independent readings. The mean value of D is found. From the known value of A_0, α_0 is found from $\alpha_0 = \frac{1}{2}D - A_0$, and then μ_g is found by (7).

When μ is to be found for a liquid, a drop is placed on the face AB, and the auxiliary prism is placed in position so as to spread

---

* The apparatus of § 36 may be used if a prism table be fitted to the shaft of the divided circle. A vertical rod carrying a cross-bar fits into the socket in the shaft. Two vertical rods fixed to the cross-bar support the prism table. The lens and needle take the place of the goniometer.

out the liquid into a very thin film. The prisms are then placed on the table and are held together by rubber bands. Two or three pairs of readings for Δ are made and the mean is found. Then θ_0 is found from $\theta_0 = \frac{1}{2}\Delta - A_0$, and μ is found by (10).

The method fails unless the index of the liquid be lower than that of *each* prism. Thus, no boundary between brightness and darkness can be obtained when aniline ($\mu = 1\cdot59$) is placed between a pair of crown glass prisms ($\mu_g = 1\cdot53$).

Since the index of a liquid diminishes with rise of temperature, the temperature of the air near the prisms should be noted. The value of μ obtained for water with sodium light may be compared with that given by $\mu = 1\cdot3330 - 0\cdot00008\,(t - 20)$, where $t°$ is the Centigrade temperature.

89. Practical example. Mr W. E. Aylwin used prism of §84, for which
$$A_0 = \tfrac{1}{2}\,(A + B) = 44° \,58'\cdot6.$$

Index of prism. The table had two pointers. Readings made with auxiliary prism are marked with *.

Rays emerging from AC		Rays emerging from BC		D	
Pointer 1	Pointer 2	Pointer 1	Pointer 2	Pointer 1	Pointer 2
149°·6	329°·9	72°·7	252°·9	76°·9	77°·0
158 ·9	338 ·9	81 ·0	261 ·5	77 ·9	77 ·4
260 ·0*	79 ·9*	337 ·0*	156 ·9*	77 ·0	77 ·0
264 ·5*	84 ·4*	341 ·5*	161 ·6*	77 ·0	77 ·2

Mean 77° 10'·5. Hence $a_0 = \frac{1}{2}D - A_0 = -6° \, 23'\cdot4$. Then, by (7),
$$\mu_g{}^2 = 2\cdot3414, \qquad \mu_g = 1\cdot5301.$$

Index of water. The mean Δ from three sets of readings was $138° \, 42'\cdot0$. Hence $\theta_0 = \frac{1}{2}\Delta - A_0 = 24° \, 22'\cdot4$. By (10), $\mu = 1\cdot3334$.

EXPERIMENT 16. **Determination of pyramidal error of prism.**

90. Introduction. Let CGH (Fig. 62) be a plane triangle. Through CG, CH draw planes at right angles to the plane CGH, and let these planes intersect in CO. Through O draw the plane OGH. Then $OCGH$ is a tetrahedron. If

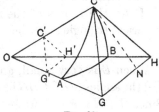

FIG. 62

we take a plane $C'G'H'$ parallel to CGH, the part of the

tetrahedron between these planes approximates to a prism when OC is large compared with the sides of CGH; when OC is infinite, a true prism is formed. Opticians fashion blocks of glass which are bounded by two unpolished ends CGH, $C'G'H'$ and by three polished faces lying in the planes OHC, OCG, OGH. When OC is not infinite, the "prism" is said to have pyramidal error.

With O as centre and OC as radius, describe a sphere and let the planes through O cut it in the "great circles" BC, CA, AB. Since OC is perpendicular to CG, CH, these lines touch the arcs CA, CB at C. Let the angles between the pairs of planes which intersect in OA, OB, OC be A, B, C. Then A, B, C are the angles of intersection of the arcs at A, B, C. Let the angles which the edges OA, OB, OC make with the faces OBC, OCA, OAB be ϵ_1, ϵ_2, ϵ_3 respectively.

In Fig. 62, let CN be perpendicular to GH, and join ON. Since OCG, OCN, CNG are right angles,

$$ON^2 + GN^2 = OC^2 + CN^2 + CG^2 - CN^2 = OC^2 + CG^2 = OG^2.$$

Hence GH is normal to the plane OCN, for it is perpendicular to ON as well as to CN. Thus the plane OCN cuts the plane OGH at right angles, and, therefore, the angle $CON = \epsilon_3$, and also $CN = OC \tan \epsilon_3$.

When the angles A, B, C are measured by a spectrometer, the prism must be levelled (§ 54) afresh for each edge, when there is pyramidal error. The sum $A + B + C$ of the angles so found is not π, but is greater than π. In spherical trigonometry it is shown that

$$A + B + C = \pi + E, \qquad \dots\dots\dots\dots\dots(1)$$

where E is the "spherical excess." We may take E as measuring the pyramidal error.

If OC, the radius of the sphere, be r, and if the area of the spherical triangle ABC be S, then, by spherical trigonometry,

$$E = S/r^2. \qquad \dots\dots\dots\dots\dots\dots(2)$$

When E is very small, an approximate value is easily found. Let the angles CGH, CHG be G, H. Then $GH = CN (\cot G + \cot H)$. Since $G + H = \pi - C$, we easily find $GH = CN \sin C/(\sin G \sin H)$. Hence

Area $CGH = \tfrac{1}{2} CN . GH = \tfrac{1}{2} CN^2 \sin C/(\sin G \sin H)$.

When CO/CN is very large, the angles G, H are nearly equal to A, B, the area CGH is nearly equal to S, and we may write $CN = OC \tan \epsilon_3 = r\epsilon_3$. Then

$$E = \epsilon_3{}^2 \sin C/(2 \sin A \sin B). \quad \ldots\ldots\ldots\ldots(3)$$

Interchanging the letters, we have

$$E = \epsilon_1{}^2 \sin A/(2 \sin B \sin C) = \epsilon_2{}^2 \sin B/(2 \sin C \sin A)....(4)$$

Hence $\qquad \epsilon_1 \sin A = \epsilon_2 \sin B = \epsilon_3 \sin C. \quad \ldots\ldots\ldots\ldots(5)$

The *nominal* values of A, B, C may be used in (3), (4) since ϵ_1, ϵ_2, ϵ_3 are, by supposition, small.

If the nominal values be $A = B = \frac{1}{4}\pi$, $C = \frac{1}{2}\pi$, then $E = \epsilon_3{}^2$. If $\epsilon_3 = 1°$ or ·01745 radian, $\epsilon_3{}^2 = ·0003045$ radian or $1'\ 3''$. By (5), $\epsilon_1 = \epsilon_2 = \epsilon_3 \sqrt{2} = 1°\ 24'\cdot8$. If A, B, C were accurately measured, we should find $A + B + C = \pi + E = 180°\ 1'\ 3''$.

91. Method. Let Fig. 63 represent a "prism" in which C is a right angle, and let the plane ABC be a principal plane for the faces which meet in CC'. Let P be a point on the face $ABB'A'$. In the experiment, this face is approximately vertical and the edge CC' is approximately hori-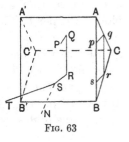
zontal. Let a ray PQ normal to $ABB'A'$ be reflected first at Q on $ACC'A'$ and then at R on $BCC'B'$, and let the pro-jections of PQ, QR, RS on the principal plane ABC be pq, qr, rs. Then, by § 18, $Aqp = Cqr$, and $Crq = Brs$. Since $C = \frac{1}{2}\pi$, rs is parallel to pq by § 79. Further, by § 18, each part of $PQRS$ is inclined at the

Fig. 63

same angle to the principal plane ABC. If a point travel along the ray, and if its distance from pq increase as it moves from P to Q, its distance from rs will increase as it moves from R to S.

The argument is unaffected if the ray from P normal to $ABB'A'$ fall first on $BCC'B'$ instead of first on $ACC'A'$.

Since PQ is normal to the face $ABB'A'$, and since pq is its projection on ABC, the angle between the co-planar lines PQ and pq is ϵ_3. The angle between the co-planar lines RS, rs is also ϵ_3. Hence the projection of RS on the plane $PQqp$ makes an angle $2\epsilon_3$ with PQ.

If the ray RS pass from glass to air at S and be refracted along ST, the angle θ between ST and SN, the outward normal, will be $2\mu\epsilon_3$, since ϵ_3 is small. If the face $ABB'A'$ be vertical, and the edge CC' be horizontal, PQ and its projection pq on the vertical plane ABC are horizontal. Hence sr, which is parallel to pq, is horizontal. Since sr is the projection of SR on ABC, SR itself is horizontal. The normal SN is horizontal and thus the plane TSN is horizontal.

The measurement of ϵ_3 is made, on the lines of EXPERIMENT 14, by the auto-collimating goniometer (§ 71). The prism is set so that the face $ABB'A'$ is approximately vertical and the edge CC' approximately horizontal. The goniometer is first set to such a height that the rays from its small reflecting prism fall on the *lower* half of $ABB'A'$, this case corresponding to Fig. 63. The image of the vertical wire of the goniometer can be made to coincide with the wire itself in two positions of the arm. In one position, the rays fall normally on $ABB'A'$; in the other position, a ray TS from the goniometer passes along $TSRQP$, is reflected at P and retraces its path. The angle θ between the two positions is obtained from the goniometer readings. If a piece of wet blotting paper be placed on $ACC'A'$, the image corresponding to TS will become faint. If T be to the left of N, then Qq is greater than Pq, and the point in which the plane $ABB'A'$ intersects the line CC' is to the right in Fig. 63.

The goniometer is then raised so that the rays fall on the upper half of $ABB'A'$, and the observations are repeated. The mean value of θ is used in the equation

$$\epsilon_3 = \theta/2\mu. \quad\dots\dots\dots\dots\dots\dots(6)$$

The index μ may be found on a drawing board as in § 83. If A, B be nominally 45°, $E = \epsilon_3{}^2$.

If C be not exactly 90°, the two images will not be at the same height. The difference corresponds to γ in Figs. 57, 58.

92. Practical example. G. F. C. Searle used the prism of § 84; $\mu = 1\cdot54$. Pivot to scale of goniometer, $l = 40$ cm. The mean readings are given.
Goniometer low. For normal, 11·580; other image, 11·387. Difference ·193 cm.
Goniometer high. For normal, 10·467; other image, 10·283. Difference ·184 cm.
Mean difference = ·188 cm. Hence $\theta = \cdot188/40 = \cdot0047$. By (6),

$$\epsilon_3 = \theta/2\mu = \cdot0047/3\cdot08 = \cdot001526 \text{ radian} = 5'\cdot2.$$

The zero of the goniometer scale is to the left. Hence the readings show that T (Fig. 63) is to the left of N. Hence $ABB'A'$ meets CC' to the right of C.

By (3), since, nominally, $C = 90°$, $A = B = 45°$, we have

$$E = \epsilon_3{}^2 = (\cdot001526)^2 = 2\cdot33 \times 10^{-6} \text{ radian} = \cdot48''.$$

Then, accurately, $A + B + C = 180°\ 0'\ 0''\cdot48$.

EXPERIMENT 17. Determination of angles of a nearly equilateral prism.

93. Method. The three angles of a nearly equilateral prism, with *three* polished faces, can be found by a modification of EXPERIMENT 14.

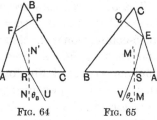

Fig. 64   Fig. 65

In Fig. 64, ABC is a section of the prism by a principal plane. Let the ray PF, which is normal to BC, be reflected at F, be refracted at R, and emerge along RU. Let the angle URN, which RU makes with the normal $N'RN$, be denoted by θ_B, and be counted positive when ARU is *obtuse*. Since $BPF = \frac{1}{2}\pi$, $AFR = BFP = \frac{1}{2}\pi - B$. Hence

$$FRN' = FAR + AFR - \tfrac{1}{2}\pi = A - B;$$

the angle FRA is acute when $A > B$. Thus, if μ be the index of the prism, $\sin\theta_B = \mu \sin(A - B)$. When $A - B$ is small,

$$\theta_B = \mu\,(A - B). \quad\quad\quad\quad\quad\quad\quad(1)$$

Similarly, in Fig. 65, let the ray QE which is normal to BC, be reflected at E, refracted at S, and emerge along SV, making the angle VSM or θ_C with the normal. Let θ_C be positive when ASV is *obtuse*. Then

$$\theta_C = \mu\,(A - C). \quad\quad\quad\quad\quad\quad\quad(2)$$

By (1), (2), $\theta_B + \theta_C = \mu\,(2A - B - C)$. But $B + C = \pi - A$; hence

$$3A - \pi = (\theta_B + \theta_C)/\mu \quad\quad\quad\quad\quad(3)$$

If the letters A, B, C be moved on one step in cyclical order in Figs. 64, 65, the base of the triangle in Fig. 64 becomes BA and in Fig. 65 it becomes CB. The angles corresponding to URN and VSM in Figs. 64, 65, will be denoted by ϕ_C, ϕ_A.

When the letters are moved on an additional step, the angles corresponding to URN and VSM are ψ_A, ψ_B. Then

$$\phi_C = \mu\,(B - C), \qquad \phi_A = \mu\,(B - A),$$
$$\psi_A = \mu\,(C - A), \qquad \psi_B = \mu\,(C - B).$$

Hence
$$3B - \pi = (\phi_C + \phi_A)/\mu, \quad \ldots\ldots\ldots\ldots\ldots(4)$$
$$3C - \pi = (\psi_A + \psi_B)/\mu. \quad \ldots\ldots\ldots\ldots\ldots(5)$$

Thus, if μ be known, (3), (4), (5) give the small "errors" in the angles of the prism.

The student is recommended to draw the necessary four figures to complete the six.

The measurements are made just as in § 83. In Fig. 64, the face AC is turned towards the goniometer, and, for θ_B, the auto-collimation is effected by reflexion first at AC for the normal and then at BC; a screen is used to prevent rays from the goniometer from falling on the right half of AC. The screen is then moved to prevent rays from falling on the left half of AC, and, for ψ_B, the auto-collimation is effected by reflexion at AC for the normal and then at AB.

The minimum deviation D, for the angle A, is measured by a spectrometer or, with sufficient accuracy, on a drawing board; the index is taken as $\mu_0 = 2 \sin \frac{1}{2}\,(D + \frac{1}{3}\pi)$, since $\sin \frac{1}{6}\pi = \frac{1}{2}$. If A_0, the value of A obtained from (3) on using μ_0 for μ, differ seriously from $60°$, we may recalculate μ by the equation

$$\mu = \sin \tfrac{1}{2}\,(D + A_0)/\sin \tfrac{1}{2}A_0,$$

and then use this value for μ in (3), (4), (5).

The equations
$$\theta_B = -\,\phi_A, \qquad \phi_C = -\,\psi_B, \qquad \psi_A = -\,\theta_C,$$

which follow from previous equations, may be used to test the consistency of the measurements.

The prism has been assumed free from pyramidal error. Unless this error be small, it will not be possible to set the prism so that the images of the goniometer wire corresponding (1) to NR and (2) to UR, in Fig. 64, can *both* be brought into view by merely adjusting the arm of the goniometer.

94. Practical example. The prism used by G. F. C. Searle and W. A. Wooster had $D = 44°\ 52'$ for angle A for sodium light. Hence

$$\mu_0 = 2 \sin \tfrac{1}{2}\,(D + \tfrac{1}{3}\pi) = 1\cdot585.$$

The values of θ_B, θ_C, ... obtained from the goniometer readings were

$$\theta_B = -\cdot00700, \qquad \phi_C = \cdot04475, \qquad \psi_A = -\cdot03700 \text{ radian},$$
$$\theta_C = \cdot03475, \qquad \phi_A = \cdot00775, \qquad \psi_B = -\cdot04425 \text{ radian}.$$

Hence $3A_0 - \pi = (\theta_B + \theta_C)/1\cdot585 = \cdot017508$ radian $= 1° \; 0' \; 11''$.

Similarly, with $\mu_0 = 1\cdot585$, $3B_0 - \pi = 1° \; 53' \; 52''$, $3C_0 - \pi = -2° \; 56' \; 13''$.
Hence

$$A_0 = 60° \; 20' \; 4'', \qquad B_0 = 60° \; 37' \; 57'', \qquad C_0 = 59° \; 1' \; 16''.$$

On re-calculating μ, with $A = 60° \; 20' \; 4''$, we find $\mu = 1\cdot581$. Then re-calculating A, B, C, using $\mu = 1\cdot581$,

$$A = 60° \; 20' \; 7'', \qquad B = 60° \; 38' \; 3'', \qquad C = 59° \; 1' \; 6''.$$

Taking the sum, $A + B + C = 179° \; 59' \; 16''$.

Experiment 18. Range finding by prism of constant deviation.

95. Design of prism of constant deviation. Let AB, CD

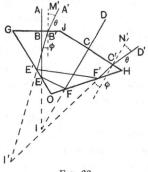

FIG. 66

(Fig. 66) be given as initial and final rays, the two intersecting in I. On AB, CD take points E, F, and join EF. If we convert the ray ABE into the ray EF, and EF into FCD, we shall have the desired result. A plane mirror GEO, passing through E and perpendicular to the plane AID, will convert ABE into EF, if $GEB = OEF$. Similarly a mirror OFH, perpendicular to the plane AID, will convert EF into FCD, if $OFE = HFC$.

Let planes GBJ, HCJ be drawn perpendicular to AB, CD, let the space enclosed by $OGJH$ be filled with homogeneous glass, and let the faces OG, OH be silvered. The ray AB is not bent by refraction at B, since the incidence is normal. The ray BE gives rise to FC which passes out along CD without change of direction.

Let a ray $A'B'$ in the plane AID, i.e. in a principal plane of the mirrors, become $B'E'$ after refraction, the angles of incidence and refraction at B' being θ, ϕ. The ray $B'E'$ gives rise to $E'F'$, $F'C'$, and the angle $B'I'C'$ between $E'B'$ and $F'C'$ is, by § 79, equal to BIC. Hence, the angle between $F'C'$ and the normal to HJ is ϕ, and, therefore, the emergent ray $C'D'$ makes with the normal $C'N'$ at C an angle θ, equal to that between $A'B'$ and the normal $B'M'$. The angle between the rays $A'B'$, $C'D'$ equals that between the rays AB, CD, and hence the deviation is independent of the angle of incidence of the initial ray on GJ.

If $A'B'$ be the *projection* on the principal plane of the part $A''B''$ of a ray $A''B''E''F''C''D''$, which is *not* in a principal plane, the projection of $C''D''$ is parallel to, though not coincident with, $C'D'$, and the inclination of $C''D''$ to the principal plane equals that of $A''B''$ (see §§ 18, 20).

For a given initial ray, the *position* of the final ray, but not its *direction*, depends upon the index of the glass. Thus, if $A'B'$ be a ray of white light, the emergent rays of different colours will have different but parallel paths. If these rays fall on an achromatic lens properly placed, they will meet in a single point at the principal focus.

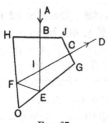

In an alternative design, we take E, F beyond I, the intersection of AB, CD, as in Fig. 67. The mirrors OEG, OFH are placed so that $GEA = FEO$, $HFC = EFO$, and the section of the prism is completed by drawing HBJ, GCJ perpendicular to AB, CD respectively.

FIG. 67

The prisms used in naval range finders give a constant deviation of $90°$ and are made on the plan of Fig. 68. The prism is symmetrical about OJ, and hence $IFE = IEF = 45°$, and $GOH = 45°$. The narrower part of GOH is optically useless, and the prism is therefore terminated at KL; this surface is not polished. The prism has then *five* faces and is sometimes called a pentagonal prism, although the ray meets only *four* faces. The faces GK, HL are silvered *.

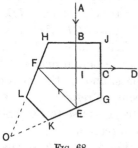

FIG. 68

96. Theory of method. The geometry of the problem, when the constant deviation is a right angle, is shown in Fig. 69. Let

* These prisms can be obtained from Messrs Barr and Stroud, Ltd., Anniesland, Glasgow, Scotland.

P be the inaccessible point, and $PQ = d$ the unknown distance.
If the prism be placed at Q, the ray PQ becomes QO, where
$PQO = 90°$. The prism is shown greatly
magnified. Select a point O on QO.
If the prism be at any other point R
on the semicircle PQO, the ray PR
becomes RO. Let $QPR = \theta$; then
$QOR = \theta$. The angle θ at O can be
measured. Draw RS, RT perpendicular to QP, QO, and let $RS =$
$TQ = x$, $RT = y$. Since $PS = x \cot \theta$,
we have

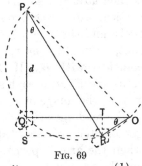

FIG. 69

$$d = PQ = x \cot \theta - y. \quad \ldots\ldots\ldots\ldots(1)$$

97. Experimental details. The distant object P may be a
vertical rod placed in front of a white card. The prism is placed
on a carriage C (Fig. 70) on an optical bench BB. The carriage

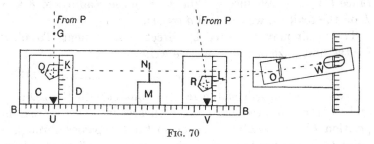

FIG. 70

has a scale D at right angles to BB, and a transparent face of
the prism is placed in contact with D. The carriage index is
first set to some definite reading U on BB, and the right angle
edge of the prism is placed at a definite point K on D. An
auxiliary carriage M, bearing a plate pierced by a small hole N,
is placed so that QN is small, and the plate is adjusted so that
a line through N parallel to BB falls on the centre of the face
of the prism. Then QN is made large, and the observer looks
through N and adjusts the direction of BB so that he sees P
by reflexion in the prism; BB is now perpendicular to PQ. If
the prism stand on a small levelling platform which rests on
the carriage C and can slide along the scale D, the principal
plane of the prism is easily set horizontal.

A goniometer, placed at the proper end of BB, is adjusted so that its pivot O lies approximately on the line QN. The arm is then adjusted so that the image of P coincides with W, the vertical wire of the goniometer*. The carriage is then moved to V, and the edge of the prism is set to L on the scale D, so that, on readjusting the goniometer arm, the image of P again coincides with W. If the passage from U to V be made by steps and if, after each step, both prism and goniometer be set so that the image of P is visible, the image will not be "lost."

The readings of U and V on BB, of K and L on D, and of the goniometer are recorded, as well as the reading of the latter when the arm is perpendicular to its scale. If the index wire cut the edge of the scale at distances h_1, h_2 cm. from its centre in the positive direction, and if the distance from the centre of the pivot to the edge of the scale be l cm.,

$$\cot \theta = (l + h_2 h_1/l)/(h_2 - h_1). \quad \ldots\ldots\ldots\ldots(2)$$

Since $UV = x$, and since y is known from the readings of K and L on the scale D, we can find d by (1).

The result may be tested by direct measurement. To allow for the finite size of the prism, we write

$$d = z + \tfrac{1}{2}c, \quad \ldots\ldots\ldots\ldots\ldots\ldots(3)$$

where c is the length of the face in contact with the scale D, and z is the distance from P to the nearer face of the prism in position U. A plumb-line is set so that it passes through a point G on QP near Q and marks out a point G' on the floor. The horizontal distance between the face of the prism and the plumb-line is measured. The horizontal distance of P from a point H' on the floor is similarly measured, and $G'H'$ is measured by steel scales or by a *reliable* tape.

A prism giving a constant deviation of 90° may be used for another purpose. The arm of the goniometer is kept parallel to the bench BB (Fig. 70), and light is sent from W to O either by the auto-collimating prism or otherwise. If, now, the carriage

---

* Since PQ is finite, the image of P is not exactly in the focal plane of the goniometer lens. If, however, a converging lens of about 2 dioptres be *fixed* to the goniometer arm a few centimetres in front of W in such a position that the image of P coincides with W when the carriage C is midway between U and V, the focusing will be satisfactory over the whole range of the carriage.

be moved along the bench, the beam of parallel rays, which issues from the prism Q and corresponds to X, the point on the wire W defining the line of collimation of the goniometer, will remain at right angles to that line, in spite of any lack of straightness of the bench. If the beam issuing from Q be nearly horizontal, the *projection* of that beam on a horizontal plane will suffer only negligible (second order) variations of direction as the carriage moves along the bench. The device thus furnishes the equivalent of a parallel beam of great width.

98. Practical example. G. F. C. Searle and T. Bradbury used prism for which $c = 2.8$ cm. By direct measurement,

$$d = z + \tfrac{1}{2} c = 1406.1 + 1.4 = 1407.5 \text{ cm.}$$

Central reading of goniometer was 10 cm., and l was 40 cm. The mean goniometer readings for positions U, V gave $h_1 = -1.460$, $h_2 = 1.387$ cm., and, by (2),

$$\cot \theta = (40 - 1.387 \times 1.460/40)/(1.387 + 1.460) = 14.032.$$

The readings for U, V gave $x = 100$ cm.; those for K, L gave $y = 1.3$ cm. By (1), $d = PQ = x \cot \theta - y = 1403.2 - 1.3 = 1401.9$ cm.

CHAPTER VI

SPHERICAL MIRRORS

99. Image formed by spherical mirror*. If a luminous point P be at C, the centre of curvature of the mirror, any ray falling on the mirror is reflected back along its path; hence *all* the reflected rays meet in C, which is thus self-conjugate. When P has any other position and the angle of incidence is more than infinitesimal, the reflected rays do not meet accurately in a single point, and focal lines are formed. In this Chapter only rays falling nearly normally on the mirror will be considered.

Let the centre of curvature of the mirror AA' be at C (Fig. 71) and let r be its radius; AA' is called the aperture of the mirror. We count r *positive* when the mirror is *concave*. If PC cut the mirror in H, then CH is called the axis, and H the pole, of the mirror with respect to P. Let a ray PM

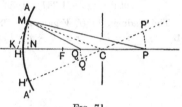

FIG. 71

meet the mirror in M, and let its direction after reflexion cut the axis in Q. Let $HP = u$, $HQ = v$, and let u, v be positive when the rays actually pass through P, Q. By the law of reflexion, $PMC = QMC$, since MC is the normal. Hence $MQH - MCH = MCH - MPH$, or $MPH + MQH = 2MCH$, an *exact* equation. Draw $MN = z$, perpendicular to HC. When z/r is small, the small angles MQH, MCH, MPH are z/v, z/r, z/u respectively. Then z divides out, and we have

$$1/u + 1/v = 2/r = 1/f. \quad \dots\dots\dots\dots\dots(1)$$

Since z does not appear in (1), *all* rays from P striking the mirror very near H meet again in Q, which is thus the image of P. If the luminous point be at Q, its image will be at P.

---

* Many mirrors are formed of glass silvered at the back, the surfaces being spherical but not necessarily concentric. Such mirrors are considered in EXPERIMENT 24.

If P be at infinity, $1/u = 0$, and then Q is at F, where $HF = \frac{1}{2}r$. The point F is the focus of the mirror, and HF is the focal length f, which is counted positive when the mirror is concave. If P be placed at F, the reflected rays will be parallel to the axis.

If P describe an arc PP' of a circle about C, the pole corresponding to P moves along the mirror to H'. If the rays from P' be restricted to fall on a *small* area surrounding H', they will meet again in Q' on $H'C$, where $H'Q' = HQ$. The restriction may be effected by placing a "stop" at C, i.e. a plate pierced by a small circular hole. If P' lie between the mirror and the stop, the rays from P' can, indeed, strike the mirror at considerable distances from the corresponding pole, but only those striking the mirror near H' penetrate the stop, and these form at Q' an aberrationless image of P'.

For simplicity, we shall now suppose that CH is the normal from C to the plane of the circular rim of the mirror, and shall call CH the principal axis, and H the vertex of the mirror. We further suppose that the angle $P'HC$ is very small and that only rays nearly parallel to the axis are utilised. We may then replace the arcs PP', QQ' by short straight lines normal to the axis, the two lines being in one plane. Let $PP' = h$, $QQ' = k$. Since $P'CQ'$ is straight, $h/k = CP/CQ$. But the ray from P' which is reflected at H passes through Q', and the angles of incidence and reflexion are equal. Hence $h/k = HP/HQ = u/v$.

Let $CP = p$, $CQ = q$, and let p, q be *positive* when P, Q lie on the *same* side of C as F. Then $u = r - p$, $v = r - q$, and (1) becomes, on reduction,

$$1/p + 1/q = 2/r. \quad \dots\dots\dots\dots\dots(2)$$

In every case of reflexion, if the luminous point move along the axis, its image moves in the *opposite* direction.

An *exact* expression is easily found for the distance from C (Fig. 71) of the point Q in which the ray reflected at M cuts the axis. Let $MCH = \theta$, $PMC = QMC = \phi$. Then

$$\frac{1}{CQ} - \frac{1}{CP} = \frac{\sin(\theta + \phi)}{r\sin\phi} - \frac{\sin(\theta - \phi)}{r\sin\phi} = \frac{2\cos\theta}{r} = \frac{2}{CK}, \dots(3)$$

where K is the point in which the tangent to HA at M cuts the axis.

EXPERIMENT 19. **Measurement of radius and focal length of concave mirror.**

100. First method. The mirror AA' (Fig. 72) is fixed with
its principal axis horizontal,
and a pin is adjusted so that
its *tip* is self-conjugate. Then
the *tip* is at C, the centre of
curvature of the mirror. The
distance of C from any point
of the mirror is r. The image of
the pin is inverted.

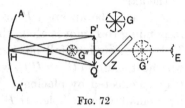

FIG. 72

If the surface of AA' be of glass, the image of CP' will be
weak unless CP' be bright. If the mirror be formed of a lens
with a concave unsilvered face, the back surface should be
coated with black varnish to diminish reflexion at that
surface. If CP' be a small flame, the inverted image is
easily seen, but accurate measurements cannot be made. A
good plan is to use a glass plate Z to reflect light on to the
mirror from a flame G. Let G' be the image of G by reflexion
at Z, and G'' the image of G' by reflexion at the mirror. An eye
at E sees the image of the flame at G'', and this image forms
a bright background against which both the pin CP' and its
image CQ' stand out as dark spikes. The coincidence of image
and object is tested by the parallax method by moving the eye
from side to side. The image G'' should subtend a considerable
angle at C; thus Z should be close to C, and G as near Z as is
convenient and safe. Since the image is formed by reflexion
and not by refraction, a white flame may be used. If G be an
electric lamp, a sheet of ground glass is placed between G and Z.
The reflector Z should be of tested (§ 31) plate glass.

The distance HC may be measured by the device shown in
Fig. 129. If the mirror and the pin be mounted on carriages on
an optical bench, the reading of the pin carriage is taken when
the tip of the pin is at C. The carriage is then set so that a
rod of known length touches H and the tip of the pin at the
same time, and the carriage reading is taken again. From
these readings and the length of the rod, the radius HC is
found.

101. Second method. We now use a *pair* of conjugate points. If the pin Q (Fig. 73) lie between C and F, an inverted image will be formed beyond C; a second pin is set at P so that its tip coincides with the image of the tip of Q. Then $HP = u$ and $HQ = v$ are measured, as in § 100, and $1/f$ is found by (1), § 99. Observations may be made with a series of values of v; the

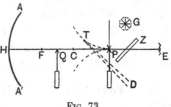

FIG. 73

values should be spread over a considerable range. The mean value of $1/f$ is used to give f.

If $1/v$ be plotted against $1/u$, the points will lie on a straight line cutting either axis at $1/f$ from the origin.

When an unsilvered mirror is used, a sloped glass plate Z and a flame G may be used as in § 100.

When Q is not coincident with C, the reflected rays do not pass accurately through a point at P but touch a caustic surface (Fig. 73); the caustic is not drawn to scale but is exaggerated to show the effect. To an eye which moves through a small distance across the axis near E the image of Q appears to be at P. When the eye is near D, the rays appear to come from the neighbourhood of T. For large movements of the eye, the parallax test thus appears to show that the pin at P is further from H than the image of Q. The observer must, therefore, be content to restrict himself to rays nearly parallel to the axis PCH.

102. Third method. We can find f by actually finding the position of the focus F. A glass plate Z, which has been tested for flatness of surfaces, § 31, reflects the rays from a distant object on to the mirror. After reflexion at Z, these rays still form a parallel beam, and thus, after reflexion at the mirror, come to a focus at F, or at least in the focal plane, and an

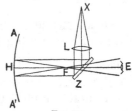

FIG. 74

eye at E looking towards the mirror sees an image of the object at F. A pin is then adjusted so that its tip coincides at F with the image. The plate Z is placed so that F lies between H and Z; then both pin and image will be viewed through Z. The distance HF is measured. Instead of a distant object, a collimator may be used, as in Fig. 74. The rays from X, the intersection of the cross-wires, form a parallel beam after passing through the lens L, and an image of X is formed at F. The goniometer of § 71 may be used as the collimator.

103. Fourth method. If a pin be placed with its tip at F (Fig. 75), the rays from F, after reflexion at the mirror, form a parallel beam. If this beam be received on the lens L of a goniometer (§ 71), an image of the pin will be formed on the cross-wire at X. The glass plate Z reflects light from the flame G towards the mirror.

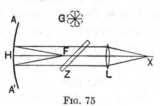

FIG. 75

104. Practical example. Mr H. W. B. Skinner used the front (concave) face of a lens.

First method. The mean reading of the pin carriage when tip of pin coincided with its image, was 76·20 cm. Reading for 30 cm. rod, 85·89 cm. The bench readings increased in the direction F to H. Hence

$$f = \tfrac{1}{2}(85{\cdot}89 + 30 - 76{\cdot}20) = 19{\cdot}85 \text{ cm.}$$

Second method. The results, over a rather small range, were

u	v	u^{-1}	v^{-1}	f^{-1}
47·65 cm.	34·00 cm.	·020986 cm.$^{-1}$	·029412 cm.$^{-1}$	·050398 cm.$^{-1}$
46·59	34·50	·021464	·028986	·050450
45·78	35·00	·021844	·028571	·050415

The mean value of f^{-1} is ·05042 cm.$^{-1}$. Hence $f = 19{\cdot}83$ cm.

Third and fourth methods. Each gave $f = 19{\cdot}86$ cm.

EXPERIMENT 20. **Measurement of radius of convex mirror by use of converging lens.**

105. Method*. The mirror M (Fig. 76), a converging lens S and a pin P are mounted on carriages on an optical bench. For accurate results, the lens must be reasonably free from spherical aberration under the conditions of the experiment.

---

* A mirror of finite thickness and silvered at the back can be treated by the methods of EXPERIMENT 24.

The system in Fig. 76 consists of two plano-convex lenses L_1, L_2, mounted coaxially in a tube; the plane faces A, B of the

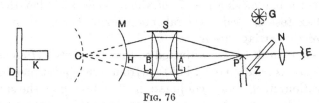

FIG. 76

lenses are turned outwards. The distance AP is such that rays from P, after passing through L_1, form a parallel beam. This parallel beam falls on L_2 and is brought to a focus by L_2. All the rays from P, which pass through S, will not meet *accurately* in a point, but the aberration is small—much smaller than when a single double-convex lens is used.

The line joining the tip P of the pin to C, the centre of curvature of the mirror, should be parallel to the optical bench. The adjustment is facilitated by a disk D fitted with a shank K. The shank is fixed to a carriage and, by a set-square, the face of D is made perpendicular to the length of the bench. Lycopodium is placed on D, which is then brought into contact with the mirror. The point of contact, H, is the centre of the patch of lycopodium left on the mirror. A pointed rod, e.g. a knitting needle, is fixed to a carriage and is adjusted so that its tip touches H. If the pin be now adjusted so that its tip P touches the tip of the rod, PC is parallel to the bench.

The lens S and M are now adjusted so that an image of the lycopodium at H is formed at P, when the pin P is (approximately) at the principal focus of L_1 for rays falling first on the convex surface of L_1. If L_1, L_2 be equal lenses, HP will be a minimum. The carriages bearing P and S then remain in position.

If the mirror be of unsilvered glass, the pin is illuminated by the sloped glass plate Z and the flame G. Unless the system be achromatic, a sodium flame is used. If the axis of S coincide with PC, it will be possible, by sliding the mirror carriage along the bench, to make the tip of the inverted image of P coincide with the tip of P itself. In this case, the rays fall normally upon the mirror, and hence C, the centre of curvature, is conjugate to P with respect to the lens. To obtain coincidence, some sideways adjustment of S may be necessary; the axis of S must be parallel to the bench. The pin and its image may be observed through a magnifying lens N of about 5 cm. focal length. The parallax test is used. Several independent bench readings of the mirror carriage are made.

The mirror M is next adjusted so that the vertex H comes to the previous position of C. The rays from P now meet (neglecting aberration) in the point conjugate to P and lying on the surface of M. After reflexion, the rays return from this point and meet again in P. The same is true for any point near P in the plane through P transverse to the bench, and hence an *erect* image of the pin is formed (see § 140). Since this image is not convenient for observation, we use the lycopodium at H, and adjust the mirror so that the images of the grains near H lie in the transverse plane through P. The lycopodium should be illuminated by sodium light and, as far as convenient, should be screened against other light. Several independent settings of the mirror carriage are made.

The difference between the two mean readings of the mirror carriage gives the radius, r, of the mirror.

The auto-collimating arrangement shown in Fig. 77 gives stronger illumination than the sloped plate. A wide slit, expanding into the opening J, is cut in the plate D. The slit is covered by the reflecting prism T. When rays from G fall on T, they are reflected through the slit past the wire W. The arrangement is fixed to a carriage and takes the place of the pin P.

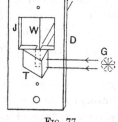

FIG. 77

If BC (Fig. 76) be large compared with the radius CH, the field of view, over which the image of P is visible, will be small, and the image will be difficult to find. If possible, the lens S should be such that BH is small compared with BC.

106. Practical example. G. F. C. Searle used the convex surface of a lens. This mirror was used in § 116. The bench readings of mirror carriage were:

C conjugate to P; 51·18, 51·18, 51·19, 51·19, 51·19. Mean 51·186 cm.

H conjugate to P; 44·61, 44·59, 44·60, 44·58, 44·58. Mean 44·592 cm.

Hence radius $= r = 51\cdot186 - 44\cdot592 = 6\cdot594$ cm.

EXPERIMENT 21. **Measurement of focal length of convex mirror.**

107. Direct method. The mirror is mounted on a carriage on an optical bench. The point H (Fig. 78) on the mirror where the tangent plane is perpendicular to the bench is marked by a small patch of lycopodium.
A glass plate Z of good quality is used as a reflector and is set with its plane vertical and inclined at about 45° to the length of the bench. A micro-scope M, with a micrometer scale at T, is mounted on a carriage so that (1) its axis is parallel to the bench, and (2) when the lycopodium at H

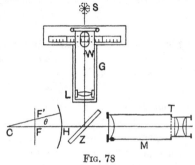

FIG. 78

is in focus, as viewed through Z, it is in the centre of the field of view. A goniometer G (§ 71) is placed with its axis approximately perpendicular to the bench. A source of light S is placed so that rays after passing through the lens L are reflected by Z on to the mirror. An image of the wire W is formed in the focal plane FF', and is viewed by the micro-scope. The goniometer arm is adjusted so that the image of W is central in the field of view. A piece of ground glass may be placed near W, between W and S. A *well illuminated distant object* may be used instead of the goniometer.

The microscope carriage is adjusted so that the image of W is focused, without parallax, on the micrometer scale, and the carriage reading is taken. The carriage is then moved back until the lycopodium at H is in focus. The difference of readings gives f directly. The plate Z remains in position throughout.

108. Goniometer method. If, when the microscope is focused on the image of the wire at F, the goniometer arm be moved, the image of W will move along the micrometer scale. When the image formed by the mirror moves from F to F', the direction of the parallel beam falling on the mirror changes from FC to $F'C$, where C is the centre of curvature, and the angle FCF' or θ equals the angle described by the arm. Since $CF = f$, we have $FF' = f.\theta$.

Let FF' cover n divisions of the micrometer scale. If, when the microscope is focused on a fine mm. scale, one micrometer division correspond to d cm., $FF' = nd$ cm. If θ correspond to x cm. on the goniometer scale and if the distance from pivot to scale be l cm., $\theta = x/l$. Hence

$$f = nld/x. \qquad \ldots\ldots\ldots\ldots\ldots\ldots\ldots(1)$$

109. Practical example. The mirror of §§ 106, 116 was used.

Direct method. The mean bench readings of microscope when focused (1) on lycopodium at H, (2) on image of wire, were 36·487, 33·200 cm. Hence

$$f = 36{\cdot}487 - 33{\cdot}200 = 3{\cdot}287 \text{ cm.}$$

Goniometer method. For goniometer, $l = 40$ cm.

Goniometer	12·0	11·5	11·0	10·5	10·0	9·5 cm.
Micrometer	1·15	1·75	2·32	2·91	3·48	4·06 divisions

When readings for 12·0, 11·5, 11·0 are subtracted from those for 10·5, 10·0, 9·5 cm., mean difference is $n = 1{\cdot}743$ divisions for $x = 1{\cdot}5$ cm. Also 0·3 cm. covered 4·38 micrometer divisions; thus $d = {\cdot}3/4{\cdot}38$ cm.

By (1), $f = nld/x = 1{\cdot}743 \times 40 \times {\cdot}3/(1{\cdot}5 \times 4{\cdot}38) = 3{\cdot}184$ cm.

EXPERIMENT 22. **Determination of radius of spherical surface by revolving table method*.**

110. Method. A table turning about a vertical axis is required. The plane of the top of the table is normal to the axis, and the top carries a straight scale against which slides a carriage bearing the spherical surface (see Fig. 80). The scale is so adjusted on the table that the straight line described by the centre of curvature of the surface, when the carriage slides along the scale, *intersects* the axis of the table. The position of the carriage relative to the table-top when the centre of curvature lies on the axis of the table will be called the *first position*. If the table be turned through any angle about its axis when the carriage is in this position, the motion merely substitutes one part of the surface for another. Hence, if rays from a point fall upon the surface, the reflected rays will be unaffected by the motion. If a vertical line be used as an object, the "image"

* "On a revolving table method of determining the curvature of spherical surfaces." By G. F. C. Searle, A. C. W. Aldis, and G. M. B. Dobson, *Philosophical Magazine*, Feb. 1911.

by reflexion will be a vertical line formed by the vertical focal lines corresponding to individual points of the object. If a microscope be used for observing the image, the adjustment of the carriage along the scale can be made with great accuracy. The microscope must be furnished with cross-wires or with a micrometer-scale. In the absence of a microscope, a telescope with cross-wires may be used, if an extra converging lens be fitted in front of the objective to allow focusing at short distances.

The carriage is now moved along the scale into a *second position* in which the axis of the table is a tangent line to the spherical surface. If the table be turned about its axis, a grain of lycopodium on the surface at the point of contact of the vertical tangent line will remain stationary. When a microscope is used, this setting can be made with great accuracy.

The radius of curvature is given by the difference of the two scale-readings of the carriage in the first and second positions.

When the straight line described by the centre of curvature, as the carriage slides along the scale, misses the axis of revolution of the table by a short distance, the error in each setting of the carriage will be very small, provided the table be turned about a mean position in which the scale is parallel to the axis of the microscope.

111. Experimental details. The adjustments of the surface

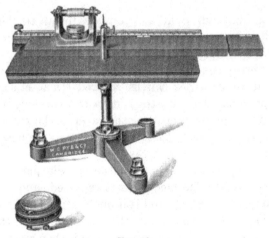

Fig. 79

to be tested are facilitated by the use of a small lathe-head to form the "carriage"; we describe the method of making the measurements when the lathe-head is used.

The table-top turns freely, but without shake, about a vertical rod carried by a tripod stand. A scale graduated in millimetres can be clamped in any position on the table by the screw seen in Fig. 79.

The lathe-head is attached by a screw to a straight-edged board resting upon the top of the table with its straight edge in contact with the scale. The spindle of the lathe-head is

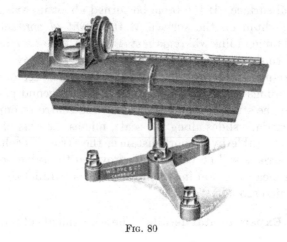

FIG. 80

screwed at one end to fit a face-plate furnished with three screws; these serve to adjust a second plate to which the lens or mirror is attached by wax. Each end of the spindle is conical, and each vertex lies on the axis of the spindle. One arm of a rod bent at right angles can be secured by a set-screw in a socket supporting the table-top, and a clip carrying a pin can be fixed in any position on the other arm of the rod by a set-screw. The bent rod can also be used to clamp the board to the table in the manner shown in Fig. 80.

The first step is to make the lathe spindle parallel to the edge of the board. The face-plate is removed, and the tip of the pin carried by the bent rod is brought into contact with the vertex of one end of the spindle, the straight edge of the board being in contact with the scale. The board is then removed from the table and is replaced so that the pin is near the other vertex. If, by sliding the board along the scale, the tip of the pin can be made to *touch* this vertex, the spindle is parallel to the edge of the board. The clamping screw enables the lathe-head to be adjusted on the board. No provision has been made for *adjusting* the spindle so that the angle between it and the plane of the top of the table may be zero. If this small angle be θ, the error in the measurement of the radius r will be $r\,(1 - \cos\theta)$ or $\frac{1}{2}r\theta^2$. With careful workmanship, the error will be negligible.

When the adjustment of the lathe-head on the board is complete, the scale is adjusted on the table-top so that the axis of the spindle *intersects* the axis of the table. The scale and the board are first adjusted roughly so that one of the conical ends is not far from the axis of the table. A microscope, with a vertical cross-wire, is then focused on the vertex when the spindle is approximately perpendicular to the microscope, the image of the vertex lying on the cross-wire. The table-top is then turned through 180°; if the image of the vertex do not again lie on the cross-wire, the difference is *halved* by moving the board along the scale, and the microscope is then moved to bring the image again to the cross-wire. The table-top is now turned so that the spindle is parallel to the microscope. If the image of the vertex do not lie on the cross-wire, the scale is moved at right angles to itself until the coincidence be obtained. The vertex of the spindle then lies on the axis of the table, and the axis of the spindle will intersect that of the table for all positions of the board along the scale.

The face-plate is now attached to the spindle, and the lens or mirror to be tested is fixed to the adjustable plate by wax. If the lens be thin, care must be taken to avoid straining it. In the case of a lens, the back surface should be smeared with lamp-black and vaseline or be coated with black varnish to stop reflexion at that surface. The spindle is rotated, and some object is observed by reflexion at the spherical surface. If the rotation cause a motion of the image, the adjusting screws are manipulated until it remain stationary. The centre of curvature then lies on the axis of the spindle. This method is (in effect) used in lens factories.

A vertical line drawn on ground glass forms a convenient

object, if well illuminated, but a vertical rod suffices. A microscope may be used to facilitate the adjustment. As the spindle
is rotated, the image of the line will move to and fro over a
definite range. The spindle is turned so that the image of the
line is as far to the right as possible. The plate carrying the
surface is then moved as nearly as possible about a *vertical*
axis by the appropriate screw or screws, so that the image
moves to the left through the proper distance, i.e. through half
the range. The adjustment is then tested by rotating the
spindle.

The preliminary adjustments are now complete. If they be
not quite perfect, the error which they cause in the radius as
found by this method will be very small, since that error
depends on the *second* powers of the errors in the adjustments.

112. Measurement of radius. The board carrying the
lathe-head is now set into the "first position" in which the
centre of curvature of the surface lies on the axis of revolution
of the table. The ground glass with the vertical line is set in
such a position that the image of the line which is observed in
the microscope is formed by rays falling *nearly normally* upon
the surface. If the image move when the table is turned about
its axis, the board carrying the lathe-head is adjusted along the
scale until the image remain at rest. The centre of curvature
then lies on the axis of the table. The scale-reading of an index
mark on the board is then recorded. The mean of several
readings is taken. The accuracy of setting is sometimes limited
by the departure of the surface from perfect sphericity.

The board is next adjusted to the "second position" in which
the axis of the table is a tangent line to the reflecting surface.
Lycopodium is placed on the surface, and then a pointed piece
of wood is held against the surface while the spindle is rotated,
so as to remove all the lycopodium but a *small* circular patch.
If the axis of the spindle be perpendicular to that of the table,
the tangent plane to the surface at the centre of this patch
will be parallel to the axis of the table. The board is then
adjusted along the scale so that the central grains of the patch
remain stationary when the table is turned about a mean
position in which the spindle is *parallel* to the microscope. When

the adjustment is complete, the scale-reading of the index mark is recorded. The mean of several readings is taken.

The difference between the two mean readings in the first and second positions gives the radius of the surface. Each setting is easily made to $\frac{1}{10}$ mm.

The following experiment tests the method. Two lenses, one convex and the other concave, are selected so that wide Newton's rings are seen by sodium light, when the marked face of the convex lens is in contact with the marked face of the concave lens. The radius of the marked convex surface is determined by the revolving table, and the radius of the marked concave surface is deduced from measurements of the rings (see § 313). The radius of this surface is also determined by the table, and the two values are compared.

113. Practical example. Lenses A, B were used. Radius R of convex surface of A found by table to be 10·81 cm. Lens A was placed on B; Newton's rings observed with sodium light and normal incidence. Hence $\phi = 0$.

No. of ring, n	1	2	3	4	5
Diameter $2\rho_n$	·78	1·23	1·51	1·75	1·96 cm.
$\rho_n{}^2 - \rho_{n-1}{}^2$	—	·2261	·1918	·1956	·1948 cm.^{2}

The mean is ·2021 cm.^{2}. If S be the radius of concave surface of B, by § 312,

$$1/R - 1/S = \lambda/(\rho_n{}^2 - \rho_{n-1}{}^2) = 5\cdot89 \times 10^{-5}/0\cdot2021 = 2\cdot914 \times 10^{-4} \text{ cm.}^{-1},$$

and hence $S = 10\cdot84$ cm. Measurements with table gave $S = 10\cdot82$ cm.

EXPERIMENT 23. **Measurement of radius of mirror by Chalmers's method.**

114. Method. The mirror AQ (Fig. 81) is mounted on a carriage K, which moves along a *straight* horizontal slide S. Let O be the centre of curvature of the mirror, OAB a horizontal line perpendicular to S. Let CD be a horizontal scale, parallel to S, so placed that OAB meets the divided part of CD; the divided face is vertical and is turned towards the mirror. Let M be a suitable microscope or telescope fitted with cross-wires or with a micrometer scale, and let ETQ, its line of collimation, cut CD in T and the mirror in Q, the line TQ being parallel to OAB and in the same horizontal plane. Let P be a point on CD, in the horizontal plane through O, such that a ray PQ, after reflexion at Q, travels along QT. The observer will then see P on the vertical cross-wire, if the microscope be properly focused.

Let the radius $OA = r$. Let $AB = h$, $BT = x$, $TP = y$, and let $AOQ = \phi$. Then $x = r \sin \phi$. Draw QN perpendicular to

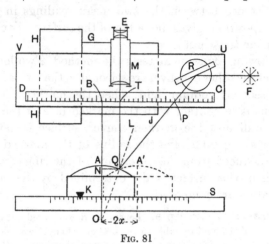

FIG. 81

OAB, and join AQ. Then $AQN = \frac{1}{2}\phi$, and thus $AN = x \tan \frac{1}{2}\phi$. Since QT is parallel to OAB, and since the radius OQL bisects TQP, we have $TQP = 2\phi$. Thus *

$$r = x/\sin \phi, \quad \dots\dots\dots\dots\dots\dots(1)$$

where $\quad \tan 2\phi = \dfrac{TP}{TQ} = \dfrac{y}{AB + AN} = \dfrac{y}{h + x \tan \frac{1}{2}\phi}. \quad \dots\dots(2)$

To find ϕ we use a method of approximation.

When x/h is small, and $y < h$, we have roughly $2\phi = y/h$. Then, as a rough approximation, $x \tan \frac{1}{2}\phi = xy/4h$. We therefore take

$$\tan 2\phi_0 = \frac{y}{h + xy/4h}, \quad \dots\dots\dots\dots\dots(3)$$

as a starting point, and find ϕ_1, ϕ_2, ... by the equations

$$\tan 2\phi_1 = \frac{y}{h + x \tan \frac{1}{2}\phi_0}, \quad \tan 2\phi_2 = \frac{y}{h + x \tan \frac{1}{2}\phi_1}, \quad \dots$$

and so on. After two or three steps, a value of ϕ will be reached which is not appreciably changed by an additional step.

* In the case of a *concave* mirror we have, in place of (2),
$$\tan 2\phi = y/(h - x \tan \frac{1}{2}\phi).$$

115. Experimental details. The mirror is mounted so that it can be moved along the *straight* slide S; the straightness of S is necessary for accuracy. The mirror is, of course, *not* of glass silvered at the back. A rod is mounted on the mirror carriage, and the carriage is moved along S; if the distance between the scale CD and the tip of the rod remain constant, CD is parallel to S. The microscope is adjusted so that its axis is perpendicular to CD. If CD be a glass scale, the microscope may be placed as in Fig. 81, the observations being taken through the scale. The scale CD is fixed to a board UV, which is carried by a block GJ, sliding against the guide HH. The slide S and the guide HH may, conveniently, be attached to a metal "surface plate" or to a sheet of plate glass. The block GJ and the mirror carriage K can then slide on the plate. The scale CD may be made parallel to S by adjusting HH. The slide S may be an engineer's steel rule fixed to a strip of wood with its divided face horizontal, so that K is guided by its straight edge.

The microscope (Fig. 23) may rest in V's; its objective is about a centimetre from CD. The block GJ is moved along the guide HH until the image of a point P on the scale CD be focused on the cross-wire, the mirror being in a position such as that shown in Fig. 81. The block GJ should now be secured in position. The final adjustment is made by sliding the microscope in its V's.

The height of the mirror is adjusted so that the image of the dividing line at P crosses the micrometer scale. Since the reflexion takes place at an unsilvered surface, good illumination is necessary. A plane reflector R, on a base fitted with levelling screws, stands upon the board UV, and directs light from the flame F past P towards the point Q on the mirror.

When the reading of the carriage K corresponding to P has been found, K is moved along S until the image of a point P', such that $P'T = TP$ approximately, come to the cross-wire; it will then also be in focus. This second position of the mirror is indicated in Fig. 81. The reflector R and the flame F must, of course, be moved to the other side of M. The difference between the two readings of K is $2x$ and that between the readings of P and P' is $2y$.

The distance $AB = h$ is measured by an adjustable distance

piece (Fig. 130) which may rest in two **V**'s to secure that it lie along the line OAB.

116. Practical example. In an experiment by Mr E. H. Taylor, line of collimation cut glass scale at 12 cm. The readings of P were 8, 7, ... and of P' were 16, 17, ... Thus $2y$ was 8, 10, ... cm. Distance $h = 10\cdot62$ cm. For $2y = 16$ cm., x was $2\cdot030$ cm. Then $xy/4h = 0\cdot38$ cm., $2\phi_0 = 36° 1' 39''$, $2\phi_1 = 36° 10' 20''$, $2\phi_2 = 36° 10' 8''$, $2\phi_3 = 36° 10' 8''$. It was unnecessary to go beyond ϕ_1. By (1),

$$r = x/\sin \phi = 2\cdot030/(\sin 18° 5' 4'') = 6\cdot540 \text{ cm.}$$

EXPERIMENT 24. **Determination of focal length of thick mirror.**

117. Theory of thick mirror*. The mirror may be formed of a piece of glass bounded by two spherical surfaces and silvered at the back, or may consist of any number of thick or thin lenses arranged coaxally along an axis, the reflexion taking place at a plane or spherical silvered surface situated behind the last of the lenses.

In Fig. 82, AC is the axis of the spherical surfaces. Let $P'X$ be an incident ray parallel to the axis. When this ray emerges

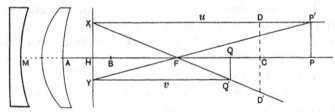

FIG. 82

from the system after reflexion at the silvered surface M, it will not, in general, be parallel to the axis, but will intersect the incident ray in some point X and cut the axis in some point F. Draw a plane XH cutting the axis at right angles in H. Then, if we reverse the ray XF, we have the two incident rays $P'X$ and FX giving rise to the emergent rays XF and XP', and thus X is its own image. From the properties of refractions and reflexions at spherical surfaces, it follows that *every* point near H in the plane XH is self-conjugate.

* The student may draw the figure for the case in which H lies between F and P—corresponding to a convex mirror.

The point conjugate to F is at infinity, since the ray XP' is parallel to the axis, and hence *every* ray which passes through F on incidence and makes a small angle with the axis is parallel to the axis on emergence.

The point F is called the Focus, and H the Principal or Unit Point of the system. The plane HX is the Principal or Unit Plane.

We can now find the image of P'. Let the incident ray $P'F$ cut the principal plane in Y. Since Y coincides with its own image, the emergent ray passes through Y. This ray YQ' is parallel to the axis, since the ray $P'FY$ passes through F. The image of P' is therefore at Q' where XF and YQ' intersect. If P and Q be the feet of the perpendiculars from P' and Q' upon the axis, then P and Q are conjugate points.

Let u and v be the distances of P and Q from the principal plane; let $HF = f$, and let u, v and f be counted positive when, in each case, a point moving in the direction of the reflected light reaches H before it reaches P, Q and F respectively. Then

$$f/u = YH/YX, \qquad f/v = XH/XY.$$

Hence, by addition, $\quad f/u + f/v = 1,$

or $\hspace{3cm} 1/u + 1/v = 1/f. \dotfill (1)$

Thus the system may be replaced by an ideal mirror of focal length f, if the surface of the latter cut the axis at right angles in H.

If we put $u = 2f$, we find from (1) that $u = v = 2f$, and then P and its image Q coincide; we may say that P is self-conjugate. If C (Fig. 82) be the corresponding point, the focus F is midway between C and H.

There are thus two points on the *axis* which are self-conjugate, viz. H and C, and if we use merely a *point* as an object we cannot determine by experiment which point is H and which is C. But the image of a *finite* object in the transverse plane through C lies in that plane and is *inverted*, for if this plane cut $P'X$ in D and XF in D', D and D' are on *opposite* sides of C. Since $HF = FC$ and since XD is parallel to the axis, $CD' = CD$. The point C corresponds to the centre of the ideal mirror and may be called the "centre" of the optical system.

On the other hand, the image of an object in the principal plane HX is erect.

The focal length f can be found by measuring CF. The point C is found by adjusting a pin so that its tip coincides with the tip of its *inverted* image; F can be found by a small screen, for distant objects will be focused at F.

It is better to use a telescope with cross-wires or a goniometer. This is focused for infinity and is turned to view the image, seen by reflexion in the mirror system, of a grain of lycopodium on that face of a glass plate which is turned towards the mirror. When the image is focused upon the cross-wire, the object is in the focal plane (see § 103).

Since the position of C can be readily found, we transform (1) so that we can use C as an origin of measurement. Let p and q be the distances of P and Q from C, and be positive when P and Q lie on the *same* side of C as F. Then $u = 2f - p$, $v = 2f - q$.

Hence, by (1), $1/(2f - p) + 1/(2f - q) = 1/f$,

or $$1/p + 1/q = 1/f. \quad \dots\dots\dots\dots\dots\dots(2)$$

Thus, if we find two conjugate points P and Q and measure their distances from the centre C, we can find f. The positions of F and H follow at once, since $CF = f$, and $CH = 2f$.

118. Gauss's method. The method corresponds for mirrors to Gauss's method (§ 215) of finding the focal length of lenses. The centre C (Fig. 82) is found by adjusting a pin so that its tip is self-conjugate. The distance of the tip of the pin from the vertex A is measured. Grains of lycopodium are then placed on the surface; let one grain be at A. The system will form an image of this grain at some point B (Fig. 82). The distance AB is measured by a microscope attached to a carriage sliding on an optical bench, the axis of the microscope being parallel to the direction of motion of the carriage. The microscope is first focused on A and the carriage is then moved until B, the image of A, be in focus; AB equals the distance through which the carriage has been moved.

The microscope should be fitted with cross-wires or with a micrometer scale; the absence of parallax between the image of the lycopodium and the wire or scale is a test of adjustment.

To distinguish between A and its image, an object is held against the surface. Binocular vision will decide whether B is in front of A or behind it.

The best illumination of the grains is obtained if a strong beam of light be made to fall upon them in a direction nearly tangential to the surface of the mirror.

Let $AB = x$, and let x be positive when B lies *in front* of A. If $AC = p$, $BC = p - x$; then, by (2), $1/f = 1/p + 1/(p - x)$.

Hence $$f = p\,(p - x)/(2p - x). \quad\ldots\ldots\ldots\ldots\ldots(3)$$
Thus f is nearly equal to $\frac{1}{2}p$ when x/p is small.

If the distance of H from A be h, and if h be positive when H lies *in front of* A, we have
$$h = p - 2f = px/(2p - x). \quad\ldots\ldots\ldots\ldots\ldots(4)$$
Thus h is nearly equal to $\frac{1}{2}x$ when x/p is small.

119. Practical example. Mr Heath used a glass concave mirror silvered at the back. Distance of centre C from vertex $A = p = 19\cdot67$ cm. The image of A was $\cdot955$ cm. *behind* A; hence $x = -0\cdot955$ cm.

By (3), $f = p\,(p - x)/(2p - x) = 19\cdot67 \times 20\cdot625/40\cdot295 = 10\cdot07$ cm.

By (4), $h = px/(2p - x) = -19\cdot67 \times 0\cdot955/40\cdot295 = -0\cdot47$ cm.

Hence the system is equivalent to an ideal concave mirror of focal length $10\cdot07$ cm., or of radius $20\cdot14$ cm., placed with its reflecting surface $0\cdot47$ cm. behind the front surface of the glass.

120. Micrometer method. Let K (Fig. 83) be a point on the surface near the vertex A. Let FK cut the principal plane HZ in Z. Then, since the ray FZ passes through the focus F, the corresponding emergent ray is parallel to the axis and, since Z is its own image, this emergent ray passes through Z. Hence the emergent ray is ZP, parallel to AF. Since the incident ray FK passes through K, the emergent ray ZP passes through the image of K, i.e. the image of K lies somewhere on ZP. Hence, an object AK, of height a, in contact with the surface, has its image of height b, where $b = HZ$. Since $CA = p$ and $CF = f$, we have $AF = p - f$, and thus $a/b = AF/HF = (p - f)/f$.

Hence $$f = bp/(a + b). \quad\ldots\ldots\ldots\ldots\ldots\ldots(5)$$

In this method we are not concerned with the *position* of the image of K, but this image is at the point in which CK cuts ZP, since the incident ray CK becomes the emergent ray KC.

The measurements are made by aid of a micrometer micro-scope. The eye-piece is adjusted so that the micrometer divisions

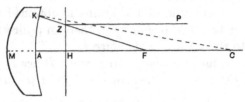

FIG. 83

are clearly seen. The points A, K may be represented by two grains of lycopodium. If the surface be *convex*, a glass scale may be placed with its divided face *in contact* with the vertex A. Since the faces of the scale are parallel, it has no effect upon the number of micrometer divisions covered by either AK or its image. The microscope is focused upon the two points so that there is no parallax between their images and the micro-meter scale, and the micrometer reading is taken for each point. Let the difference of the readings be m. Then the microscope is focused upon the images of K and A, and the difference of micrometer readings is again found; let this be n. Then, since $a/b = m/n$, we have, by (5),

$$f = np/(m + n). \quad \dots\dots\dots\dots\dots\dots(6)$$

The distance $p = AC$ is found, by aid of a microscope, as in Gauss's method, § 118.

The microscope may be mounted on a sliding carriage, as in Gauss's method. If a glass scale be used to provide the points A and K, it will not affect x (§ 118), since it displaces both K and its image through equal distances parallel to the axis.

121. Practical example. G. F. C. Searle used a double-convex lens placed in front of a plane mirror. A glass scale provided the points to be observed. When the glass scale was seen directly, the mean number ($7m$) of divisions corresponding to 7 mm. on the glass scale was 60·88. When the image of the scale was viewed, 4 mm. covered 55·48 or $4n$ divisions. Hence $m = 8·70$, $n = 13·87$. The distance $CA = p = 20·08$ cm. By (6), (4),

$f = 13·87 \times 20·08/22·57 = 12·34$ cm., $h = 20·08 - 24·68 = -4·60$ cm.

The system is equivalent to an ideal concave mirror of focal length 12·34 cm. with its surface 4·60 cm. behind the front surface of the glass.

122. Goniometer method. Let GF (Fig. 84) be an object of length s in the focal plane. Then, if GZ, parallel to the axis, cut the principal plane HZ in Z, the ray GZ will give rise to

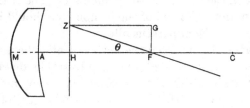

FIG. 84

the emergent ray ZF. If ZFH be θ, then, for small angles, $\theta = ZH/HF = s/f$. Or

$$f = s/\theta. \quad\quad\quad\quad\quad\quad (7)$$

Since G is in the focal plane, all the rays starting from G which emerge from the system are, on emergence, parallel to ZF; the aperture of the system is supposed so small that aberration is not conspicuous. If these emergent rays be received by a goniometer (§ 71), which has been focused for "infinity," an image of G will be formed on the wire, if the arm of the goniometer be properly directed. Thus, θ is the angle between the two positions of the arm in which (1) the image of F and (2) the image of G is focused on the wire.

The apparatus is arranged as in Fig. 85. The mirror system AM is set up with its axis horizontal. A glass scale S is attached

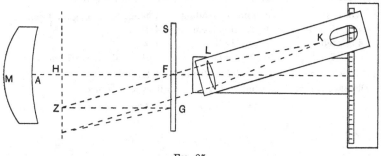

FIG. 85

to a sliding carriage moving parallel to the axis AF. The *divided* face is turned *towards* the mirror, and the plane of this face is

perpendicular to AF. The goniometer is then set with its axis
along AF, and S is adjusted by its carriage so that the image of
a nearly central division line is seen sharply focused (without
parallax) upon the wire K of the goniometer. The scale requires
good illumination. The end of the goniometer should be as close
to the scale S as the apparatus allows.

The distance AF between the vertex A and the divided face
of the scale S is measured by an adjustable distance piece
(§ 206). Or the reading of the carriage is taken, and then the
carriage is adjusted so that a rod of known length touches both
the vertex A and the glass scale.

The goniometer arm is then turned so that the cross-wire K
coincides with the images of a number of division lines on S in
turn, and the reading of the arm is taken in each case; the mean
distance, d cm., on the goniometer scale corresponding to s cm.
on the glass scale is found. The distance, l cm., from the centre
of the pivot to the edge of the goniometer scale is measured.
Then, if the small angle θ correspond to s cm. on the glass
scale, $\theta = d/l$ radians. Then, by (7),

$$f = s/\theta = sl/d. \quad\ldots\ldots\ldots\ldots\ldots\ldots(8)$$

The thickness of the scale S has no influence on the result.
The (parallel) rays emerging from the mirror system pass
through S, but their *directions* are unchanged by the passage.

123. Practical example. G. F. C. Searle used system of § 119.
Distance from centre of pivot to edge of scale of goniometer = l = 40 cm. The
glass scale was set in focal plane with divided face towards mirror.

Glass scale	Goniometer reading	Change of reading for 1 cm.
4·0, 5·0 cm.	6·58, 10·59 cm.	4·01 cm.
4·5, 5·5	8·59, 12·59	4·00
5·0, 6·0	10·59, 14·60	4·01 Mean 4·007 cm.

By (8), since $s = 1$ cm., $f = sl/d = 40/4·007 = 9·98$ cm.

Distance AF of scale from vertex $= f + h = 9·59$ cm. Hence

$$h = 9·59 - 9·98 = -0·39 \text{ cm.}$$

CHAPTER VII

THIN LENSES

124. Refraction at a spherical surface. Let C (Fig. 86) be the centre of curvature of the surface AX, let P be a luminous point and PX an incident ray. Let PC meet the surface in A and let the distance of X from CA be h. Let the media to the left and right of AX have indices μ_1 and μ respectively. Let $CA = r$, $AP = u$. Let RXZ be the direction of the ray after refraction; since RXZ is coplanar with PX, CX, the ray meets the axis in R. Let $AR = w$. Since CXN is the normal at X, $\mu_1 \sin PXN = \mu \sin RXN$. But

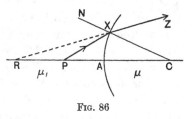

FIG. 86

$$PXN = XPA + XCA = h/u + h/r,$$

when the angles are small. Similarly $RXN = h/w + h/r$. Hence

$$\mu_1 (1/u + 1/r) = \mu (1/w + 1/r),$$

or
$$\mu_1/u - \mu/w = (\mu - \mu_1)/r. \quad\dots\dots\dots\dots\dots(1)$$

Since h does not appear in (1), *all* the rays from P which make very small angles with PC are refracted so that their directions pass through the single point R.

If P move so as to increase the angle PXN, R moves so as to increase the angle ZXC. Hence, if P move along PC, R will move along PC in the *same* direction as P.

Unless the angle PXN be small, those refracted rays which have suffered refraction at points very near X will not pass through a single point but will pass through two focal lines.

By a second application of (1), we can find the image of a point formed by rays refracted through a lens of any thickness separating two media. Lenses of finite thickness are considered in Chapter IX. For the present, we assume the thickness of the lens to be very small compared with the radii of curvature of its faces. Such lenses are best treated by the method of deviation.

125. Deviation method. As the standard lens we take a double-convex lens of refractive index μ relative to air. Let C, D (Fig. 87) be the centres of curvature of the faces AX, BY. Let the axis CD cut the faces in A, B, and let $AC = r$, $BD = s$. The radii r, s are counted *positive* when the corresponding faces are *convex*.

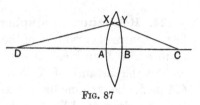

Fig. 87

The reciprocals $1/r$ and $1/s$ are the curvatures of the faces. Let the distance of either X or Y from the axis be h, and let the acute angle between CX and DY be i. Then $i = XCA + YDB$. But $\sin XCA = h/r$, $\sin YDB = h/s$. When the greatest value of h is so small compared with r and s that the angles may be replaced by their sines,

$$i = h/r + h/s = h\,(1/r + 1/s). \quad\ldots\ldots\ldots\ldots(2)$$

Let O (Fig. 88) be the mid-point of AB. Let P be a luminous point on the axis and let $OP = u$. Let a ray PX strike the lens at X, at distance h from the axis. The ray will be refracted at X, and, if XPA be a small angle, the inclination to the axis of the ray in the glass will also be small. Hence, if the refracted ray

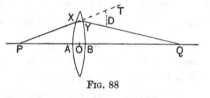

Fig. 88

strike the second face at Y, the distance of Y from the axis may be taken as equal to h, since the lens is thin. The ray emerges at Y and meets the axis in Q, where $OQ = v$. The angle between the faces at X and Y is, to our order of approximation, $h\,(1/r + 1/s)$, and the deviation, D, is the same as that due to a prism of this small angle. Thus, by § 27,

$$D = (\mu - 1)\,i = (\mu - 1)\,h\,(1/r + 1/s). \quad\ldots\ldots\ldots(3)$$

But D is also the sum of the angles XPA, YQB and thus, since those small angles are h/u and h/v,

$$1/u + 1/v = (\mu - 1)\,(1/r + 1/s). \quad\ldots\ldots\ldots\ldots(4)$$

For a given value of u, v is independent of the position of X,
and hence *all* the rays from P which strike the lens meet again
in the single point Q—the image of P. Approximations have
been made in the work, and thus (4) is only accurate for those
rays whose paths in the lens are (1) very near the axis and (2)
make very small angles with the axis. These conditions ensure
that the paths on either side of the lens are nearly parallel to
the axis.

If we write $\qquad (\mu - 1)(1/r + 1/s) = 1/f,$ ....(5)

then f is called the focal length of the lens. The reciprocal $1/f$
is called the "power" of the lens. Equation (4) now becomes

$$1/u + 1/v = (\mu - 1)(1/r + 1/s) = 1/f. \qquad ....(6)$$

If P be at infinity, $1/u = 0$, and then $v = f$. If Q be now at
F_2, then F_2 is one principal focus. If Q be at infinity, then
$u = f$, and if P be now at F_1, then F_1 is the other principal
focus. Hence $OF_1 = OF_2 = f$. The principal foci are thus at
equal distances from O, whether the radii r, s be equal or not.
The focal length of a thin lens of given index μ depends only
upon the *sum* of the curvatures, viz. upon $1/r + 1/s$.

Practical lens designers, lens manufacturers, opticians and
oculists assign a positive focal length to a thin converging lens.
Their example is followed in this book.

If OP (Fig. 89) be less than f, the ray is not bent sufficiently
to cut the axis after passing through
the lens. If, however, the *direction* of
the emergent ray be continued *backwards*, it will meet the axis at Q,
as in Fig. 89; the image is virtual.

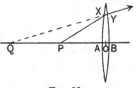

If X be at unit distance from the
axis, the deviation of the ray is $1/AP$

FIG. 89

$- 1/AQ$, and this equals $1/f$. We obtain this result if we put
$v = -AQ$ in (6).

The value of u is positive when the object is real, i.e. when the
rays actually diverge from P before they meet the lens. The
value of v is positive when the image is real, i.e. when Q and the
source of light are on opposite sides of the lens. The radii r
and s have positive values when the surfaces are convex to the
air. And the value of f is positive when a ray is bent towards

the axis. Thus the focal lengths of converging and diverging lenses have positive and negative values respectively.

Equation (6) is symmetrical with respect to P and Q, and thus P and Q are said to be conjugate. If P be an actual luminous point so placed that its image Q is real, it is true that a luminous point placed at Q will have its image at P.

But, if P be real and Q virtual, it is not true that a luminous point at Q has its image at P. Fig. 89 indicates that a luminous point at Q has its image either further from the lens than Q and on the same side of the lens or on the opposite side of the lens. The cases in which (a) P is virtual and Q real and (b) both P and Q are virtual call for similar warnings.

In general, we must distinguish between (1) the geometrical straight lines (extending to infinity in both directions) which we call the *directions* of the rays and (2) the rays themselves. With this distinction in mind, we may state the relation between image and object as follows:

Let the *directions* of the rays falling on the A-face of the lens pass through a point P which may be on either the A-side or the B-side of the lens. Then the *directions* of the rays after emergence from the B-face will pass through a point Q, which may be on the A-side or on the B-side of the lens according to circumstances. If the rays be *reversed*, so that the *directions* of the rays incident on the B-face pass through Q, the *directions* of the rays after emergence from the A-face will pass through P.

For a given value of XA or h (Fig. 88) the angle D is constant. Hence, if P move along the axis, Q moves along the axis in the *same* direction as P.

The meaning of (6) is simply that the deviation of a ray meeting the lens at unit distance from the axis equals the reciprocal of THE focal length. The expression "THE focal length" implies that there is the same medium on either side of the lens.

When the lens, of index μ, is *surrounded* by a medium of index μ', the relative index is μ/μ'. The focal length is then given by

$$\frac{1}{f} = \left(\frac{\mu}{\mu'} - 1\right)\left(\frac{1}{r} + \frac{1}{s}\right) = \frac{\mu - \mu'}{\mu'}\left(\frac{1}{r} + \frac{1}{s}\right). \quad \ldots\ldots(6a)$$

126. Newton's formula. If we write $u = f + p$, $v = f + q$, so that p, q are the distances of object and image from the corresponding foci, equation (6) becomes

$$1/(f + p) + 1/(f + q) = 1/f.$$

On simplification, we obtain Newton's formula

$$pq = f^2. \qquad \dots\dots\dots\dots\dots\dots\dots(7)$$

127. Thin lens separating two media. In Fig. 90, let the media to left and right of the lens have indices μ_1, μ_2; the index of the lens is μ. The ray $PXYQ$ is bent just as if the lens were replaced by a prism of small angle i, where i is the angle between the tangent planes at X, Y. If we write θ_1, ϕ_1 for θ, ϕ, and θ_2, ϕ_2 for θ', ϕ', equations (16), (15) of § 26 become $D = \theta_1 + \theta_2 - i$, $i = \phi_1 + \phi_2$.

When all the angles are "small," equations (18a) of § 26 give $\phi_1 = \mu_1 \theta_1/\mu$, $\theta_2 = \mu\phi_2/\mu_2$. Thus, since $\phi_2 = i - \phi_1$, we have

$$D = \theta_1 + \frac{\mu}{\mu_2}\left(i - \frac{\mu_1\theta_1}{\mu}\right) - i = \theta_1\frac{\mu_2 - \mu_1}{\mu_2} + i\frac{\mu - \mu_2}{\mu_2}. \quad \dots(8)$$

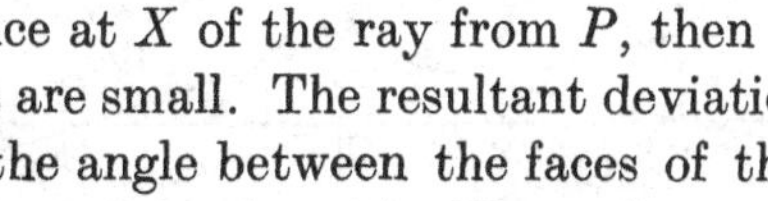

FIG. 90

If X (Fig. 90) be at unit distance from the axis and θ_1 be the angle of incidence at X of the ray from P, then $\theta_1 = 1/u + 1/r$, when the angles are small. The resultant deviation of the ray is $1/u + 1/v$, and the angle between the faces of the lens at unit distance from the axis is $1/r + 1/s$. Hence, by (8),

$$\frac{1}{u} + \frac{1}{v} = \left(\frac{1}{u} + \frac{1}{r}\right)\frac{\mu_2 - \mu_1}{\mu_2} + \left(\frac{1}{r} + \frac{1}{s}\right)\frac{\mu - \mu_2}{\mu_2},$$

or

$$\frac{\mu_1}{u} + \frac{\mu_2}{v} = \frac{\mu - \mu_1}{r} + \frac{\mu - \mu_2}{s} = \frac{1}{g}, \quad \dots\dots\dots\dots(9)$$

where g is used as an abbreviation.

Since the lens is very thin, we may consider that u and v are measured from O, the mid-point of AB.

If, when v is infinite, $u = f_1$, and when u is infinite, $v = f_2$, then f_1, f_2 are the two focal lengths of the system of lens and media. Thus

$$\mu_1/f_1 = \mu_2/f_2 = 1/g. \qquad \ldots\ldots\ldots\ldots(10)$$

The focal lengths are thus proportional to the indices of the two media in contact with the lens. Since $\mu_1 = f_1/g$, $\mu_2 = f_2/g$, equation (9) becomes

$$f_1/u + f_2/v = 1. \qquad \ldots\ldots\ldots\ldots(11)$$

If $u = f_1 + p$, $v = f_2 + q$, (11) becomes, on reduction,

$$pq = f_1 f_2, \qquad \ldots\ldots\ldots\ldots(12)$$

which is Newton's formula, when f_1 is not equal to f_2.

128. The magnification. If, in Fig. 86, P describe a small arc about C as centre, R describes an arc about the same centre, and in the same plane, and *all* the rays from P in any position on its arc will, after refraction at the face AX, pass through the corresponding position of R. For the small arcs we may substitute two short straight lines perpendicular to CPR. A second application of this result for refraction at the second face of the lens shows that, if P (Fig. 88 or Fig. 90) move along a short straight line intersecting the axis at right angles, Q moves along a similar line, the two lines being co-planar.

Let QQ' (Fig. 91) be the image of PP'. Then the ratio QQ'/PP' is called the magnification and is denoted by m. The magnification is *positive* when QQ' is erect.

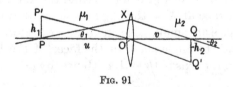

FIG. 91

Since, very close to O, the surfaces of the lens are parallel, the angles $P'OP$ and $Q'OQ$ are, by § 22, related in the same way as if the lens were removed and the ray $P'O$ were refracted at an interface between the two media of indices μ_1 and μ_2; the axis POQ is normal to the interface. Thus, if $PP' = h_1$, $QQ' = -h_2$,

$$\mu_1 h_1/u = -\mu_2 h_2/v, \qquad \ldots\ldots\ldots\ldots(13)$$

and $$m = h_2/h_1 = -(\mu_1 v)/(\mu_2 u). \qquad \ldots\ldots\ldots(13a)$$

Equation (9) determines the position of the image and (13) shows its size.

When $\mu_1 = \mu_2$, the rays $P'O$, OQ' are merely parts of the same straight line, and then $PP'/u = QQ'/v$, or

$$m = -\,QQ'/PP' = -\,v/u. \quad\quad\quad\dots\dots\dots\dots(14)$$

If $XPO = \theta_1$, $XQO = -\theta_2$, we have (Fig. 91), $u\theta_1 = -v\theta_2$, and then (13) becomes

$$\mu_1 h_1 \theta_1 = \mu_2 h_2 \theta_2, \quad\quad\quad\dots\dots\dots\dots\dots\dots(14a)$$

which is an example of Helmholtz's theorem. (See §§ 182, 189.)

When the angles $P'OP$, $Q'OQ$ are finite, the rays from P' do not meet in a single point at Q', but pass through two focal lines. See Chapter XI.

129. The elongation. Let the points P, R (Fig. 92) on the axis have Q, S for their images. Let $PO = u$, $QO = v$, $RO = u'$, $SO = v'$. Then the ratio

$$e = QS/PR = (v' - v)/(u - u')$$

is called the "elongation" of QS relative to PR. If μ_1, μ_2 be the indices of the media to the left and right of the lens, we have, by (9), $\mu_1/u + \mu_2/v = \mu_1/u' + \mu_2/v'$, and thus

$$e = (v' - v)/(u - u') = \mu_1 vv'/\mu_2 uu'. \quad\quad\dots\dots\dots(15)$$

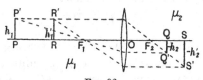

FIG. 92

If QQ' ($-h_2$) be the image of PP' (h_1), and SS' ($-h_2'$) be that of RR' (h_1'), we have, by (13),

$$v/u = -\,\mu_2 h_2/\mu_1 h_1, \quad\quad v'/u' = -\,\mu_2 h_2'/\mu_1 h_1',$$

and thus $e = (v' - v)/(u - u') = \mu_2 h_2 h_2'/\mu_1 h_1 h_1'. \quad\dots\dots\dots(16)$

When $\mu_2 = \mu_1$, as when the lens is in air,

$$e = (v' - v)/(u - u') = h_2 h_2'/h_1 h_1', \quad\quad\dots\dots\dots(17)$$

and now the elongation is the product of the magnifications $m = -\,h_2/h_1$ and $m' = -\,h_2'/h_1'$ at Q and S.

130. Combination of two thin lenses in contact in air.
Let the focal lengths be f_1, f_2. If a ray such as PX (Fig. 93)
meet the first lens at unit
distance from the axis, it will,
to our approximation, meet the
second lens also at unit distance
from the axis, since the lenses
are thin and in contact. The

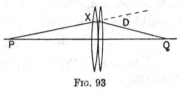

FIG. 93

lenses will bend the ray by $1/f_1$ and $1/f_2$ respectively. The total
deviation, D, is thus $1/f_1 + 1/f_2$ radians towards the axis. A
single thin lens of focal length f, produces the same deviation, if

$$1/f = 1/f_1 + 1/f_2. \qquad \dots\dots\dots\dots\dots(18)$$

If $f_2 = -f_1$, then f is infinite, and the combination acts as a
thin plate of glass with parallel faces.

131. Power of a lens. In many cases, it is more convenient
to use the *reciprocal* of the focal length than to use the focal
length itself. This reciprocal is called the "Power" of the lens
—a term due to Herschel—and is denoted by F. Thus

$$F = 1/f. \qquad \dots\dots\dots\dots\dots(19)$$

By § 130, if F be the power of the combination formed by two
thin lenses, of powers F_1, F_2, in *contact*,

$$F = F_1 + F_2. \qquad \dots\dots\dots\dots\dots(20)$$

In technical optics, the power of a converging lens having a
focal length of one metre is taken as the unit and is called a
Dioptre; such a lens is specified as "$+ 1D$." A diverging lens
of 25 cm. focal length would be specified as "$- 4D$." Spectacle
lenses are made to definite powers, such as $+ \cdot25D$, $+ \cdot5D$,
$+ 1D, \dots$ and $- \cdot25D, - \cdot5D, -1D, \dots$.

**132. Application of principle of least time to find pri-
mary image.** Let AKB (Fig. 94) be a thin lens. Let S be a
luminous point on the axis of the lens and let T be its image.
Let the radii of the faces AK, BK be a, b cm. and let the index
of the lens be μ. The radii are positive when the faces are
convex. Let the focal length be f cm., counted positive when

the lens is converging. Capital letters will be used to denote the power of the lens and the reciprocals of distances. Thus

$$1/f = F, \quad \dots\dots\dots\dots\dots\dots(21)$$

where F cm.$^{-1}$ or $100\,F$ dioptres is the power of the lens.

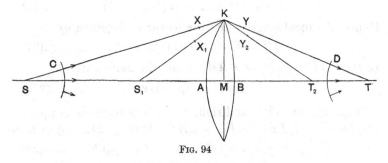

FIG. 94

When spherical aberration is negligible, any spherical wave front C, which expands from S as centre, becomes, after passing through the lens, a spherical wave front D, contracting to T as centre. Hence the time taken by light in passing from S to T is the same for every ray.

Let M be the point in which the plane of the edge of the lens intersects the axis. Let $SM = u, MT = v, KM = h$. The position of T is found by equating the optical length of the path from S to T for the ray which starts along SK to that for the ray which starts along SA. If X, Y be points on SK, TK such that $SX = SA$ and $TY = TB$, the optical length of the path from X to Y is equal to that of the path from A to B. Since the speed of light in glass is only μ^{-1} times its speed in air, the optical length of AB is μAB. Hence the optical equation is

$$XK + KY = \mu AB. \quad \dots\dots\dots\dots(22)$$

Since the distance of M from the centre of curvature of AK is $(a^2 - h^2)^{\frac{1}{2}}$, we have $AM = a\,\{1 - \sqrt{1 - h^2/a^2}\}$, and similarly for BM. Expanding as far as h^2, we find $AM = \frac{1}{2}h^2/a$, $BM = \frac{1}{2}h^2/b$. Hence

$$AB = \tfrac{1}{2}h^2\,(1/a + 1/b). \quad \dots\dots\dots\dots(23)$$

Now, as far as h^2,

$$SK = \sqrt{u^2 + h^2} = u + \tfrac{1}{2}h^2/u, \qquad TK = \sqrt{v^2 + h^2} = v + \tfrac{1}{2}h^2/v,$$
$$SX = SM - AM = u - \tfrac{1}{2}h^2/a, \qquad TY = TM - BM = v - \tfrac{1}{2}h^2/b.$$

Hence $\qquad XK = SK - SX = \tfrac{1}{2}h^2\,(1/u + 1/a),$ ....(24)

$\qquad\qquad\qquad KY = TK - TY = \tfrac{1}{2}h^2\,(1/v + 1/b).$ ....(25)

Multiplying (22) by $2/h^2$, we have by (23), (24) and (25),

$\qquad\qquad (1/u + 1/a) + (1/v + 1/b) = \mu\,(1/a + 1/b),$

or $\qquad\qquad\qquad 1/u + 1/v = (\mu - 1)\,(1/a + 1/b).$ ....(26)

Hence, the focal length f and the power F are given by

$$1/f = F = (\mu - 1)\,(1/a + 1/b).\qquad\text{.............(27)}$$

In terms of the thickness $AB = c$ and the half-width h,

$$1/f = F = 2\,(\mu - 1)\,c/h^2.\qquad\text{...............(27a)}$$

If the media to left and right of the lens have indices μ_1, μ_2, (22) becomes $\mu_1 XK + \mu_2 KY = \mu AB$. By (23), (24), (25) we have

$$\mu_1/u + \mu_2/v = (\mu - \mu_1)/a + (\mu - \mu_2)/b,\qquad\text{.........(28)}$$

which agrees with the result (9) found in § 127.

133. Secondary image. Some of the light from S (Fig. 94) which falls on BK is reflected and strikes AK. Here some light is reflected and then some of this reflected light is refracted at BK and passes out of the lens, forming a secondary image of S at T_2. Let $MT_2 = v_2$. If Y_2 lie on T_2K and if $T_2Y_2 = T_2B$, the optical equation is

$$XK + KY_2 = 3\mu AB,\qquad\text{...................(29)}$$

since the light which moves along the axis has traversed AB three times. This equation only differs from equation (22), corresponding to the primary image, by having 3μ in place of μ and v_2 in place of v. Hence the method of § 132 gives

$$1/u + 1/v_2 = (3\mu - 1)\,(1/a + 1/b).\qquad\text{.........(30)}$$

If f_2 be the focal length, F_2 the power for the secondary image,

$$F_2 = \frac{1}{f_2} = (3\mu - 1)\left(\frac{1}{a} + \frac{1}{b}\right) = \frac{3\mu - 1}{\mu - 1}\cdot\frac{1}{f},\qquad\text{.........(31)}$$

and $\qquad\qquad\qquad F_2 = F\,(3\mu - 1)/(\mu - 1).$ ....(32)

The symmetry of (30) shows that, if T_2 be the secondary image of S, then S is the secondary image of T_2. If $\mu = 1\cdot 5$ we have $f_2 = \tfrac{1}{7}f$.

If we measure f and f_2, we can find $\mu - 1$ from the equation

$$\mu - 1 = 2f_2/(f - 3f_2) = 2F/(F_2 - 3F).\qquad\text{......(33)}$$

134. Images by once reflected rays. Two images of S are formed by rays which suffer a single reflexion. One image is formed by reflexion at AK. With this we are not here concerned. If AK be concave, and if S be placed at the centre of curvature of AK, then S coincides with its own image.

A second image, S_1 (Fig. 94), is formed by rays which have suffered one reflexion at BK and two refractions at AK. Let $S_1M = u_1$, and let u_1 be positive when the image, S_1, is real.

If X_1 lie on S_1K and if $S_1X_1 = S_1A$, the optical equation is

$$XK + KX_1 = 2\mu AB. \quad\quad\dots\dots\dots\dots(34)$$

Multiplying (34) by $2/h^2$ and using § 132, we have

$$(1/u + 1/a) + (1/u_1 + 1/a) = 2\mu\,(1/a + 1/b),$$

or $1/u + 1/u_1 = 2\,(\mu - 1)\,(1/a + 1/b) + 2/b = 2/f + 2/b.$

If S be adjusted so that the image S_1 coincides with S, and if p denote the common value of u and u_1 in this case,

$$P = 1/p = \tfrac{1}{2}\,(1/u + 1/u_1) = 1/f + 1/b. \quad\dots\dots(35)$$

Similarly, if an object at a distance q from M on the other side of the lens coincide with its image formed by rays reflected once at AK and refracted twice at BK,

$$Q = 1/q = 1/f + 1/a. \quad\quad\dots\dots\dots\dots(36)$$

Adding (35) and (36) we have

$$1/p + 1/q = 2/f + (1/a + 1/b) = \{2 + 1/(\mu - 1)\}/f. \quad\dots(37)$$

Thus, if p, q and f be known, we can find $\mu - 1$ from the equation

$$\mu - 1 = \frac{1/f}{1/p + 1/q - 2/f} = \frac{F}{P + Q - 2F}. \quad\dots\dots(38)$$

If we equate the values of $\mu - 1$ given by (33) and (38), we obtain

$$2\,(P + Q) = F + F_2. \quad\quad\dots\dots\dots\dots(39)$$

If the lens be a meniscus, and if BK be so strongly concave that $1/f + 1/b$ is negative, p is negative and then it will be impossible to make a real object coincide with its image. Since, however, BK is concave, the radius b can be found directly by making an object on the B-side of the lens coincide with its image by reflexion at BK. By (35), $1/p - 1/f = 1/b$, and then (38) becomes

$$\mu - 1 = \frac{1/f}{1/b + 1/q - 1/f}. \quad\quad\dots\dots\dots(40)$$

Equation (40) may also be obtained by using (36) and writing
$1/q - 1/f$ for $1/a$ in (27), § 132.

135. Secondary powers for n thin lenses in contact.
The method of § 133 is easily extended. Let P (Fig. 95) be the
object and Q a secondary image. Let the lenses be $A_1 K_1 B_1$,
$A_2 K_2 B_2 \ldots A_n K_n B_n$. Let $\mu_1, \ldots \mu_n$ be the indices and $F_1, \ldots F_n$
the powers of the lenses. Let the radii of the surfaces be
$a_1, b_1 \ldots a_n, b_n$, the radii being positive when the surfaces are
convex. Let the distance of the edges $K_1 \ldots K_n$ from the axis
be h. Let X, Y be points on PK_1, QK_n such that $PX = PA_1$,
$QY = QB_n$. Let the planes of the edges of the first and last
lenses cut the axis in M_1, M_n; let $PM_1 = u$, and let $QM_n = v$.

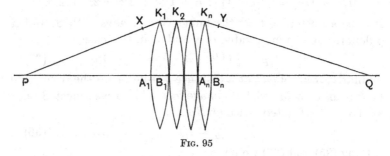

FIG. 95

Since $F_1 = (\mu_1 - 1)(1/a_1 + 1/b_1)$ and $A_1 B_1 = \frac{1}{2} h^2 (1/a_1 + 1/b_1)$,
we have $\qquad A_1 B_1 = \frac{1}{2} h^2 F_1 / (\mu_1 - 1),$ ....(41)
and similarly for the other lenses.

Let

$$W_{12} = -2(1/b_1 + 1/a_2), \quad W_{23} = -2(1/b_2 + 1/a_3). \quad ......(42)$$

Then, by putting $\mu = 1$ in (31), we see that W_{12}, W_{23}, ... are the
secondary powers of the air spaces between the lenses, for the
air space between the first and second lenses is a lens of unit
index and has radii $-b_1$, $-a_2$, the faces being concave in the
case of Fig. 95. [An optical system formed by air enclosed between
two spherical soap films is a close approximation to such a lens.
The films may be formed on the ends of a very short cylinder
furnished with a side tube. If a slight pressure be supplied, so that
the films bulge outwards, the secondary image is easily seen.
The *primary* focal length of this system is infinite.]

Since $K_1 K_2 = \frac{1}{2} h^2 (1/b_1 + 1/a_2)$, we have

$$K_1 K_2 = -\tfrac{1}{2} h^2 . \tfrac{1}{2} W_{12}, \quad K_2 K_3 = -\tfrac{1}{2} h^2 . \tfrac{1}{2} W_{23} . \quad \ldots\ldots(43)$$

Further, as in § 132,

$$XK_1 = \tfrac{1}{2} h^2 (1/u + 1/a_1), \quad YK_n = \tfrac{1}{2} h^2 (1/v + 1/b_n). \quad \ldots(44)$$

The optical length of the path from X to Y equals that from A_1 to B_n (see § 132) and either ray traverses every space once and some spaces thrice. The whole optical length is thus equal to the optical length of the path of a ray which passes through the system without reflexions together with twice the sum of the optical thicknesses of the parts which are traversed thrice. Every space is traversed once by the ray, and some spaces are traversed *twice* in addition, making thrice for those spaces.

Let Σ denote summation for *all* the quantities of any type and let S denote summation for only those quantities which refer to spaces through which a ray has passed *three* times. Then if the optical lengths of the paths XY, $A_1 B_n$ be $[XY]$, $[A_1 B_n]$, we have, by (43), (44),

$$[XY] = \tfrac{1}{2} h^2 \left(\frac{1}{u} + \frac{1}{a_1}\right) - \Sigma \tfrac{1}{2} h^2 . \tfrac{1}{2} W - 2\mathsf{S} \tfrac{1}{2} h^2 . \tfrac{1}{2} W + \tfrac{1}{2} h^2 \left(\frac{1}{v} + \frac{1}{b_n}\right).$$

By (41), we have

$$[A_1 B_n] = \Sigma \mu AB + 2\mathsf{S} \mu AB$$
$$= \Sigma \tfrac{1}{2} h^2 . \mu F/(\mu - 1) + 2\mathsf{S} \tfrac{1}{2} h^2 . \mu F/(\mu - 1).$$

Equating $[XY]$ to $[A_1 B_n]$, removing the factor $\frac{1}{2} h^2$ and using (42), we find

$$\frac{1}{u} + \frac{1}{v} + \Sigma \left(\frac{1}{a} + \frac{1}{b}\right) - \mathsf{S} W = \Sigma \frac{\mu F}{\mu - 1} + \mathsf{S} \frac{2\mu F}{\mu - 1} . \quad \ldots(45)$$

Since for any lens $1/a + 1/b = F/(\mu - 1)$,

we have $\Sigma \mu F/(\mu - 1) - \Sigma (1/a + 1/b) = \Sigma F.$

Hence by (45), $\dfrac{1}{u} + \dfrac{1}{v} = \Sigma F + \mathsf{S} \dfrac{2\mu F}{\mu - 1} + \mathsf{S} W$

$$= \Sigma F - \mathsf{S} F + \mathsf{S}\left(F + \frac{2\mu F}{\mu - 1}\right) + \mathsf{S} W. \quad \ldots\ldots(46)$$

For any lens, we have, by (32),

$F + 2\mu F/(\mu - 1) = F(3\mu - 1)/(\mu - 1) = $ secondary power of lens.

Since $\Sigma F - \mathsf{S}F$ is the sum of the primary powers of those lenses which are traversed *only* once, the result (46) can be expressed as follows:

Secondary power of system = {*Sum of primary powers of lenses traversed* only *once*} + {*Sum of secondary powers of lenses traversed three times*} + {*Sum of secondary powers of air spaces traversed three times*}.

If the lenses be so chosen that the primary power of each is positive and the value of every W is positive, *all* the secondary powers of the system will be positive.

The rays which suffer their first reflexion at the last surface can be reflected a second time at any one of the $2n - 1$ surfaces in front of the last surface. The rays which suffer their first reflexion at the last surface but one can be reflected a second time at any one of the $2n - 2$ surfaces in front of that surface, and so on. Hence there are

$$(2n - 1) + (2n - 2) + \ldots + 2 + 1 = n(2n - 1)$$

secondary powers. Thus with 1, 2 or 3 lenses there will be 1, 6 or 15 secondary images respectively.

CHAPTER VIII

EXPERIMENTS WITH THIN LENSES

EXPERIMENT 25. **Measurement of focal length of converging lens by use of distant object.**

136. Method. The image of a very distant object P is observed when (1) the A-face, as in Fig. 96, (2) the B-face, as in Fig. 97, is turned towards P; the axis of the lens is directed to P in each case. The image of P is (1) at F_2, and (2) at F_1, in the two cases. The distances F_2B, F_1A are measured. When the

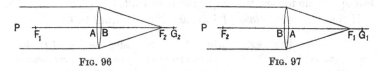

FIG. 96 FIG. 97

thickness of the lens is negligible, the mean of F_2B and F_1A may be taken as equal to f, the focal length.

The image may be received on a screen of *fine* ground glass*, the ground side facing the lens. If the lens and the screen be mounted on carriages on an optical bench (§ 207), the bench readings of the carriages are taken; then the screen carriage is moved so that a distance rod of length l fits between lens and screen. If d be the difference between the index readings of the screen, then $f = l + d$ when $f > l$, and $f = l - d$ when $f < l$.

We may also use a sharp pin, and adjust the tip by the parallax method to coincide with the image of P.

A microscope attached to a carriage may be used. The A-face of the lens is turned towards P, and the microscope is focused on lycopodium on the B-face. The carriage is then moved back until P be in focus; the difference of readings gives F_2B. The distance F_1A is found in the same way.

137. Corrections for finite distance and for thickness. When P is at a finite but large distance, we use Newton's

---

* A good surface may be made by grinding the plate with fine carborundum and water on a sheet of glass.

formula, § 126. If, when the light from P enters the lens by the B-face, the image of P be at G_1 (Fig. 97) instead of at the focus F_1, then $AG_1 > AF_1$, and $F_1G_1 = f^2/PF_2$. Hence $AF_1 = AG_1 - f^2/PF_2$. In the small term f^2/PF_2 we use AG_1 in place of f; then

$$AF_1 = AG_1 - AG_1^2/PF_2. \quad\dots\dots\dots\dots(1)$$

If, when the light from P enters the lens by the A-face, the image be at G_2 (Fig. 96),

$$BF_2 = BG_2 - BG_2^2/PF_1. \quad\dots\dots\dots\dots(2)$$

A correction for the (small) thickness c of the lens is required. By (30), § 198,

$$f = \tfrac{1}{2}(AF_1 + BF_2) + \tfrac{1}{2}c/\mu. \quad\dots\dots\dots\dots(3)$$

For spectacle lenses, we may take $\mu = 1\cdot52$; then $\tfrac{1}{2}c/\mu = c/3\cdot04$. If the A-face be *plane*, BF_2 is accurately equal to f.

138. Practical example. Miss Slater tested the lens of thickness $c = \cdot35$ cm. used in §§ 141, 146. The distance of object from nearer focus was about 6500 cm. Image received on ground glass; $AG_1 = 28\cdot35$, $BG_2 = 28\cdot30$ cm. By (1) and (2),

$$AF_1 = AG_1 - AG_1^2/PF_2 = 28\cdot35 - \cdot12 = 28\cdot23.$$
$$BF_2 = BG_2 - BG_2^2/PF_1 = 28\cdot18 \text{ cm.}$$

Taking $\mu = 1\cdot52$, $\tfrac{1}{2}c/\mu = \cdot12$ cm. By (3), § 137,

$$f = \tfrac{1}{2}(28\cdot23 + 28\cdot18) + \cdot12 = 28\cdot32 \text{ cm.}$$

When a pin was used in place of ground glass, result was $f = 28\cdot50$ cm.

EXPERIMENT 26. **Measurement of focal length of converging lens by use of plane mirror.**

139. First method. The lens is placed near a plane mirror M (Fig. 98); its axis AB is ideally normal to M. Let the B-face be towards M. If a luminous point be placed at F_1, the rays, after passing through the lens, are reflected by M along their own paths, if F_1AB be normal to M, and come to a focus again at F_1. Hence F_1 is its own image or is self-conjugate. If a pin be placed with its tip P at F_1, the image of P coincides with P itself. If AB be not quite normal to M, the self-conjugate point is not exactly at F_1, but is in the focal plane close to F_1; the distance of this point from A may be taken as

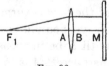

FIG. 98

equal to AF. The measurement may be repeated with the A-face towards M. If the thickness AB be negligible, the mean of AF_1 and BF_2 is taken as equal to f. A correction for the (small) thickness may be applied as in (3), § 137.

The position of the self-conjugate point does not depend upon the distance BM, since the rays between B and M are parallel, but BM is best small.

Let the axis F_1AB (Fig. 99) cut M in C, and let D be the point conjugate to C with respect to the lens when C is a *real* object point. If $BC < f$, B and D lie on opposite sides of M. Let PP' be an object in the focal plane through F_1. A ray

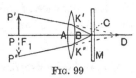

FIG. 99

$P'K'$, which is directed to D, is bent at K' and strikes M at C. Here it is reflected along CK'', and, after refraction at K'', is directed along $DK''P''$, and meets the focal plane in P''. Since the angles $K'CB$, $K''CB$ are equal, $P''P = P'P$. Thus PP'' is the image of PP'; the image is in the focal plane through F_1, is inverted and is of the same size as the object.

140. Second method. If the distance from lens to mirror be greater than f, there is a second position of the pin in which its tip P is self-conjugate. The necessary condition is that the plane of the mirror should pass through the point conjugate to P with respect to the lens. The image is now erect as may be seen from Fig. 100.

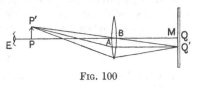

FIG. 100

Let QQ' be the real inverted image of PP' formed by the lens. Then the rays from P' which fall on the lens meet in Q'. If the face of M pass through Q', the reflected rays also pass through Q'. If M be turned so that some reflected rays strike the lens, they come to a focus in P', and P' is self-conjugate. A convex or concave mirror may be used instead of M without affecting the result. Thus the system of lens and mirror produces an erect image of PP' coincident with PP' itself. The rays returning from M are intercepted by the pin and its image would be invisible if the lens were achromatic. If the lens be not achromatic, P will be self-conjugate for one colour, but not for another. If the pin be well illuminated on the side nearer AB, an eye near E sees a coloured halo surrounding it, the colour changing with the position of the pin. A more convenient method is to use a sloped plate, as in Fig. 72.

Generally, the mirror will be of glass of thickness t and index μ', silvered at the back. If M in Fig. 100 now denote the point in which AB cuts the front surface of the glass, and if v be the effective distance of the reflecting surface from B, $v = BM + t/\mu'$. Putting $AP = u$, and neglecting the thickness of the lens,

$$1/f = 1/u + 1/v. \quad\quad\quad\quad\ldots\ldots\ldots\ldots\ldots\ldots(1)$$

If u and v be *nearly equal*, we correct for the (small) thickness of the lens by writing $u = AP + \tfrac{1}{2}c/\mu$, $v = BM + t/\mu' + \tfrac{1}{2}c/\mu$, where c is the thickness and μ the index of the lens, for, when u and v are *nearly equal*, the calculated value of f practically depends only on their sum. This approximation does not require that the faces of the lens should have nearly equal radii nor even that both faces should be convex.

A grain of lycopodium G on the surface of M has its image at a distance $2t/\mu'$ behind G. If a microscope be focused on G and then on the image, the displacement is $2t/\mu'$.

141. Practical example. Miss Slater tested the lens of §§ 138, 146; for this lens $c = \cdot35$ cm.; $\tfrac{1}{2}c/\mu = \cdot12$ cm.

First method. When tip of pin coincided with tip of inverted image, $AF_1 = 28\cdot22$, $BF_2 = 28\cdot15$ cm. By (3), § 137,

$$f = \tfrac{1}{2}(28\cdot22 + 28\cdot15) + \cdot12 = 28\cdot30 \text{ cm.}$$

Second method. By microscope on bench, $2t/\mu'$ for plane mirror was found to be $\cdot96$ cm. Hence $t/\mu' = \cdot48$ cm.

When pin P coincided with erect image, $AP = 57\cdot85$, $BM = 55\cdot17$ cm., where M denotes the front surface of mirror. If Q be image of P formed by lens, $BQ = BM + t/\mu' = 55\cdot65$ cm. Since lens is thin and AP, BQ are *nearly equal*, we may write $u = AP + \tfrac{1}{2}c/\mu = 57\cdot97$ cm., $v = BQ + \cdot12 = 55\cdot77$ cm. Then

$$1/f = 1/u + 1/v = 1/57\cdot97 + 1/55\cdot77 = \cdot03518 \text{ cm.}^{-1}, \quad f = 28\cdot42 \text{ cm.}$$

EXPERIMENT 27. **Measurement of focal length of thin converging lens by conjugate points.**

142. Introduction. The lens is fixed in a holder on a carriage on an optical bench (§ 207). The object and the image

locator are fixed to two other carriages. The object should lie
on the axis of the lens, which should be parallel to the bench.
The lens, whose centre is Q, is first set near the centre of the
bench with its axis as nearly parallel to the bench as can be
judged by eye; for many purposes this is sufficient. The object
point P, viz. the tip of a pin, or the intersection of cross-wires,
is set close to the lens and is adjusted by eye so that OP is
parallel to the bench. Then P is moved to its end of the bench
and the image locator is adjusted so that its locating *point* T,
viz. the tip of a pin or the intersection of cross-lines on a screen,
coincides with Q the image of P. Then the locator is moved to
its end of the bench, and P is adjusted to coincide with the
image of T. After this process has been repeated two or three
times, the line PO will be very nearly parallel to the bench.
(See §§ 208, 210.) If a pin be used as locator, it should be
mounted as described in § 207 (Fig. 131), so that it can be moved
slightly at right angles to the bench.

When the lens can be moved transversely, its adjustment is
easy. A pointer is fixed to an additional carriage; the P- and
T-carriages are brought up in turn, and P and T are set to
touch the tip of the pointer. The line PT is then parallel to the
bench for all positions of the carriages on the bench. The lens
is now placed at a distance of about $2f$ from P, and the T-
carriage is set so that T is in the plane conjugate to P. The lens
is then adjusted so that Q coincides with T.

To set the axis of the lens parallel to the bench, the locator,
which may conveniently be a pin, is placed so that an image
of it is formed by internal reflexion in the lens, as in Boys's
method (§ 157), and the lens is adjusted in angular position so
that T is self-conjugate. Unless the lens fittings be specially
designed, the previous settings of P and T may now need a
little readjustment.

If a simple lens be used with white light, the images due to
the various parts of the spectrum will not be coincident. If a
sodium flame be placed behind the object, the difficulty will
be avoided.

143. First method. The lens, the object P and the locator
are arranged on the optical bench as in § 142. If the distance

of P (Fig. 101) from O, the centre of the lens, be greater than f, the focal length, a real inverted image of P will be formed at Q. If $OP = u_0$, $OQ = v_0$, then approximately

$$1/f = 1/u_0 + 1/v_0. \quad \ldots(1)$$

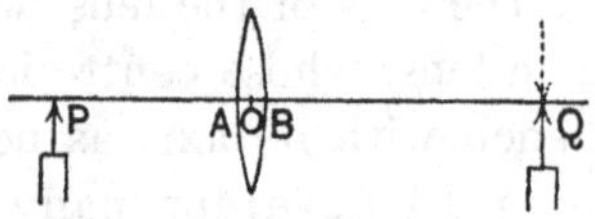

FIG. 101

Unless we know the ratio of the radius of one face of the lens to that of the other, we cannot satisfactorily correct u_0 and v_0 for the thickness, if we use (1) without restrictions on the value of u_0. Hence this method is unsuited to accurate work.

The object may be a pair of cross-wires or a piece of perforated zinc with a flame behind it. In this case, unless the illumination of the room be strong, the image can be received on a white screen or on a piece of fine ground glass, the ground side facing the lens. A simpler plan is to use a pin as object, with a fairly bright background, e.g. white paper, or, better, a sodium flame, and to locate the image by making the tip of a locating pin coincide at Q with the image of the tip of the object pin.

The values of u_0 and v_0 are deduced from the readings of the carriages and the thickness, c, of the lens; a rod of known length is used (§ 211). A series of values of u_0 is taken, v_0 is found in each case, and f is found from the mean value of $1/f$.

The focal length may also be found graphically. If $1/v_0$ be plotted against $1/u_0$, the straight line passing most evenly among the points will cut either axis at approximately $1/f$ from the origin.

Since the least distance between image and object is $4f$, a bench more than $4f$ in length is required. If this be not available, we fall back on § 136 or § 139.

144. Second method. Let O (Fig. 102) be the centre of the lens AB. Let P be an object and Q its real image. Let $PQ = l$, $PO = x$, where $x < l$, since O is between P and Q; hence $OQ = l - x$. Then, if the thickness of the lens be negligible, we have $1/x + 1/(l - x) = 1/f$. Hence

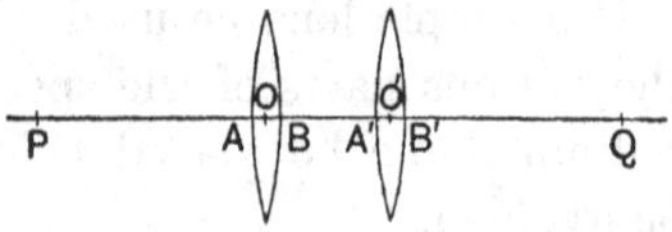

FIG. 102

$$f = x(l - x)/l. \quad \ldots\ldots\ldots\ldots\ldots\ldots(2)$$

Now let the lens be moved to the position O', where $O'Q = x$. Then $O'P = l - x$. The two distances have thus been interchanged, and an image of P is again formed at Q. If $OO' = a$, we have $l - a = 2x$; then $x = \frac{1}{2}(l - a)$, and $l - x = \frac{1}{2}(l + a)$. Hence, by (2),

$$f = (l^2 - a^2)/4l. \quad\quad\quad\quad\quad (3)$$

The whole distance PQ or l is measured and also OO' or a, the displacement of the lens. Measurements from the lens itself are thus avoided.

The lens is attached to a carriage; the two readings of this carriage give OO'. The object P and the screen or pin by which the image is located are borne by other carriages. The distance $PQ = l$ is deduced from the readings of the P and Q carriages by aid of a standard rod. See § 211.

If the length of the object be h, the length of the image is $h(l - x)/x$ in the first case, and $hx/(l - x)$ in the second.

When the lens is of finite thickness, the conjugate distances must be measured from the principal or unit planes and not from O. If the distance between these planes be t, we write $PQ - t$ for l in (3). In the case of a thin lens, t is positive; it will suffice (§ 198) to take $t = (\mu - 1)c/\mu$, where c is the thickness. When $\mu = 1\cdot5$, $t = \frac{1}{3}c$.

145. Third method. If we solve (3) for l, we obtain, since l is positive, $l = 2f + (4f^2 + a^2)^{\frac{1}{2}}$. Thus the distance l between a real object and its real image has its least value when $a = 0$; then $l = 4f$.

The lens is placed so that OP is greater than f, and Q is found. The lens is then moved to the second position for which Q is conjugate to P. The lens is next put midway between these two positions, and the new position of Q is found. Repetitions of this process lead rapidly to the minimum distance between P and Q. When the adjustment is nearly complete, a considerable motion of the lens is required to cause an appreciable displacement of the image, though, of course, the size of the image changes. When this state has been reached, the lens may be left in what seems the best position and Q may be moved to get the best possible focusing. The distance l between P and Q is now measured. Then $f = \frac{1}{4}l$.

When the distance between the principal planes is t, the least distance between image and object is $4f + t$. We then have $f = \frac{1}{4}l - \frac{1}{4}t$. Taking $t = (\mu - 1) c/\mu$, where c is the (small) thickness and μ the index of the lens (§ 198), we have

$$f = \tfrac{1}{4}l - \tfrac{1}{4} (\mu - 1) c/\mu. \quad \ldots\ldots\ldots\ldots(4)$$

If $\mu = 1\cdot5$, we have $\frac{1}{4} (\mu - 1) c/\mu = \frac{1}{12}c$, and thus the correction for thickness is very small.

146. Practical example. The lens of (assumed) index $\mu = 1\cdot52$ and thickness $c = \cdot35$ cm., used in §§ 138, 141, was tested.

First method. Miss Slater, using a 30 cm. rod, found
$u_0 = $ (bench reading of P) $- 103\cdot52$ cm., $v_0 = 103\cdot44$ cm. $-$ (bench reading of Q), where u_0, v_0 are distances from centre of lens.

u_0	v_0	$1/u_0$	$1/v_0$	$1/u_0 + 1/v_0$
40·93 cm.	93·44 cm.	·02443 cm.$^{-1}$	·01070 cm.$^{-1}$	·03513 cm.$^{-1}$
49·98	66·07	·02001	·01513	·03514
60·98	53·44	·01640	·01871	·03511
76·38	45·44	·01309	·02201	·03510

If the mean $\cdot03512$ cm.$^{-1}$ be taken as $1/f$, we have $f = 28\cdot47$ cm.

Second method. Miss Laverick measured PQ and a. To allow (§ 144) for distance between principal planes, we write $l = PQ - \cdot12$. Then f is given by (3). Mean is $f = 28\cdot47$ cm.

PQ	a	l	f
132·81 cm.	50·00 cm.	132·69 cm.	28·46 cm.
126·63	39·84	126·51	28·49
117·29	19·96	117·17	28·47

Third method. Miss Laverick found minimum value of PQ to be 114·13 cm. Then $4f = 114\cdot13 - \cdot12 = 114\cdot01$ cm., and $f = 28\cdot50$ cm.

EXPERIMENT 28. **Measurement of focal length of diverging lens.**

147. Introduction. In testing spectacle lenses, opticians use the method of neutralisation. A converging lens of focal length f' is placed in contact with a diverging lens of focal length f. If the thicknesses be negligible, the power of the combination is $1/f' + 1/f$. Spectacle lenses are made to nominally definite powers, on the dioptre scale, such as $\pm \cdot5D$, $\pm 1\cdot0D$, $\pm 1\cdot5D, \ldots$, and sufficient accuracy is reached when a converging lens is found such that, if the combination be moved across the line of sight, there is little or no apparent movement of objects seen through the combination. The system acts as a lens of zero power and thus $f = -f'$. The method is not, however, applicable to intermediate powers and, moreover, neglects the thickness of the lenses in the general case.

If each lens have one *plane* face and if, when the curved surfaces are in contact, the combination have zero power, then $f = -f'$ accurately.

The methods of §§ 149, 150 do not require exact equality between f and $-f'$.

148. Combination of converging with diverging lens. Let P (Fig. 103) be such an object point on the axis of a converging lens CD that its image Q is real. Between CD and Q place a diverging lens AB, so that the rays from P fall first on the A-face. Let the image of Q formed by AB be at R.

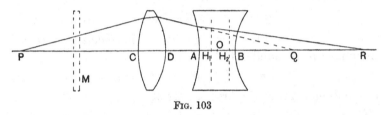

FIG. 103

Then, if R be a real image, and if we neglect the thickness $c = AB$ and measure from O, the mid-point of AB,

$$1/f = 1/OR - 1/OQ. \quad \dots\dots\dots\dots(1)$$

Thus, we can find f, if we measure the distance of the image R from the B-face and know the distance from B of the point Q, where the image of P is formed when AB is absent.

It is convenient to have CP nominally infinite or very large; in this case a moderate movement of CD relative to P has no appreciable effect on the distance DQ, which can then be treated as constant.

When the thickness is not negligible, we must replace (1) by

$$1/f = 1/H_2R - 1/H_1Q, \quad \dots\dots\dots\dots(2)$$

where H_1, H_2 are the principal or unit points. The distances $AH_1 = x_1$, $BH_2 = x_2$, can be calculated by § 196.

149. First method. When the power of AB is numerically less than that of CD, the lenses may be placed in *contact*.

The first step is to measure the distance $DQ = d$. We may either use a *distant* point for P or may place a plane mirror at M and then make Q self-conjugate with respect to the system of CD and M. A collimator adjusted for infinity may be used as an alternative. The lens AB is now placed in *contact* with CD, so that the A-face touches the D-face. Then $AQ = DQ$. We suppose that the D-face is more curved than the A-face, so that the lenses touch in a point; contact will be indicated by the formation of Newton's rings. The distance $BR = k$ of the image from B is then measured. When a plane mirror is used at M, R is adjusted to be self-conjugate. If we neglect the thickness c of AB,

$$1/f = 1/k - 1/d. \quad \text{......................(3)}$$

For accurate work, (3) is replaced by

$$1/f = 1/(k + x_2) - 1/(d - x_1). \quad \text{...............(4)}$$

The lens AB is now reversed so that B is in *contact* with D, and the image R' is found. If $R'A = k'$,

$$1/f = 1/(k' + x_1) - 1/(d - x_2). \quad \text{...............(5)}$$

Expanding as far as the first powers of x_1, x_2, we have

$$1/f = 1/k - 1/d - x_2/k^2 - x_1/d^2 = 1/k' - 1/d - x_1/k'^2 - x_2/d^2. \quad (6)$$

In practice, k' is nearly equal to k. Replacing the small term x_1/k'^2 by x_1/k^2, taking the average, and (§ 198) writing $\frac{1}{2}c/\mu$ for $\frac{1}{2}(x_1 + x_2)$, we have

$$1/f = \frac{1}{2}(1/k + 1/k') - 1/d - (\frac{1}{2}c/\mu)(1/k^2 + 1/d^2). \quad \text{...(7)}$$

The measurements are made on an optical bench. A pin attached to a carriage may be used to locate Q and R. The distances d, k, k' are deduced from the bench readings of the pin carriage by aid of a rod of known length.

The lenses may be held in a suitable spring clamp. The pressure should be slight to avoid distorting the lenses at the point of contact where the Newton's rings are formed. If the eye be moved across the axis, the image will appear to *jump* a short distance as the eye crosses the axis, when the pressure is considerable.

150. Second method. When the power of AB (Fig. 103) is numerically greater than that of CD, the combination of the lenses in *contact* cannot produce a real image of the real distant object P. If, however, AB be placed at such a distance from CD that Q lies between B and the focus of AB corresponding to parallel rays falling on the B-face, a real image of P will be formed at R. The focal length can then be found from (2), if AD be measured.

It is convenient to place a telescope, set for infinity, to the *right* of AB, and to adjust the distance between the lenses so that an image of P is formed on the cross-wire. A goniometer, § 71, may be used in place of the telescope. As explained in § 148, it is convenient if P be a distant point. We may, however, use a plane mirror M and locate Q by making a pin self-conjugate with respect to CD and M. The device of Fig. 77 may be used. The distance $DQ = d$ is measured before AB is put into place. If the mirror be used, the auto-collimating fitting (§ 71) may be attached to the goniometer when the distance AD is to be adjusted.

When a telescope set for infinity is used, R is at infinity and (2) becomes $f = - H_1 Q$. If the distance AD between the lenses, when A faces D, be l, and if we neglect the thickness $AB = c$,

$$f = - (d - l). \quad \dots (8)$$

For accurate work we replace (8) by

$$f = - (d - l - x_1). \quad \dots (9)$$

If AB be reversed, so that B faces D, and if, now, l' be the distance between the lenses, when adjusted,

$$f = - (d - l' - x_2). \quad \dots (10)$$

Taking the mean of (9) and (10), we have, by § 198,

$$f = - d + \tfrac{1}{2}(l + l') + \tfrac{1}{2}(x_1 + x_2) = - d + \tfrac{1}{2}(l + l') + \tfrac{1}{2}c/\mu. \quad (11)$$

151. Practical example. G. F. C. Searle and Miss Laverick tested a diverging spectacle lens of nominal power $-3D$. The thickness was $c = \cdot083$ cm. Assuming $\mu = 1\cdot52$, $x_1 + x_2 = c/\mu = \cdot054$ cm.

First method. The power of CD was nominally $4D$. A plane mirror was used. When Q was self-conjugate with respect to CD and mirror, $d = QD = 25\cdot03$ cm.

When A touched D, and R was self-conjugate with respect to lenses and mirror, $k = BR = 105\cdot61$ cm.

When B touched D, R' was self-conjugate when $k' = AR' = 105\cdot11$ cm.

The readings were somewhat irregular; the lenses were distorted at point of contact. By (7),

$$\frac{1}{f} = \frac{1}{2}\left(\frac{1}{105\cdot61} + \frac{1}{105\cdot11}\right) - \frac{1}{25\cdot03} - \cdot027\left(\frac{1}{105\cdot61^2} + \frac{1}{25\cdot03^2}\right) = -\cdot03051 \text{ cm.}^{-1}.$$

Hence $f = -32\cdot78$ cm.

Second method. The power of CD was nominally $2D$. A distant object (65 metres) was used, and $d = 49\cdot82$ cm. When A was towards D, $l = 17\cdot08$ cm. When B was towards D, $l' = 17\cdot11$ cm. By (11),

$$f = -49\cdot82 + \tfrac{1}{2}(17\cdot08 + 17\cdot11) + \cdot027 = -32\cdot70 \text{ cm.}$$

The corresponding power is $-\cdot03058$ cm.$^{-1}$.

Experiment 29. Measurement of power of weak lens.

152. First method. The method is suitable for either converging or diverging lenses for which $|f| > 200$ cm. Spectacle lenses can be obtained with $|f| = 400$ cm., or even $|f| = 800$ cm. An auxiliary lens S (Fig. 104) is required. This may be a simple lens; if, as is preferable, a lantern projection lens be used, the end B should be that which faces the slide in the lantern. The lens S is mounted on the optical bench with the end B to the right, and a carriage to the right of S bears a pin T.

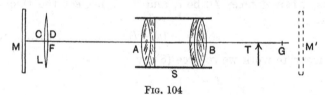

FIG. 104

The axis of S should be parallel to the length of the bench, and the tip of T should be on that axis (§ 210). By the aid of a plane mirror placed at M to the left of S, T is adjusted to coincide with its own image by reflexion at M. The bench fittings (§ 207) allow the mirror to be adjusted about horizontal and vertical axes. Then T is at the focus G of S. The mirror is now placed at M' to the *right* of S. The lens L under test, of focal length f, is placed to the *left* of S, with its D-face towards S, and is adjusted so that a scrap of paper attached to the D-face of L coincides with its image by reflexion at M'. Then the D-face of L is at the focus F of S.

The mirror is next placed to the left of L, and T is adjusted to be self-conjugate. Let $TG = y$, where y is positive when

T is to the left of G. Since the reflected rays are normal to M, T is the final image corresponding to rays parallel to the axis which fall on L from the left. After passing through L, these rays are directed to a point at a distance f to the *right* of F, when L is converging. After passing through S, the rays will meet in T. Then these two points are conjugate with respect to S. Hence, if the focal length of S be g, we have, by Newton's formula (§ 190), $fy = g^2$. Thus

$$1/f = y/g^2. \quad\ldots\ldots\ldots\ldots\ldots\ldots\ldots(1)$$

Hence $1/f$, the power of L, is proportional to y, the displacement of T.

The mirror M is now removed, and the lens L is moved away from S through a considerable distance* such as 50 or 100 cm., so that $DF = k$, and the pin T is adjusted so that it coincides with the image of the paper fixed to L at D. Since D is to the left of F, its image is to the right of G. If $TG = w$ in this adjustment, we have, by Newton's formula,

$$g^2 = kw, \quad\ldots\ldots\ldots\ldots\ldots\ldots\ldots(2)$$

and thus

$$1/f = y/kw. \quad\ldots\ldots\ldots\ldots\ldots\ldots(3)$$

To avoid chromatic errors, sodium light should be used. A sloped glass plate (§ 100), or an auto-collimating device (Fig. 77), may be used.

The quantity f, as found by (1), is not strictly the focal length of L, but is the distance from the D-face (Fig. 104) of the focus for parallel rays falling first on the C-face. If L be a simple lens and if the C-face be plane the error vanishes.

153. Second method. In view of the great focal length of L, any reasonably good mirror suffices for locating F. The mirror will also serve for locating G, if the displacement of T from G prove to be at least 2 or 3 cm. To avoid depending on the perfection of the mirror for *small* values of GT, we modify the method. The point G is first located and the lens L is placed with its D-face at F, the mirror being used in each case. Then L is moved through distance k (50 cm.) away from S, and w is found; then (2) gives g^2. The lens L is replaced at F, and T is adjusted

---

* This distance should be large if we are to retain the advantage of using a lens S which is corrected for parallel rays entering by the A-face.

to the image of an object P at any distance from F exceeding
100 cm. If $PF = l$, and if x be the distance to the right of F of
the image of P formed by L, $1/x + 1/l = 1/f$ and also $1/x = z/g^2$,
where z is the distance of T from G. The lens L is now removed;
let the distance from G of the image of P be v. We count z and
v positive when T is to the *left* of G. Then $1/l = -v/g^2$. Hence

$$1/f = 1/x + 1/l = (z - v)/g^2 = (z - v)/kw. \quad \dots\dots(4)$$

It will be seen that the small distance $z - v$ is independent of
any error in the assumed position of G.

154. Practical example. First method used. Zero of bench was on
F side of S. Mean of four bench readings, when T was at G, was 143·975 cm.
When weak lens L was placed between M and S, and T was self-conjugate,
mean of four readings was 143·013 cm. Hence $y = \cdot962$ cm.

Distance of L from F was increased by 100 cm.; thus $k = 100$ cm. When T
coincided with image of paper on D-face, T was at 146·72 cm. Hence

$$w = 146\cdot72 - 143\cdot975 = 2\cdot745 \text{ cm. and } g^2 = kw = 274\cdot5 \text{ cm.}^2.$$

Then $1/f = y/g^2 = \cdot962/274\cdot5$ cm.$^{-1}$; $\quad f = 285\cdot3$ cm.

EXPERIMENT 30. **Test of relation between elongation and
magnification.**

155. Method. A converging lens L (Fig. 105) and scales
P, Q are fixed to carriages sliding on a bench. The scale Q, on
which the image of P is received, is of glass; the scale P may be
of glass or white celluloid. The divided faces of the scales are
towards L.

The scale P is placed in any
position which gives a real
image, and Q is adjusted so
that the image of P is focused

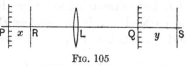

FIG. 105

on Q. The distance n_2 cm. on Q covered by the image of n_1 cm. on
P is observed, and the magnification $m = -n_2/n_1$ is found. The
magnification is negative, since the image is inverted. The object
scale is then moved to R, the other scale is moved to S so that
the image is again in focus and the magnification m' is found.
The displacements of the scales, viz. $x = PR$, $y = QS$, are
obtained from the bench readings of their carriages. Then the
elongation $e = y/x$ and mm' are found. By § 129, $e = y/x = mm'$,
since the lens is in air. The measurements may be repeated for
a number of pairs of positions of P.

156. Practical example. Mr F. L. Dawney compared e with the product mm'. The table shows that in the first *pair* of readings the P-scale was moved through $53 - 17$ or 36 cm. along bench, and the Q-scale was moved through $134\cdot93 - 130\cdot55$ or $4\cdot38$ cm. in the same direction. Hence $e = 4\cdot38/36 = \cdot1217$. In first set of readings, $n_1 = 8$ cm. on P-scale; image covered $8\cdot16 - 6\cdot24$ cm. on glass Q-scale, or $n_2 = 1\cdot92$ cm.; hence $m = -1\cdot92/8 = -\cdot2400$. The values of m' are marked *.

| Readings for | | n_1 | Readings for n_2 | m | e | mm' |
P	Q					
cm.	cm.					
17	130·55	8	8·16 – 6·24	– ·2400		
53	134·93	6	8·64 – 5·50	– ·5233*	·1217	·1256
26	131·50	6	7·98 – 6·35	– ·2717		
62	137·32	5	8·40 – 5·46	– ·5880*	·1617	·1598

EXPERIMENT 31. **Measurement of radii of curvature of thin lens by Boys's method.**

157. Converging lens. Since the lens is converging, one face at least must be convex; we assume the A-face to be convex. If K (Fig. 106) be a luminous point on the axis of the lens AB, a ray KJ will enter the lens by the B-face at J and meet the A-face at G. Most of the light will pass out through the A-face, but some will be reflected. If JG be normal to the A-face at G, the reflected ray will retrace its course, and an image of K will be

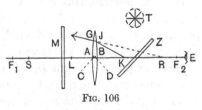

FIG. 106

formed at K; let $BK = k$. Let R be the centre of curvature of AG, and $AR = r$. Then R and K are conjugate with respect to the lens. If the lens be treated as "thin" and if f be the focal length,

$$1/r = 1/k - 1/f. \quad\quad\quad (1)$$

If the B-face be convex with radius s, and if the luminous point be placed at L on the A-side of the lens, the point will be self-conjugate when $AL = l$. When the lens is "thin,"

$$1/s = 1/l - 1/f. \quad\quad\quad (2)$$

If the B-face be concave, so that its radius s is negative, the radius r of the convex A-face can still be found by Boys's method. If the B-face be so strongly concave that $1/f + 1/s$ is negative, it will be impossible to make a real object coincide with its image, and Boys's method fails for the B-face. Since, however, the B-face is concave, there is a second self-conjugate point, viz. at S, the centre of curvature of that face; the point S is now on the same side of the lens as R. One image can be distinguished from the other, for S and R must be further from the lens than R and K respectively, since the lens is converging. Hence S is the more distant of the two self-conjugate points.

158. Lens of finite thickness. If, in Fig. 106, F_1, F_2 be the foci and $F_1 A = U, F_2 B = V$, we have, by Newton's formula (§ 190), $F_1 R . F_2 K = f^2$, or $(U + r) (V - k) = f^2$. Hence

$$r = f^2/(V - k) - U. \quad\quad\dots\dots\dots\dots(3)$$

By § 198, when the lens is formed of a single piece of glass of small thickness $c, f = \frac{1}{2} (U + V) + \frac{1}{2}c/\mu$.

By § 215, $f^2 = U (V + b)$, where b is the distance from B of C, the image of A. Then (3) becomes

$$r = U (V + b)/(V - k) - U = U (b + k)/(V - k). \quad\dots(4)$$

For further details see § 245.

159. Experimental details. The foci are found by aid of a plane mirror (§ 139) or otherwise. When the thickness of the lens is entirely neglected, the distance of either focus from the lens may be taken as the focal length. For the "luminous point" K, the *tip* of a pin may be used. The image of the pin is inverted. Since the reflexions take place at unsilvered surfaces, good illumination is necessary. We may use a plate of glass, Z, to reflect light from the flame T to the lens. To an eye at E, the image of the pin will stand out dark against a bright background. The device of Fig. 77 may be used. A black surface M should be placed behind the lens.

160. Diverging lens. Since one face at least must be concave, we suppose that the A-face is concave. The radius r of this face is found by placing a pin at R, its centre of curvature. If the B-face be *concave*, its radius s can be found in the same way. If the B-face be *convex*, s is positive and can be found by Boys's method if $s < |f|$, or $(\mu - 1) s < \mu |r|$, where $|f|$,

$| r |$ are the numerical values. When s exceeds $| f |$, Boys's method fails for the B-face. The radius of the B-face can be found if an auxiliary converging lens H be placed on the A-side of AB and if a pin P be adjusted so as to coincide with its image formed by light which has passed twice through H and has been reflected at the B-face. If Q be conjugate to P with respect to $\dot{H}$, then S is conjugate to Q with respect to the lens AB, the virtual object Q giving rise to the virtual image S. Hence S can be located and s found.

161. Practical example. G. F. C. Searle tested lenses of § 172. An optical bench was used.

Thin lens. With A towards mirror, BF_2 (Fig. 106) was 25·08 cm. Using rays reflected at A, pin was self-conjugate at K, when $k = BK = 12\cdot82$ cm. Treating BF_2 as f, we have, by (1),

$$r = \{1/12\cdot82 - 1/25\cdot08\}^{-1} = \{\cdot03813\}^{-1} = 26\cdot23 \text{ cm.}$$

Correction for thickness. With B towards mirror, $AF_1 = U = 24\cdot99$ cm. As above, $V = 25\cdot08$ cm. Thickness $= c = \cdot23$ cm. Assuming $\mu = 1\cdot52$, $c/\mu = \cdot15$ cm. By § 158, $f = \frac{1}{2}(U + V + c/\mu) = 25\cdot11$ cm. By (3), § 158,

$$r = 25\cdot11^2/(25\cdot08 - 12\cdot82) - 24\cdot99 = 26\cdot44 \text{ cm.}$$

Thick lens. Thickness $= 1\cdot24$ cm. By plane mirror, $U = 17\cdot52$, $V = 17\cdot52$ cm. A pin was self-conjugate at K when $k = BK = 8\cdot72$ cm. A microscope, used in Gauss's method, § 215, gave $b = \cdot84$ cm. By (4), § 158,

$$r = 17\cdot52 \times 9\cdot56/8\cdot80 = 19\cdot03 \text{ cm.}$$

In § 172, when there was water ($\mu_1 = 1\cdot3331$) between the A-face of this lens and the mirror, d_1 was 25·66 cm. By (5), § 170,

$$r = \cdot3331 \times 17\cdot52 \times 26\cdot50/8\cdot14 = 19\cdot00 \text{ cm.}$$

EXPERIMENT 32. Determination of index of lens by flare spot method.

162. Method. If a luminous point be placed on or near the axis of a lens, an image is formed by rays which pass through the lens without suffering reflexion. This is the ordinary primary image used in photography. A comparatively faint secondary image is formed by rays which have been twice reflected and still fainter images are formed by rays which have been reflected 4, 6, ... times (see § 133).

A spectacle lens may be used; its primary focal length should be considerable—say, a metre,—so that its thickness may be neglected. Its surfaces should be distinguished as A and B (Fig. 107). The primary focal length f is found by aid of a *very*

distant object (500 metres) or of a *good* plane mirror (see
EXPERIMENTS 25, 26).

The secondary focal length f_2 is found on a bench. One
carriage bears the lens holder. The lens may be fitted into a
frame. A second carriage bears a plate pierced by a small hole
and furnished with cross-wires S. The hole is illuminated by a
flame. The third carriage bears a pin T which is adjustable in
a plane perpendicular to the length of the bench (§ 207); T is
adjusted to coincide with the secondary image of S.

If the centre of S and the tip of T be brought in turn to an
auxiliary pointer, as in § 142, the line ST will be parallel to the
bench. The lens is then adjusted so as to bring the secondary
image of S close to T.

The secondary focal length is found by the minimum distance
method, § 145. The minimum distance between the cross-wires
and the secondary image is
$4f_2 + t_2$, where t_2 is the distance
between the principal points
corresponding to the secondary image. When c, the thickness of the lens, is *small*, $t_2 =
(\mu - 3)\,c/\mu$. The lens being
"thin," t_2 is negligible and

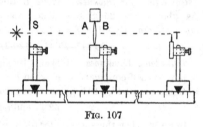

FIG. 107

hence the minimum distance may be taken as $4f_2$. The minimum distance is roughly $4f/7$.

The refractive index is now found (see § 133) by

$$\mu = 1 + 2f_2/(f - 3f_2). \quad \ldots\ldots\ldots\ldots\ldots(1)$$

The index can also be found if we use the image formed by
rays which have suffered a single reflexion. A well-illuminated
pin is made to coincide with its own image, as in Boys's method,
§ 157, when the A-face of the lens is towards the pin. The
distance of pin from lens is the length p of § 134. The distance
q, when the B-face is towards the pin, is found in like manner.

Then $$\mu = 1 + (1/f)/(1/p + 1/q - 2/f). \quad \ldots\ldots\ldots(2)$$

If one face, which we will take to be the B-face, be so strongly
concave that Boys's method fails, equation (40) of § 134 is used.
The images may be distinguished as in § 157.

163. Practical example. Mr J. L. Barritt used a double-convex spectacle lens. Primary focal length (found by plane mirror) $= f = 101\cdot85$ cm. Power $= F = 1/f = \cdot009818$ cm.$^{-1} = \cdot9818$ dioptre.

Minimum distance between object and secondary image $= 4f_2 = 59\cdot62$ cm. Hence $f_2 = 14\cdot905$ cm. and $F_2 = 1/f_2 = \cdot06709$ cm.$^{-1}$. By (1),

$$\mu = 1 + 2f_2/(f - 3f_2) = 1 + 29\cdot81/57\cdot135 = 1\cdot5218.$$

Distance from lens of pin and coincident image formed by light reflected at B-face $= p = 51\cdot90$ cm. Hence $P = 1/p = \cdot019268$ cm.$^{-1}$.

Distance from lens of pin and image formed by light reflected at A-face $= q = 52\cdot38$ cm. Hence $Q = 1/q = \cdot019091$ cm.$^{-1}$. By (2),

$$\mu = 1 + F/(P + Q - 2F) = 1 + \cdot009818/\cdot018723 = 1\cdot5244.$$

Also $2(P + Q) = \cdot07672$ cm.$^{-1}$ and $F + F_2 = \cdot07691$ cm.$^{-1}$. By (39), § 134, these quantities should be equal.

EXPERIMENT 33.   **Flare spots with two lenses.**

164. Secondary powers for two thin lenses in contact. Let AKB, CLD (Fig. 108) be the lenses. Let the radii of the faces BK, CL be b, c, and be positive when those faces are convex.

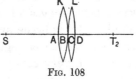

FIG. 108

Let m and m_2 be the primary and secondary focal lengths of AKB; let $M = m^{-1}$ and $M_2 = m_2^{-1}$ be the corresponding powers. Let n, n_2, N, N_2 be the primary and secondary focal lengths and powers of CLD.

Let W be the secondary power of the air space KBL. Then, by (42), § 135, $W = -2(1/b + 1/c)$.

In the case in which rays have been reflected at D and C, let f_{DC} be the secondary focal length of the system and F_{DC} the corresponding power. Thus $F_{DC} = 1/f_{DC} = 1/u + 1/v_{DC}$, where u and v_{DC} are the distances from the system of a real object S and its secondary image T_{DC}.

Extending this notation to the five other pairs of faces, and applying the result stated in § 135, we obtain the six secondary powers:

$$F_{DC} = M + N_2, \quad F_{DB} = M + N_2 + W, \quad F_{DA} = M_2 + N_2 + W,$$
$$F_{CB} = M + N + W, \quad F_{CA} = M_2 + N + W, \quad F_{BA} = M_2 + N.$$

165. Experimental details. The six secondary images are readily observed if the system consist of two converging meniscus spectacle lenses placed in contact, their *concave* surfaces facing each other. Then M, N, M_2, N_2 are positive and, since b and c are negative, W is positive. Hence the six secondary powers F_{DC}, ... are positive, and it will be possible to obtain six real secondary images of a real object. To avoid coincident images, M and N should be unequal. The primary and secondary focal lengths of each lens are found as in EXPERIMENT 32.

The radius of curvature of the *concave* surface of each lens is found by making a pin coincide with its image formed by reflexion at that surface. The image required is further from the lens than that formed by reflexion, as in Boys's method, § 157, at the *convex* surface. The distance of the pin from the surface equals the radius. The distance is measured by the appliance of Fig. 129.

The lenses are then fixed to a frame, as in Fig. 109, unless a special holder be available; the frame is mounted on a carriage on the bench, as in Fig. 107. When S is sufficiently far from the lenses, the six real secondary images of S will be easily seen on looking through the system towards S, provided the eye be far enough from the lenses. If the images be very small, they may be increased in size by bringing the lenses nearer to S. If, however, the distance be made too small, some or all of the six secondary images will become virtual. When the lenses are coaxal, the six secondary images will lie on a straight line. The secondary focal lengths are found by the minimum distance method, beginning with the shortest.

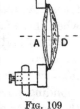

FIG. 109

It is impossible to tell by the appearance of any particular secondary image which two surfaces have acted as reflectors for that image. But, if we calculate, by § 164, the six secondary powers from M, N, M_2, N_2 and W, we shall be able to find among the observed values of the secondary powers one which will agree with any selected one of the calculated values, and so we can decide which two surfaces acted as reflectors in that case.

166. Practical example. G. F. C. Searle used two meniscus spectacle lenses nominally of $1D$ and $\cdot 5D$. The primary powers were found from distances of object and image from lens, the secondary powers by minimum distance method. For AKB, $M = \cdot 01028$, $M_2 = \cdot 07082$ cm.$^{-1}$. For CLD, $N = \cdot 00484$, $N_2 = \cdot 03317$ cm.$^{-1}$.

Radius b of $AKB = -46.44$, radius c of $CLD = -35.18$ cm. The signs are *negative*, since surfaces are *concave*. Hence $W = -2\,(1/b + 1/c) = .09992$ cm.$^{-1}$.

The six secondary powers of system were found by minimum distance method. The table gives these **observed** powers, the secondary powers calculated by § 164, and the pairs of reflecting surfaces.

Reflectors	Calculated power	Observed	Reflectors	Calculated power	Observed
D, A	·20391 cm.$^{-1}$	·20555 cm.$^{-1}$	C, B	·11504 cm.$^{-1}$	·11507 cm.$^{-1}$
C, A	·17558	·17528	B, A	·07566	·07516
D, B	·14337	·14331	D, C	·04345	·04354

EXPERIMENT 34. **Test of formula for thin lens separating two media.**

167. Method. When a thin lens separates two media, the distances u, v of a real object P and its real image Q from the lens, by (9), (10), § 127, satisfy

$$\mu_1/u + \mu_2/v = \mu_1/f_1 = \mu_2/f_2, \quad\ldots\ldots\ldots\ldots(1)$$

where μ_1 and μ_2 are the indices of the media in which P and Q are, and f_1, f_2 are the two focal lengths.

A tank T, shown in vertical section in Fig. 110, contains water. The lens L is fixed by sealing wax into a recess turned

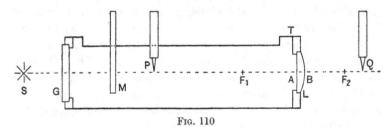

FIG. 110

in the plate forming one end of the tank. A plate G, of tested plane glass, closes the other end. The tank may be a brass tube 40 cm. long, 5 cm. in diameter, with end plates of stout brass. A slot 2·5 cm. wide is cut in the tube for the greater part of its length. The lens may be plano-convex or concavo-convex; in either case the convex surface faces outwards. The lens and plate are "edged" to fit their recesses.

The tank may be supported on a board overhanging an optical bench (§ 207), so as to allow the carriages free movement along the bench. The "object" is a needle P carried by an arm attached to a carriage. The image locator is a needle Q attached to a second carriage. A plane mirror M is fixed to a third carriage. The distance, $PA = u_0$, of P from the nearer face A of L, and the distance $QB = v_0$ are deduced from the bench readings of the carriages by the aid of a rod of known length.

The tank is placed parallel to the bench, and the tips of P and Q are set as nearly as possible on the axis of L.

The mirror M is placed in the tank, and its normal is so directed that, when the Q-carriage is in the proper position on the bench, the tip of the (inverted) image of Q coincides with the tip of Q. Then Q is at the focus F_2. An alternative method is to use a distant object P_∞ visible through G, either directly or by reflexion at a plane mirror, and to set Q on the image of P_∞. The needle P is now put into the tank, and Q is set on the image of P. A series of positions of P is taken, and the corresponding positions of Q are found.

The focus F_1 may be found by making P coincide with its image when a plane mirror is placed outside the tank near L. Or we may use a distant object Q_∞ visible through L, either directly or by reflexion at a plane mirror. Or, again, a collimator adjusted for infinity may be placed to the right of L, and then P is adjusted so that its image coincides with the wire in the focal plane of the collimator. A goniometer (§ 71) may be used.

Since u_0 is always large, the chromatic aberration is considerable, unless monochromatic light be used. When Q is set on the image of P, a sodium flame is placed at S. The device of Fig. 77 may be used throughout in place of a needle at Q; when F_2 is found by use of the mirror M, the wire is illuminated by a sodium flame and the reflecting prism.

On account of the finite thickness c of the lens, u_0, v_0 are not necessarily equal to the u, v of (1). To apply corrections, we write $u = u_0 + x_1$, $v = v_0 + x_2$, where the distances x_1, x_2 can be found from (21), (22), § 195, on putting $\mu_1 = \frac{4}{3}$ and $\mu_2 = 1$ in those equations. In calculating x_1, x_2, approximate values of the curvatures r^{-1}, s^{-1} of the faces A, B, of the index μ of the lens and of the focal lengths f_1, f_2 may be used. The radii may be found by Boys's method, § 157. To find μ, the lens is placed on a horizontal plane mirror in a trough. If the distance of a pin from the lens, when self-conjugate, be f when the lens is dry and h when the lens is *just* covered with water, then, approximately, $\mu = (\tfrac{4}{3}h - f)/(h - f)$, since $\mu_1 = \tfrac{4}{3}$. See (10), § 171.

If $1/v$ be plotted against $1/u$, the points will lie on a straight line cutting the axis of $1/u$ where $1/u = 1/f_1$ and the axis of $1/v$ where $1/v = 1/f_2$. Then $\mu_1/\mu_2 = f_1/f_2$. A better method is to calculate $\mu_1/u + \mu_2/v$ for each position of P. With careful work, this value will be nearly constant.

If the temperature be near 15° C., we may take $\mu_1 = \tfrac{4}{3}$ and tabulate $4/(3u) + 1/v$, as in § 168.

168. Practical example. The lens used by Miss Laverick had $c = \cdot37$ cm., $\mu = 1\cdot52$, $r = -37\cdot73$ cm. (concave), $s = 6\cdot66$ cm. With water ($\mu_1 = \tfrac{4}{3}$) in tank, distances of foci from lens were $AF_1 = 17\cdot79$, $BF_2 = 13\cdot55$ cm. Using these approximate values for f_1, f_2 in (21), (22), § 195, and putting $\mu = 1\cdot52$, $\mu_1 = \tfrac{4}{3}$, $\mu_2 = 1$, we have $x_1 = \cdot338$, $x_2 = -\cdot02$ cm.

Hence, we may write
$$u = u_0 + \cdot34, \qquad v = v_0 - \cdot02 \text{ cm.}$$

u	v	$1/u$	$1/v$	$4/3u + 1/v$
∞ cm.	13·53 cm.	·00000 cm.$^{-1}$	·07391 cm.$^{-1}$	·07391 cm.$^{-1}$
34·93	27·47	·02863	·03640	·07457
28·93	35·27	·03457	·02835	·07444
18·13	∞	·05516	·00000	·07355

EXPERIMENT 35. **Determination of index of a liquid or of a thin lens by use of liquid of known index.**

169. Method for liquid. A converging lens L, of focal length f, is placed on a horizontal plane mirror M (Fig. 111), silvered on its lower surface. If the tip Q of a pin be on the vertical ABX, the image of Q will coincide with Q itself, or Q will be self-conjugate, when it is at the principal focus of L (see § 139). The image of the *pin* is inverted. If L be thin, QB may be taken as equal to f.

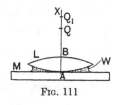

FIG. 111

If a little liquid W, of index μ_1, be placed between L and the upper face of M, it will be held in place by capillarity and will form a thin plano-concave lens, if the A-face of the glass lens be convex. If the radius of this face be r, the focal length of the liquid lens is accurately $- r/(\mu_1 - 1)$. By § 130, the focal length, f_1, of the combination is given by

$$1/f_1 = 1/f - (\mu_1 - 1)/r. \quad \ldots\ldots\ldots\ldots(1)$$

If $(\mu_1 - 1)/r < 1/f$, then f_1 is positive and greater than f. Hence Q_1, the self-conjugate position of the pin, is higher than before. The distance $Q_1 B$ is taken as f_1.

The radius r can be found by (1), if the index μ_1 of the liquid be known.

The liquid (μ_1) is replaced by a second liquid (μ_2), and the pin is self-conjugate at Q_2; $Q_2 B$ is taken as f_2, the focal length of the combination. Then,

$$1/f_2 = 1/f - (\mu_2 - 1)/r. \quad\ldots\ldots\ldots\ldots\ldots(2)$$

From (1), (2),

$$(\mu_1 - 1)/r = 1/f - 1/f_1, \qquad (\mu_2 - 1)/r = 1/f - 1/f_2.$$

Hence
$$\frac{\mu_2 - 1}{\mu_1 - 1} = \frac{1/f - 1/f_2}{1/f - 1/f_1} = \frac{f_1 (f_2 - f)}{f_2 (f_1 - f)}. \quad\ldots\ldots\ldots(3)$$

Thus, if μ_1 be known, we can find μ_2. If the first liquid be water at $t°$ C., $\mu_1 = 1\cdot3330 - 0\cdot00008 (t - 20)$, for sodium light. For the second liquid amyl alcohol or aniline may be used.

170. Correction for thickness of lens. If a ray in the liquid of index μ_1 be normal to M, it will, if it pass out of the liquid into *air* by the curved face of the liquid lens, proceed as if it came from P_1 below A, where $AP_1 = r/(\mu_1 - 1)$; the thickness of the liquid or of the mirror is immaterial. If it then traverse L, it will meet AX in Q_1, where Q_1 is conjugate to P_1 with respect to L. If F, Q be the lower and upper foci of L, then, by Newton's formula (§ 190), $P_1 F . Q_1 Q = f^2$. Let $AF = U$, $BQ = V$. Then $P_1 F + U = r/(\mu_1 - 1)$, and $Q_1 Q = d_1 - V$, where $d_1 = Q_1 B$. Hence,

$$r/(\mu_1 - 1) = U + f^2/(d_1 - V). \quad\ldots\ldots\ldots\ldots(4)$$

If b be the distance from B, measured towards A, of the image of A, we have, § 215, $f^2 = U (V + b)$, and then (4) becomes

$$r/(\mu_1 - 1) = U (d_1 + b)/(d_1 - V). \quad\ldots\ldots\ldots\ldots(5)$$

Similarly, $\quad r/(\mu_2 - 1) = U (d_2 + b)/(d_2 - V). \quad\ldots\ldots\ldots\ldots(6)$

From (5), (6),

$$\mu_2 = 1 + \frac{(d_1 + b)(d_2 - V)}{(d_2 + b)(d_1 - V)} (\mu_1 - 1). \quad\ldots\ldots\ldots\ldots(7)$$

If we use the value of r given, in Boys's method, by (4), § 158, we can use (5) to find μ_1. Or, if μ_1 be known, (5) gives r.

171. Methods for thin lens. *First method.* Let the convex faces A, B of the lens L have radii r, s, and let the index be μ. If, when liquid of index μ_1 is between L and M, the resultant focal length is f_1 or g_1 according as A or B touches M, then f_1 is given by (1) and

$$1/g_1 = 1/f - (\mu_1 - 1)/s. \quad \dots \dots \dots (8)$$

By (1), (8),

$$2/f - 1/f_1 - 1/g_1 = (\mu_1 - 1)(1/r + 1/s) = (\mu_1 - 1)/\{f(\mu - 1)\}.$$

Hence

$$\mu = 1 + \frac{(\mu_1 - 1)/f}{2/f - 1/f_1 - 1/g_1}. \quad \dots \dots \dots (9)$$

Second method. If the mirror and lens be placed in a dish and the liquid *just* cover the lens, there is a combination of L and of two liquid lenses of powers $-(\mu_1 - 1)/r$, $-(\mu_1 - 1)/s$. If the resultant focal length be h,

$$1/h = 1/f - (\mu_1 - 1)(1/r + 1/s) = 1/f - (\mu_1 - 1)/\{f(\mu - 1)\}.$$

Hence

$$\mu = 1 + (\mu_1 - 1) h/(h - f). \quad \dots \dots \dots (10)$$

In the case of a meniscus, the convex face should be uppermost.

172. Practical example. G. F. C. Searle compared amyl alcohol (μ_2) with water ($\mu_1 = 1\cdot3331$).

Thin lens. Thickness $= c = \cdot23$ cm.; $\mu = 1\cdot52$ (assumed). A-face on mirror. Distances from B of pin, when self-conjugate: Air, 25·06; (1) Water, 36·67; (2) Amyl alcohol, 40·93. Treating these distances as focal lengths, we find, by (3),

$$\mu_2 = 1 + \frac{36\cdot67\,(40\cdot93 - 25\cdot06)}{40\cdot93\,(36\cdot67 - 25\cdot06)} \times \cdot3331 = 1\cdot4079.$$

Correction for thickness. Distances 25·06 cm.,... are strictly V, d_1, d_2 (§ 170) and not f, f_1, f_2. Approximately $b = c/\mu = \cdot15$ cm. Hence $d_1 + b = 36\cdot67 + \cdot15 = 36\cdot82$, $d_2 + b = 41\cdot08$ cm. By (7),

$$\mu_2 = 1 + \cdot3331 \times 36\cdot82 \times 15\cdot87/(41\cdot08 \times 11\cdot61) = 1\cdot4081.$$

Thick lens. Thickness $= c = 1\cdot24$ cm.; A-face on mirror. Distances from B of self-conjugate pin: Air, $V = 17\cdot52$; Water, $d_1 = 25\cdot66$; Amyl alcohol, $d_2 = 28\cdot50$.

Gauss's method, § 215, gave $b = \cdot84$ cm. By (7),

$$\mu_2 = 1 + \frac{(25\cdot66 + \cdot84)(28\cdot50 - 17\cdot52)}{(28\cdot50 + \cdot84)(25\cdot66 - 17\cdot52)} \times \cdot3331 = 1\cdot4058.$$

EXPERIMENT 36. **Determination of index of lens by immersion in liquid of known index.**

173. Method. Let the focal length of a thin converging lens of index μ be f when the lens is in air. When the lens is in liquid of index μ', let the focal length be f'. Then, by (6), (6a), § 125, $f'/(\mu'f) = (\mu - 1)/(\mu - \mu')$. Hence

$$(\mu - 1)/(\mu' - 1) = (f'/\mu')/(f'/\mu' - f).\dots\dots(1)$$

The lens AB (Fig. 112) is placed in a tank which is closed at one end by a plate of glass DE of index μ_0 and thickness g. Since the lens is converging, one face, say the B-face, is convex; this face touches the plate at B. The axis ABX cuts the plate in B, C. Let rays, parallel to ABX, from a distant object P enter the tank by the plane glass end N. When

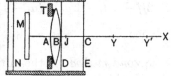

Fig. 112

the tank contains liquid, the directions in the liquid between the B-face and the D-face of the rays which emerge from the B-face pass through a point on AX at distance f' to the right of B. If there were only air to the right of the D-face, the rays in the air would converge to a point, say K, on AX at a distance f'/μ' from B. The images of K (an image of P) and B formed by the parallel plate DE are separated by the same distance, viz. f'/μ', as the points K and B themselves. Hence, if the images of P and B formed by the rays which emerge into air from the C-face be at Y' and J respectively, then $Y'J = f'/\mu'$. When the tank contains no liquid, the final images of P and B will be at Y, J, at distance f apart. The image J is virtual and $CJ = z$, where $z = g/\mu_0$. Thus $f'/\mu' = Y'J$, $f = YJ$ and then, by (1), if $YC = y$, $Y'C = y'$,

$$(\mu - 1)/(\mu' - 1) = Y'J/Y'Y = (y' + z)/(y' - y)\dots(2)$$

The tank is placed with its axis parallel to an optical bench. The lens is secured to a plate T, which is fixed to a carriage sliding on the bench. The lens *touches* the plate DE. A very distant object, or a collimator adjusted for infinity, may furnish the parallel rays. A plane mirror M, capable of adjustment about vertical and horizontal axes, may also be used. The tip of a pin placed at Y will be self-conjugate for rays reflected at M. If the device of Fig. 77 be employed, sodium light can be used. The distances $YC = y$, $Y'C = y'$ are deduced from the bench readings of the pin carriage by aid of a rod of known length.

Since J is the image of B, we can find CJ by a microscope, as in EXPERIMENT 3.

174. Correction for thickness of lens. When the lens has thickness c, the rays from the distant object P, on emerging from the B-face into the *liquid*, are directed to a point on AX at distance $f' - x_2'$ from B, where f' and x_2' are found on putting $\mu_1 = \mu_2 = \mu'$ in (20), (21), § 195. When these rays emerge into the air, they are directed to a point on AX at distance $(f' - x_2')/\mu'$ from the image of B, i.e. from J. When the tank contains no liquid, we write 1 for μ' in these expressions. Thus

$$\frac{y' + z}{y + z} = \frac{(f' - x_2')/\mu'}{f - x_2} = \frac{\mu - 1}{\mu - \mu'}\,\lambda,$$

where $\quad \lambda = \dfrac{r - (\mu - \mu')\,c/\mu}{r - (\mu - 1)\,c/\mu} \times \dfrac{r + s - (\mu - 1)\,c/\mu}{r + s - (\mu - \mu')\,c/\mu}.$ ......(3)

Then $\quad\quad\quad \mu = 1 + \dfrac{(\mu' - 1)\,(y' + z)}{(y' + z) - \lambda\,(y + z)}.$ ....(4)

When c is small compared with r, s, we may, in the terms in λ containing c, use the value of μ found by § 173, and the values of r, s found by Boys's method, § 157, neglecting c in each of the three measurements.

175. Practical example. G. F. C. Searle tested bi-convex lens; thickness $c = \cdot31$ cm. Water (15° C.) was used; $\mu' = \frac{4}{3}$. For plate DE, $g = \cdot28$ cm., $\mu_0 = 1\cdot52$ (assumed). Hence $z = g/\mu_0 = \cdot18$ cm. A plane mirror, the device of Fig. 77 and sodium light were used. Lens touched plate.

When tank contained water, Y was self-conjugate when $y' = 49\cdot05$ cm.; with no water, $y = 17\cdot96$ cm. Hence $y' + z = 49\cdot23$, $y' - y = 31\cdot09$ cm. By (2), since $\mu' - 1 = \frac{1}{3}$,

$$\mu = 1 + (\mu' - 1)\,(y' + z)/(y' - y) = 1 + \tfrac{1}{3} \times 49\cdot23/31\cdot09 = 1\cdot5278.$$

Correction for thickness. Boys's method, neglecting c, gave $r = 18\cdot7$, $s = 18\cdot8$ cm. Using $\mu = 1\cdot528$, $(\mu - \mu')\,c/\mu = \cdot04$ cm., $(\mu - 1)\,c/\mu = \cdot11$ cm. Since $y + z = 18\cdot14$, $y' + z = 49\cdot23$ cm., (3) and (4) give

$$\lambda = \frac{18\cdot70 - \cdot04}{18\cdot70 - \cdot11} \times \frac{37\cdot50 - \cdot11}{37\cdot50 - \cdot04} = 1\cdot0019, \quad\quad \mu = 1\cdot5283.$$

CHAPTER IX

COAXAL OPTICAL SYSTEMS—THICK LENSES

176. Introduction. By continued applications of the formulae relating to refraction at a single spherical surface, the optical properties of any system of coaxal lenses can be ascertained. These calculations are necessary when a photographic lens system or any other system of lenses is to be designed for a definite purpose. When, however, the system has been constructed, its optical constants can be determined by simple tests in which the system is treated as a whole, no detailed reference to the separate lenses being necessary. Before passing on to the experimental work, it will be convenient to apply the principles of geometrical optics to establish those properties of coaxal systems which are involved in the practical methods of determining the optical constants of such systems.

The diameters of the lenses will be supposed small compared with the radii of curvature of their faces, and the angles between the rays and the axis of the system will be treated as "small."

177. Approximate treatment of spherical surfaces. Let AS (Fig. 113) be a spherical surface, of radius r, its centre being at O, and let the radius OA be chosen as the axis. Draw SK perpendicular to the axis, let $SK = y$, and let $SOA = \psi$. Then $\sin \psi = y/r$. Now

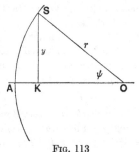

$$\sin \psi = \psi - \tfrac{1}{6}\psi^3 + \tfrac{1}{120}\psi^5 - \ldots,$$
$$\cos \psi = 1 - \tfrac{1}{2}\psi^2 + \tfrac{1}{24}\psi^4 - \ldots,$$
$$\tan \psi = \psi + \tfrac{1}{3}\psi^3 + \tfrac{2}{15}\psi^5 + \ldots.$$

Fig. 113

Thus, if ψ be so small that we may neglect ψ^2 in comparison with unity, we may use ψ instead of $\sin \psi$ or $\tan \psi$, and unity instead of $\cos \psi$.

In this case $\psi = y/r$, and thus the angle between the axis and the normal to the surface at S may be treated as proportional to y.

Since $AK = r\,(1 - \cos\psi)$, we have, to the same approximation, $AK = 0$, and thus, when we are measuring distances parallel to the axis, we may treat the part of the spherical surface, which is very near the axis, as a small portion of a plane perpendicular to the axis.

178. Refraction at a spherical surface. Suppose that the spherical surface AS (Fig. 114) separates two media of different

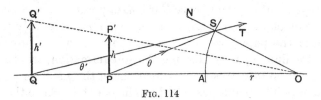

FIG. 114

refractive indices, the index of the medium on the left of AS being μ, while that of the medium on the right of AS is μ'. Let P be a point on the axis OA, and suppose that all the rays in the medium μ pass either actually or on prolongation through P. In the first case P is called a real object point and is in the medium μ, while in the second case the geometrical point P is in the medium μ' and is called a virtual object point.

Let the ray PS strike the surface at S and let ST be the refracted ray. Then, by the first law of refraction, Chapter I, § 14, ST lies in the plane through the points P, S and O, and hence the straight line ST will cut the axis OAP in some point Q at a finite or an infinite distance from O. In the latter case the ray ST is parallel to OAP.

Let $AP = u,\quad AQ = v,\quad OA = r,$

and let u and v be positive when P and Q lie to the left of A and let r be positive when A lies to the left of O.

Let the rays PS, ST make angles θ, θ' with OA, and let θ' be reckoned to have the same sign as θ when the rays PS, QS, supposed to start from P and Q and to move from left to right, deviate to the *same* side of OA.

If OSN be the normal to AS at S, we have, by Snell's law,

$$\mu \sin PSN = \mu' \sin QSN. \quad\ldots\ldots\ldots\ldots(1)$$

But $PSN = \theta + SOA,\quad QSN = \theta' + SOA,$

and if y, the perpendicular from S on OA, be so small that the angles may be used instead of their sines,

$$\theta = y/u, \qquad \theta' = y/v, \qquad SOA = y/r.$$

Hence (1) becomes

$$\mu y\,(1/u + 1/r) = \mu'y\,(1/v + 1/r),$$

or $\qquad\qquad\qquad \mu\,(u + r)/u = \mu'\,(v + r)/v.$ ....(2)

179. Conjugate points. The value of v given by equation (2) is independent of y and hence Q is a *fixed* point. Since S may be *any* point on the surface near to A, but not necessarily in the plane of the paper, we see that *all* the rays, whose directions in the medium μ passed through P before refraction, are changed by refraction into rays whose directions in the medium μ' pass through Q. The point Q is called the image of P. Since the path of a ray may be reversed, we see that, if Q be taken as the object point so that the directions of the rays in μ' before refraction pass through Q, the directions of the rays in μ after refraction will pass through P, which will thus be the image of Q. Hence the two points are said to be *conjugate*.

When Q is to the right of AS, the rays in the medium μ' actually pass through Q, and Q is then called a *real* image, while, if Q be on the left of AS, as in Fig. 114, the rays only appear to come from Q, and then Q is called a virtual image. (See § 125.)

When y becomes so large that the sines of the angles θ, θ' and SOA differ appreciably from the angles themselves, it is no longer true that all the rays, whose directions in the first medium pass through P, are so refracted that their directions in the second medium pass through a single point; in other words spherical aberration has to be taken into account.

180. Conjugate planes. About O (Fig. 114) as centre describe spheres through P and Q; let a radius $OP'Q'$ cut them in P', Q'. Then, since Q is the image of P, the image of P' is at Q'. When the angle $P'OP$ is very small, the straight lines PP', QQ' may be treated as perpendicular to OA, and hence, if planes be drawn at right angles to the axis through the conjugate points P and Q, all the points near the axis on one plane will have their images on the other. These planes are, therefore, called *conjugate planes*.

Since the straight lines QP and $Q'P'$ intersect in O, the lines PP' and QQ' both lie in one plane passing through the axis.

181. Magnification. The ratio of $Q'Q$ to $P'P$ is called the magnification of the image; it may be denoted by m. The magnification is constant for a given pair of conjugate planes, for

$$m = Q'Q/P'P = OQ/OP.$$

The magnification is reckoned positive when P' and Q' lie on the same side of the axis OPQ.

182. Helmholtz's formula for refraction. In Fig. 114 let $PP' = h$, $QQ' = h'$, and let h' be reckoned to have the same sign as h, when Q' and P' are on the *same* side of OPQ. Since a ray *crosses* the normal on refraction, all the possible cases are shown in Figs. 115, 116 and 117. From these figures it will be

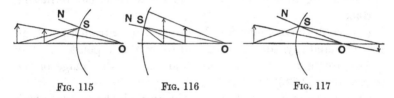

FIG. 115　　　　FIG. 116　　　　FIG. 117

seen that, when the signs of the angles are determined as in § 178, h' and θ' are either both positive or both negative.

When we know which of the two media has the greater refractive index, we can decide which of the two arrows in either Fig. 115 or Fig. 116 is to be taken as the object. Thus in Fig. 115, if μ' be greater than μ, the refracted ray is nearer to the normal than the incident ray and so the arrow which is the nearer to the refracting surface is the object. If, however, μ be greater than μ', the other arrow is the object. Fig. 116 can be discussed in a similar manner.

From Fig. 114 we have

$$h'/h = (v + r)/(u + r), \qquad \theta'/\theta = u/v.$$

Hence, by (2), we obtain Helmholtz's formula,

$$\mu h\theta = \mu'h'\theta'. \qquad \ldots\ldots\ldots\ldots\ldots\ldots(3)$$

183. Helmholtz's formula for reflexion. Let SA (Fig. 118) be a spherical reflecting surface. Let PP' be an object and

QQ' its image. Let $PA = u$, $QA = v$. Let $PP' = h$, $QQ' = - h'$.
Let $SPA = \theta$, $SQA = - \theta'$, the negative sign being used because
in Fig. 118 the ray, as it advances,
crosses the axis in the opposite direction
to the ray PS. Since a ray from P',
incident at A, passes, after reflexion,
through Q', we have $- h'/h = v/u$. Also
$u = SA/\theta$, $v = - SA/\theta'$, and hence $h\theta =$

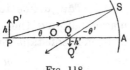

FIG. 118

$h'\theta'$. If μ be the index of the medium, this is the same for both
rays. If, for convenience, we use μ' for the ray SQ, on the
understanding that $\mu' = \mu$, we obtain $\mu h\theta = \mu' h'\theta'$, as in (3).

**184. Four properties of a spherical refracting or re-
flecting surface.** In §§ 179 to 182 it has been shown that,
when the angles of incidence and refraction are very small,
the following four results are true for a spherical refracting
surface.

(A) If all the incident rays be directed to or from a point, all
the refracted rays are also directed to or from a point. The two
points are said to be conjugate and each is the image of the
other. (See § 179.)

(B) If any radius be chosen as the axis, all the points, which
lie on a straight line intersecting the axis at right angles and are
very near to the axis, have their images on a parallel straight
line, which also intersects the axis. The two planes which are
normal to the axis and pass through a pair of conjugate points
are called conjugate planes.

(C) For a given pair of conjugate planes there is a constant
magnification.

(D) If two conjugate points on the axis be joined by a ray,
the angles which the ray makes with the axis before and after
refraction are related to the linear magnitude of a small object
in one of the conjugate planes and to that of its image in the
other conjugate plane by Helmholtz's formula $\mu h\theta = \mu' h'\theta'$.

By § 99, (A), (B), (C) apply to reflexion at a spherical surface,
and § 183 shows that (D) also applies to this case.

The four properties apply to a plane surface for either
refraction or reflexion, since a plane is merely an extreme form
of a spherical surface.

185. Foci and focal planes of a coaxal system. A coaxal optical system is formed by any number of different media separated by plane or spherical surfaces which are so placed that a single straight line, called the axis, intersects every surface at right angles; the centres of all the spherical surfaces lie, there-fore, on the axis. For such a system, the four properties (A), (B), (C), (D) of § 184 are handed on from surface to surface and thus are true for the whole system.

The media at the two ends of the system will be called the first and second media, and their indices of refraction will be denoted by μ_1 and μ_2. In the general theory μ_1 and μ_2 will be treated as unequal.

Let $F_1 F_2$ (Fig. 119) be the axis of the system, and let $A_1 R_1 R_2 A_2$ be a straight line parallel to the axis. Then a ray $A_1 R_1$, which

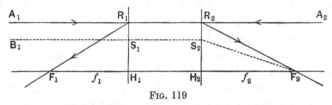

FIG. 119

moves from left to right in the first medium and coincides with the straight line $A_1 R_1 R_2 A_2$ before it suffers refraction, will emerge into the second medium along some straight line $R_2 F_2$, in the plane containing $A_1 R_1$ and the axis, and this straight line will cut the axis at the point F_2, which we shall suppose is at a finite distance from the system of spherical surfaces. By (A), § 184, *all* the rays, which move parallel to the axis from left to right in the first medium, are directed to F_2 in the second medium, since they start from a single point at infinity on the axis in the first medium. Hence the position of F_2 is independent of the distance of $A_1 R_1$ from the axis.

Similarly a ray $A_2 R_2$, which moves from right to left in the second medium and coincides with the straight line $A_2 R_2 R_1 A_1$ before it suffers refraction, and all rays parallel to $A_2 R_2$ will be directed to or from F_1, a *fixed* point on the axis, when they emerge into the first medium.

The two fixed points F_1 and F_2 are called the *First Focus* and the *Second Focus* of the system, and the planes through F_1

and F_2 perpendicular to the axis are called the first and second *Focal Planes*.

When the foci F_1, F_2 are at infinity, the system is said to be "telescopic"; it then possesses special properties, which are described in §§ 200 to 204.

186. Principal planes and principal points. Let the *straight lines* $A_1 R_1$ and $R_2 F_2$ intersect in R_2 and the *straight lines* $A_2 R_2$ and $R_1 F_1$ in R_1, and let the planes $R_1 H_1$ and $R_2 H_2$ be perpendicular to the axis, the points H_1, H_2 being on the axis.

Reversing the direction of the ray $R_1 F_1$, we see that the rays $A_1 R_1$ and $F_1 R_1$ in the first medium meet in R_1 and give rise to the rays $R_2 F_2$ and $R_2 A_2$ in the second medium. Hence, by (A), § 184, *every* ray, whose direction in the first medium passes through R_1, gives rise to a ray whose direction in the second medium passes through R_2, and therefore R_2 is the image of R_1. Thus R_1 and R_2 are conjugate points, and $R_1 H_1$ and $R_2 H_2$ are conjugate planes, every point in the plane $R_1 H_1$ having its image in the plane $R_2 H_2$. Since the distances $R_1 H_1$ and $R_2 H_2$ are equal, these conjugate planes are planes of positive unit magnification.

Since, by (C), the magnification is constant for a given pair of conjugate planes, it follows that, if S_1 be any point on $R_1 H_1$, the image of S_1 is at S_2 on the plane $R_2 H_2$, and that $S_2 H_2 = S_1 H_1$. Hence, any ray $B_1 S_1$, which is parallel to the axis in the first medium, emerges into the second medium along $S_2 F_2$, (1) because it passes through S_2, for S_2 is the image of S_1, and (2) because it must pass through the focus F_2. Thus the lines $B_1 S_1$ and $S_2 F_2$ intersect on the plane $R_2 H_2$, and hence the point H_2 is fixed and is independent of the distance between the ray $A_1 R_1$ and the axis. Since H_1 is conjugate to H_2, it is also a fixed point.

The two conjugate points H_1, H_2 are called the first and second *Principal Points* of the system, and the conjugate planes through them are called the first and second *Principal Planes*. Since the principal planes have positive unit magnification, they are sometimes called the *Unit Planes*, and the principal points are then called the *Unit Points*.

The distance $H_1 H_2$ may be denoted by t, and t will be reckoned positive when H_1 lies on the left of H_2. When t is negative, so that H_1 lies to the right of H_2, we may say that the principal points are inverted.

The reader must be careful to bear in mind that the points H_1, H_2 do *not*, as a rule, lie on the end surfaces of the coaxal system, and that the geometrical straight line R_1R_2 is *not*, as a rule, the path of a ray. The positions of the end surfaces relative to the principal planes H_1R_1, H_2R_2 depend upon the constitution of the system. Optical systems may differ greatly in this respect. When the first refracting surface lies between A_1 and R_1, the ray is refracted before it reaches R_1. If, on the other hand, R_1 lie between A_1 and the first surface, the ray actually passes through R_1.

187. Focal lengths. The distances F_1H_1 and F_2H_2 are called the first and second *Focal Lengths* of the system and will be denoted by f_1 and f_2. If F_1 lie on the left of H_1, f_1 is reckoned positive, and, if F_1 lie on the right of H_1, f_1 is reckoned negative. If F_2 lie on the right of H_2, f_2 is counted positive, and, if F_2 lie on the left of H_2, f_2 is counted negative. By § 189, f_1 and f_2 are either both positive or both negative. In the general case, the two focal lengths are unequal.

188. Formula for conjugate points. When the positions of the principal planes and of the foci are known, the image of any given point is easily found. Figs. 120 and 121 correspond to positive focal lengths; in Fig. 120 H_1H_2 or t is positive, and in Fig. 121 t is negative. Figs. 122 and 123 correspond to negative focal lengths; in Fig. 122 t is positive, and in Fig. 123 t is negative.

Let the directions of the rays in the first medium pass through the point Q_1 (Figs. 120 to 123), so that Q_1 is an object point. The ray Q_1R_1, parallel to the axis, emerges along R_2F_2, while the ray Q_1F_1, through the focus F_1, emerges along S_2Q_2 parallel to the axis. The rays R_2F_2, S_2Q_2 will intersect in a point Q_2, which is the image of Q_1. If the conjugate planes through Q_1, Q_2 meet the axis in P_1, P_2, the points P_1 and P_2 are themselves conjugate. In each of these figures, the object point is to the *left* of the first principal plane H_1R_1. The student may construct the four figures for the case when the object point is to the *right* of H_1R_1.

In any given case, we cannot tell whether the object P_1 is real or virtual and whether the image P_2 is real or virtual unless we know the positions of P_1 and P_2 relative to the end surfaces of the system.

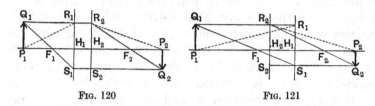

FIG. 120   FIG. 121

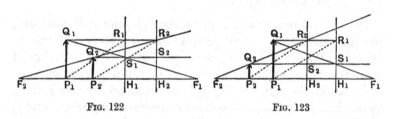

FIG. 122   FIG. 123

Let $P_1 H_1 = u$ and $P_2 H_2 = v$, and let u be positive when P_1 lies on the left of H_1, and let v be positive when P_2 lies on the right of H_2. Let $P_1 Q_1 = h_1$ and $P_2 Q_2 = h_2$. In Figs. 120, 121, Q_2 and Q_1 are on opposite sides of the axis and hence, by § 182, h_2 is negative; in Figs. 122 and 123, Q_2 and Q_1 are on the same side of the axis, and hence h_2 is positive.

Since $R_2 H_2 = R_1 H_1 = h_1$ and $S_2 H_2 = S_1 H_1 = -h_2$ in Figs. 120, 121, we have

$$\frac{f_1}{u} = \frac{H_1 F_1}{R_1 Q_1} = \frac{S_1 H_1}{S_1 R_1} = \frac{-h_2}{h_1 - h_2}, \quad \frac{f_2}{v} = \frac{H_2 F_2}{S_2 Q_2} = \frac{R_2 H_2}{R_2 S_2} = \frac{h_1}{h_1 - h_2} \quad \ldots (4)$$

Hence
$$\frac{f_1}{u} + \frac{f_2}{v} = \frac{h_1 - h_2}{h_1 - h_2} = 1. \quad \ldots\ldots\ldots\ldots (5)$$

By algebra from (4) or from Figs. 120, 121, we have

$$\frac{h_1}{h_2} = -\frac{u - f_1}{f_1}, \quad \frac{h_2}{h_1} = -\frac{v - f_2}{f_2}. \quad \ldots\ldots\ldots (5a)$$

We obtain the same results from Figs. 122, 123, if we note that $F_1 H_1 = -f_1$, $F_2 H_2 = -f_2$, $P_2 H_2 = -v$, and that h_2 is positive.

189. Ratio of focal lengths. Let the rays P_1R_1, R_2P_2 make angles θ_1, θ_2 with the axis. In Figs. 120, 121, the rays cross the axis in opposite directions and hence, by § 178, the angle θ_2 is negative. Then, since R_2 is the image of R_1 and P_2 of P_1, the incident ray P_1R_1 emerges along R_2P_2. Now, for small angles,

$$u\theta_1 = R_1H_1 = -v\theta_2, \quad\dots\dots\dots\dots\dots(6)$$

and, by (4), $f_1/f_2 = -uh_2/vh_1,$

or, by (6), $f_1/f_2 = h_2\theta_2/h_1\theta_1.$

By Helmholtz's formula, § 182, $\mu_1 h_1 \theta_1 = \mu_2 h_2 \theta_2$, and hence

$$f_1/f_2 = \mu_1/\mu_2. \quad\dots\dots\dots\dots\dots(7)$$

In Figs. 122 and 123, the rays P_1R_1, P_2R_2 cross the axis in the *same* direction and hence θ_2 is *positive*. In this case, as noted in § 188, h_2 is positive. Hence equation (7) holds good for negative as well as for positive focal lengths.

Thus the ratio of f_1 to f_2 is independent of everything except the two media at the two ends of the system. We see also that f_1 and f_2 are of the same sign. When $\mu_1 = \mu_2$, as is the case in most optical experiments, then $f_1 = f_2$.

Since $f_1/\mu_1 = f_2/\mu_2 = (f_1 + f_2)/(\mu_1 + \mu_2),$

the formula (5) connecting u and v can be written

$$\mu_1/u + \mu_2/v = (\mu_1 + \mu_2)/(f_1 + f_2). \quad\dots\dots\dots\dots(8)$$

From Figs. 120 to 123 it will be seen that the focal lengths are positive or negative according as the image of a distant object is inverted or erect. If we view a distant object through a lens system and place the eye so far from the system that *the image lies in front of the eye*, we can at once decide whether the focal lengths are positive or negative.

The *position* of the image does not decide the question, for a system may give a real image of a distant object and yet have negative focal lengths, and a system which gives a virtual image of the object may have positive focal lengths.

190. Newton's formula. Let the distances P_1F_1 and P_2F_2 (Fig. 120) be denoted by p and q respectively. Then $u = f_1 + p$, $v = f_2 + q$. Hence (5) becomes

$$f_1/(f_1 + p) + f_2/(f_2 + q) = 1.$$

Simplifying this equation, we obtain Newton's formula

$$pq = f_1 f_2. \quad \ldots\ldots\ldots\ldots\ldots\ldots(9)$$

Newton's formula can also be obtained directly from Figs. 120 to 123. For

$$P_1 F_1 / F_1 H_1 = R_1 H_1 / H_1 S_1 = F_2 H_2 / P_2 F_2$$

and hence
$$P_1 F_1 . P_2 F_2 = F_1 H_1 . F_2 H_2. \quad \ldots\ldots\ldots\ldots(9a)$$

When μ_2 differs from μ_1, Newton's formula does not enable us to determine more than the *product* $f_1 f_2$. If, however, the ratio μ_1/μ_2 be known, $f_2 = f_1 \mu_2/\mu_1$, by (7), and then the focal lengths can be found by the equations

$$f_1{}^2 = \mu_1 pq/\mu_2, \qquad f_2{}^2 = \mu_2 pq/\mu_1. \quad \ldots\ldots\ldots\ldots(10)$$

When, as is generally the case, the two final media are identical, $f_1 = f_2 = f$, and then Newton's formula takes the form

$$pq = f^2. \quad \ldots\ldots\ldots\ldots\ldots\ldots(11)$$

The focal length of such a system can, therefore, be deduced by Newton's formula when the distances of two conjugate points from the corresponding foci have been found.

Let f_1, f_2 be positive, and let the sequence of the points be $P_1\ F_1\ F_2\ P_2$, so that the distance between the object P_1 and the image P_2 is greater than the distance between the foci, as in Figs. 120, 121. Then the distance, l, between P_1 and P_2 is

$$l = p + q + f_1 + f_2 + t.$$

Here p, q and $(f_1 + f_2 + t)$ are positive. But $(p + q)^2 = 4pq + (p - q)^2$. Since $4pq = 4f_1 f_2$, the minimum value of $(p + q)^2$ occurs when $p = q = \sqrt{(f_1 f_2)}$. If d be the minimum value of l,

$$d = 2\,(f_1 f_2)^{\frac{1}{2}} + f_1 + f_2 + t = t + (f_1^{\frac{1}{2}} + f_2^{\frac{1}{2}})^2. \quad \ldots\ldots(12)$$

When $f_1 = f_2$, $\qquad\qquad d = t + 4f. \quad \ldots\ldots\ldots\ldots\ldots(13)$

191. Nodal points. The nodal points are two conjugate points on the axis, such that any ray, whose direction in the first medium passes through the first nodal point, gives rise to a ray, whose direction in the second medium passes through the second nodal point and is parallel to that of the ray in the first medium. The nodal points are of great importance in the theory of optical instruments.

Let N_1 (Figs. 124, 125) be the first nodal point, and let $R_1 N_1$

be any ray, whose direction in the first medium passes through
N_1 and cuts the first principal plane in R_1; Fig. 124 illustrates

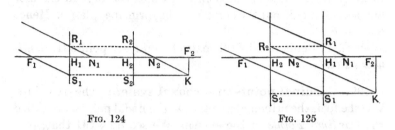

FIG. 124　　　　　　　　　　　FIG. 125

the case in which t, the distance between the principal planes,
is positive and Fig. 125 illustrates the case in which the distance
is negative*. Draw R_1R_2 parallel to the axis, cutting the second
principal plane in R_2. Then, by § 186, R_2 is the image of R_1,
and hence the direction of the emergent ray passes through R_2.
If, therefore, R_2N_2 be drawn parallel to R_1N_1, it will cut the
axis in the second nodal point N_2. From the triangles $R_1H_1N_1$,
$R_2H_2N_2$ it is clear that $H_2N_2 = H_1N_1$.

If we take a second ray, parallel to R_1N_1, which passes through
the focus F_1 and meets the first principal plane in S_1, it will
emerge parallel to the axis along S_2K and will intersect R_2N_2
at K in the second focal plane, where $F_2K = H_1S_1$. Since the
triangles N_2F_2K, $F_1H_1S_1$ are equal, $N_2F_2 = F_1H_1$.

Since $H_2N_2 = H_1N_1$, it follows that $H_2F_2 = F_1N_1$.

Thus the distance of the second focus from the second nodal
point is equal to the first focal length, and the distance of the
first focus from the first nodal point is equal to the second focal
length, while the distance between the nodal points is equal to
that between the principal points.

The distance between a nodal point and the corresponding
principal point is equal to the difference of the focal lengths.
Thus, if the focal lengths be equal, as is the case when the first
and second media are identical, the nodal points coincide with
the principal points.

The planes which cut the axis at right angles in the nodal
points are called the *Nodal Planes*.

* The construction of the two figures for positive and negative values of t,
when f_1 and f_2 are negative, is left as an exercise to the student.

If $u = f_1 + f_2$, then, by (5), $v = f_1 + f_2$. Defining θ_1, θ_2 as in § 189, $\theta_1 = y/(f_1 + f_2)$, $\theta_2 = -y/(f_1 + f_2)$, where y is the distance from the axis of the points in which the rays in the first and second media cut the corresponding principal planes. Hence $\theta_2 = -\theta_1$.

Dr G. T. Bennett calls this pair of conjugate points the *Antinodal Points*.

192. Cardinal points of a coaxal system. The six points, viz. the foci, the principal points, and the nodal points, are called the *Cardinal Points* of the system. We see by § 191 that, provided the positions of the foci be known, the nodal points can be found when the principal points are known and *vice versa*.

193. Deviation formula for a system in a single medium. When $\mu_2 = \mu_1$, then $f_1 = f_2 = f$, and (5) becomes

$$1/u + 1/v = 1/f. \quad\dots\dots\dots\dots\dots\dots(14)$$

If a ray from an object point on the axis be directed, on emergence, to an image point on the axis, the angle $(< \pi)$ between the directions of the incident and emergent parts of the ray is called the resultant deviation of the ray. If the directions of the ray on incidence and emergence cut the principal planes at a distance y from the axis, we see, by Figs. 120 to 123, that the deviation, D, is given by

$$D = y/u + y/v = y/f. \quad\dots\dots\dots\dots\dots(15)$$

Hence, the resultant deviation produced by a given system, in which $f_1 = f_2 = f$, depends only upon y and not upon the position of the object point. For the special case in which $y = 1$, the deviation is $1/f$ or is equal to the *Power* of the system. This simple result should be carefully noted, *but it must be remembered that it applies only to the case when $\mu_2 = \mu_1$*. The *general* case is considered in § 252.

The practical unit of *Power* is the *Dioptre*; this is the power of a system having each of its focal lengths 100 cm. The power is counted *positive* by practical opticians when a point which moves through the system along the axis, starting from infinity at the first end of the system, reaches the second principal plane *before* it reaches the second focus. This case is illustrated in Figs. 120 and 121.

194. Calculation of the positions of the cardinal points.
In calculating the positions of the cardinal points, we may use
the deviation formula (13) of § 25. We first consider the case in
which light moves from left to right. If we take a ray which,
before incidence on the first refracting surface, is parallel to
the axis and at a distance y from it, we can find the angle of
incidence of the ray on the first refracting surface, in terms of
the radius of the surface, and thus can calculate the deviation
due to the first refraction. The product of this deviation and
the distance between the first and second refracting surfaces,
when added, with the proper sign, to the original distance y,
gives the distance from the axis of the point where the ray
suffers the second refraction. The angle of incidence on the
second surface can then be found in terms of the radius of
that surface, and the deviation due to the second refraction
can be determined. Adding this deviation, with the proper
sign, to that due to the first refraction, we obtain the angle
between the axis and the ray after the second refraction, and
this process is continued until the ray emerges into the final
medium on the right. From Fig. 120 or Fig. 121, we see that the
angle between the axis and the ray in the final medium on the
right is equal to y/f_2, since $R_2 H_2 = y$. In the course of this work
we shall have found the distance from the axis of the point
where the ray suffers the final refraction. If this be z_2, the
distance of the second focus from the last refracting surface is
$f_2 z_2/y$, and thus the positions of the second focus and of the corresponding principal point are known with reference to the system.

Similarly, the first focus and the first principal point can
be determined by tracing a ray from right to left. The nodal
points are then found by § 191.

When the media on the right and left are identical, the nodal
points coincide with the principal points, and then $f_1 = f_2 = f$.

Rays which start from a point but strike the refracting surfaces at finite distances from the axis, or make finite angles with
the axis, do not meet *accurately* at a point after passing through
the system, and thus the paths of the emergent rays can be only
approximately found by aid of the foci and principal points.

But, if Snell's second law be used in the exact form, so that
the *sines* of the angles are used and not merely the angles

themselves, the paths of a number of selected rays can be traced through any given system by the aid of trigonometrical tables. If the object point lie on the axis, this process will give the points in which the emergent rays meet the axis. In the practical work of designing lens systems for photographic and other purposes, this method is employed, and the radii of the refracting surfaces and the other constants of the system are varied until the emergent rays pass sufficiently nearly through a single point. This point is the one which is determined by the method of foci and principal planes.

195. Cardinal points of thick lens separating two media.
Let $APQB$ (Fig. 126) be a lens of index μ and thickness c, separating a medium of index μ_1 (on the left) from a medium of index μ_2 (on the right). Let O_1, O_2 be the centres of curvature of AP, BQ, the radii r, s being positive when the faces are convex, as in Fig. 126. The

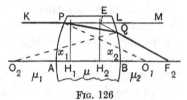

FIG. 126

axis O_1O_2 cuts the faces in A, B. Let KP be an incident ray parallel to the axis in the medium μ_1, PQ its path in the lens; the emergent ray QF_2 meets the axis in the second focus F_2. Let the straight line $KPLM$ cut the lens in P, L, and let $PA = LB = y$. The difference between the arc PA and the perpendicular from P on the axis is to be neglected. Let PQ, QF_2 make the small angles α, β with the axis, and let the direction of QF_2 cut KM in E. The plane perpendicular to the axis through E cuts the axis in H_2 and is the second principal plane. Then H_2F_2 is the second focal length, and $H_2F_2 = f_2$. Since $EF_2H_2 = \beta = y/f_2$, we have

$$1/f_2 = \beta/y. \quad \ldots\ldots\ldots\ldots\ldots\ldots(16)$$

The small angles which KP, PQ make with O_1P are y/r, $y/r - \alpha$, and, by Snell's law, for small angles, $\mu_1 y/r = \mu (y/r - \alpha)$. Hence

$$\mu\alpha = (\mu - \mu_1)\, y/r. \quad \ldots\ldots\ldots\ldots\ldots(17)$$

If $QB = z$, the small angles which PQ, QF_2 make with O_2Q are $z/s + \alpha$, $z/s + \beta$, and, by (16), $\beta = y/f_2$. By Snell's law,

$$\mu\,(z/s + \alpha) = \mu_2\,(z/s + y/f_2).$$

Thus
$$\mu_2 y/f_2 = (\mu - \mu_2) z/s + \mu a. \quad\ldots\ldots\ldots\ldots(18)$$

But $z = y - ca$, or, by (17),
$$z = y \{1 - (\mu - \mu_1) c/(\mu r)\}. \quad\ldots\ldots\ldots\ldots(19)$$

By (18), (19) and (17),
$$\frac{\mu_2}{f_2} = \frac{(\mu - \mu_2) z}{sy} + \frac{\mu a}{y} = \frac{\mu - \mu_2}{s}\left\{1 - \frac{(\mu - \mu_1) c}{\mu r}\right\} + \frac{\mu - \mu_1}{r}.$$

Since the expression on the right remains unchanged when we write μ_2, s for μ_1, r and *vice versa*, it is also equal to μ_1/f_1. Hence
$$\frac{\mu_1}{f_1} = \frac{\mu_2}{f_2} = \frac{\mu - \mu_1}{r} + \frac{\mu - \mu_2}{s} - \frac{(\mu - \mu_1)(\mu - \mu_2) c}{\mu rs} \ldots(20)$$

Thus $f_1/f_2 = \mu_1/\mu_2$, as already shown in § 189.

Let $H_2 B = x_2$, and let x_2 be positive when H_2 is to the left of B. Then $x_2/f_2 = LQ/LB = ca/y$. Hence, by (17),
$$x_2 = \frac{cf_2 a}{y} = \frac{(\mu - \mu_1) f_2 c}{\mu r}. \quad\ldots\ldots\ldots\ldots(21)$$

Similarly, if $H_1 A = x_1$, and if x_1 be positive when H_1 is to the right of A, we have
$$x_1 = \frac{(\mu - \mu_2) f_1 c}{\mu s}. \quad\ldots\ldots\ldots\ldots\ldots(22)$$

If t be the distance between the principal points, reckoned positive when H_1 is to the left of H_2, we have
$$t = c - x_1 - x_2. \quad\ldots\ldots\ldots\ldots\ldots\ldots(23)$$

When the medium on either side of the lens has index μ', we write $\mu_1 = \mu_2 = \mu'$ in the formulae.

196. Thick lens surrounded by air. Since the air is of unit refractive index, we put $\mu_1 = \mu_2 = 1$ in § 195. We then obtain
$$\frac{1}{f} = (\mu - 1)\left\{\frac{1}{r} + \frac{1}{s} - \frac{(\mu - 1) c}{\mu rs}\right\}, \quad\ldots\ldots\ldots\ldots(24)$$

or
$$f = \frac{rs}{(\mu - 1)\{r + s - (\mu - 1) c/\mu\}}. \quad\ldots\ldots\ldots(25)$$

Further,
$$x_1 = \frac{(\mu - 1) cf}{\mu s} = \frac{cr}{\mu(r + s) - (\mu - 1) c}, \quad\ldots\ldots(26)$$

$$x_2 = \frac{(\mu - 1) cf}{\mu r} = \frac{cs}{\mu(r + s) - (\mu - 1) c}, \quad\ldots\ldots(27)$$

and $\qquad t = c - x_1 - x_2 = \dfrac{(\mu - 1)(r + s - c)\,c}{\mu\,(r + s) - (\mu - 1)\,c}.$ ....(28)

197. Lens in air with one face plane. When one face of the lens is plane, the formulae of § 196 take simple forms. Thus, when the face A (Fig. 126) is plane, we have $r = \infty$, and then

$$f = s/(\mu - 1), \qquad x_1 = c/\mu, \qquad x_2 = 0, \qquad t = (\mu - 1)\,c/\mu.$$

Hence, if rays from a distant point fall normally upon the plane face of the lens, the distance of the image from the curved face is accurately equal to the focal length. This result might have been foreseen, for the ray KP (Fig. 126) will not be bent till it meets the curved surface QB, and hence the second principal plane, which is defined by the intersection of KP and the emergent ray QF_2, must pass through B.

198. Approximate results for a thin lens in air. When c, the thickness of the lens, is small in comparison with r and s, the radii of its faces, we may write, in place of (26), (27), (28),

$$x_1 = \frac{cr}{\mu\,(r + s)}, \qquad x_2 = \frac{cs}{\mu\,(r + s)}, \qquad t = \frac{(\mu - 1)\,c}{\mu}. \quad\dots(29)$$

For many lenses $\mu = \tfrac{3}{2}$ approximately, and then $t = \tfrac{1}{3}c$, nearly.

If the surfaces of the lens be both convex or both concave and if the radii be equal, so that $r = s$, we have $x_1 = x_2 = \tfrac{1}{2}c/\mu$. When $r = s$ and $\mu = \tfrac{3}{2}$, the principal points are at a distance $\tfrac{1}{6}c$ from the middle point of the lens.

By (29), $\qquad\qquad x_1 + x_2 = c/\mu,$ ....(29a)

and thus $x_1 + x_2$ is, to the accuracy implied in (29), independent of the radii of the faces.

If the distances F_1A and F_2B be denoted by U and V,

$$U = f - x_1, \qquad V = f - x_2,$$

and thus, by (29a),

$$f = \tfrac{1}{2}(U + V) + \tfrac{1}{2}(x_1 + x_2) = \tfrac{1}{2}(U + V) + \tfrac{1}{2}c/\mu. \quad\dots(30)$$

When $\mu = \tfrac{3}{2}$, $\qquad\qquad f = \tfrac{1}{2}(U + V) + \tfrac{1}{3}c.$ ....(31)

Equation (30) may be used in finding the focal length of a thin lens. The equation does not require us to know the radii of the faces of the lens.

199. Cardinal points of a compound system in air. Let
the compound system be formed of two optical systems S and S'
arranged along a common axis. In the case of each system, the
surrounding medium is air of unit refractive index. Let f be the
focal length and h, k (Fig. 127) the first and second principal
points of S, the second principal point being the one determined
by a ray moving from left to right, as in §§ 186, 188. Let f' be
the focal length and h', k' the principal points of S'. Let $hk = t$
and $h'k' = t'$, the distances t and t' being positive when the
second principal point of the corresponding system is to the

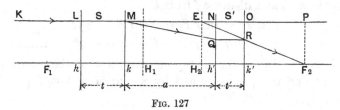

FIG. 127

right of the first principal point. Further, let $kh' = a$, the distance being positive when k is to the left of h'.

We will now trace the path of a ray, which moves from left
to right and before incidence on the first system is parallel to
the axis and at a distance y from it.

Let KL be the incident ray, and let its direction cut the principal planes of S and S' in L, M, N, O and the second focal
plane of the compound system in P as in the Figure.

By § 193, the deviation due to the passage of the ray through
the first system is y/f and hence, if the direction of the ray, on
emerging from S, meet the first principal plane of S' in Q, we have
$NQ = ay/f$, and therefore $h'Q = y(1 - a/f)$. The deviation due
to the second system is $h'Q/f'$, or $(1 - a/f)y/f'$, and thus the
resultant deviation is $y/f + (1 - a/f)y/f'$. But, if F be the focal
length of the compound system, the resultant deviation is y/F,
and thus

$$\frac{1}{F} = \frac{1}{f} + \frac{1}{f'} - \frac{a}{ff'}, \qquad F = \frac{ff'}{f + f' - a}. \quad\ldots\ldots\ldots(32)$$

If the distance between the second focus of S and the first
focus of S' be b, and if b be counted positive when the first focus
of S' is to the right (in Fig. 127) of the second focus of S,

we have $b = a - f - f'$. The expression for F now takes the simple form

$$F = - ff'/b. \quad \dots\dots\dots\dots\dots\dots(33)$$

Let F_2 be the second focus of the compound system, R a point on $k'O$ such that $k'R = h'Q$, and let F_2R meet KP in E. Then RF_2 is the emergent ray and, if EH_2 be drawn perpendicular to the axis, H_2 is the second principal point of the compound system, while H_2F_2 is the focal length.

Let H_2k' be denoted by x_2 and be reckoned positive when H_2 is to the *left* of k'. Then, since the straight line MQ, if continued, cuts the axis at a distance f from k,

$$x_2/F = OR/PF_2 = NQ/PF_2 = a/f.$$

Hence $\qquad x_2 = Fa/f = af'/(f + f' - a) = - af'/b. \quad \dots\dots(34)$

Similarly, if H_1 be the first principal point of the compound system, and if H_1h be denoted by x_1 and be reckoned positive when H_1 is to the *right* of h,

$$x_1 = Fa/f' = af/(f + f' - a) = - af/b. \dots\dots\dots(35)$$

Finally, if the distance H_1H_2 be denoted by T and be reckoned positive when H_1 is to the *left* of H_2, $T = t + t' + a - x_1 - x_2$, or

$$T = t + t' - a^2/(f + f' - a) = t + t' + a^2/b. \quad \dots(36)$$

200. Telescopic systems. A telescopic system, or, briefly, a "telescope," is a coaxal system such that any ray, which is parallel to the axis before entering the system, is again parallel to the axis after it leaves the system. There are no foci, unless points at an infinite distance from the system be called foci, and thus the system cannot be said to have focal lengths.

An observer, who can see distinctly an object at an infinite distance, will still be able to see it distinctly if he view it through a "telescope."

201. Transverse magnification of a telescope. Let XY (Fig. 128) be the axis, and let an incident ray PP', which is

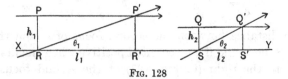

FIG. 128

parallel to the axis, give rise to the emergent ray QQ', which, by hypothesis, is also parallel to the axis. The lenses are *not* shown in Fig. 128.

Let P be a point on PP' and Q be the image of P. Then, if PR and QS be drawn perpendicular to the axis, S is the image of R. Let $PR = h_1$ and $QS = h_2$. Now, by § 181, the magnitudes of the successive images, formed by the successive refracting surfaces, are proportional to the magnitude of the object, and hence, for a given position of R, the ratio of h_2 to h_1 is constant for all values of h_1. But, since the image of P' lies on QQ' at a point which may be denoted by Q', it follows that, for a given value of h_1, the ratio of h_2 to h_1 is constant for all positions of R. Combining these results, we see that the ratio of h_2 to h_1 is constant for a given telescope, being independent both of the position of R and also of the magnitude of h_1. If, therefore, we write

$$h_2/h_1 = m, \quad \dots\dots\dots\dots\dots\dots(37)$$

the quantity m, which we may call the *transverse magnification*, is constant for a given telescope. As in § 181, we reckon m positive when P and Q lie on the same side of the axis and negative when they lie on opposite sides. The *transverse magnification* must be carefully distinguished from the *magnifying power*.

202. Magnifying power of a telescope. Since Q' is the image of P' and S is the image of R, the incident ray RP' gives rise to the emergent ray SQ'. Let θ_1, θ_2 be the angles which the rays RP' and SQ' make with the axis, the two angles having the same sign when a point, moving from left to right along the incident and emergent rays, crosses the axis at R and S in the same direction. Then, if μ_1 and μ_2 be the refractive indices of the media at the two ends of the system, we have, by Helmholtz's formula (§ 182)

$$\mu_1 h_1 \theta_1 = \mu_2 h_2 \theta_2.$$

Hence, by (37), $\quad \theta_2/\theta_1 = \mu_1 h_1/(\mu_2 h_2) = \mu_1/(\mu_2 m)\dots\dots\dots\dots(38)$

Thus *all* rays parallel to RP' on incidence are parallel to SQ' on emergence.

If the rays RR' and RP' come from two infinitely distant object points subtending an angle θ_1 at R, the emergent rays will enter an eye placed at S as if they came from two infinitely distant points subtending an angle θ_2 at S. But, since the object points are infinitely distant, the angle they subtend at S is equal to θ_1, the angle they subtend at R. Hence θ_2/θ_1 is the ratio of the apparent angular distance between two very distant object points, when viewed through the telescope, to the actual angular distance which they subtend at the observer's station. The ratio θ_2/θ_1 is, therefore, called the *magnifying power* of the telescope. Thus, by (38),

Magnifying power $\times$ Transverse magnification $= \mu_1/\mu_2$.   (39)

203. Longitudinal magnification. If $P'R'$, $Q'S'$ be drawn perpendicular to the axis, the point S' is the image of R'. By Helmholtz's formula, h_2 and θ_2 have the same sign and hence R' and S' lie either both to the right or both to the left of R and S respectively. Thus, if an object point move along the axis, the image moves along the axis *in the same direction* as the object. This result is true not merely for a "telescope" but also for *any* coaxal system, for it follows from the corresponding result stated in § 124.

Let $RR' = l_1$ and $SS' = l_2$; then $l_1\theta_1 = h_1$, $l_2\theta_2 = h_2$, and hence, by (37) and (38),

$$l_2/l_1 = h_2\theta_1/(h_1\theta_2) = \mu_2 m^2/\mu_1 \dots\dots\dots\dots(40)$$

The ratio l_2/l_1 is called the longitudinal magnification or the elongation. We see that it is constant for a given telescope.

204. Formulae when $\mu_1 = \mu_2$. In many cases, air forms the medium at each end of the system, and thus $\mu_2 = \mu_1$. Then

$$h_2/h_1 = m, \qquad \theta_2/\theta_1 = 1/m, \qquad l_2/l_1 = m^2.$$

Hence, in this case,

Magnifying power $\times$ Transverse magnification $= 1$.

Longitudinal magnification $=$ (Transverse magnification)^2.

CHAPTER X

EXPERIMENTS WITH COAXAL SYSTEMS

205. Notes on practical work with lens systems. *In all experiments with lens systems, the student should leave the apparatus in adjustment until he have obtained the following information:*

(1) The marks—say A and B—which distinguish one end of the system from the other.

(2) The distances of the A-surface and of the B-surface from the corresponding foci. The distances are, of course, measured from the points in which the axis intersects the optical surfaces.

(3) The thickness of the lens system, i.e. the distance AB between the two points just mentioned.

When the two focal lengths have been found, the data of (1), (2) and (3) will enable the student to make a diagram to scale showing the positions of the cardinal points (§ 192) relative to the system. Such a diagram should be made in every case. Care must be taken to obtain the data required under each of the three headings, for the diagram cannot be made if the information be incomplete.

206. Measuring rods. The measurements, in general, are best made on an optical bench. When a bench is not used, the distances of pins or scales from the surfaces of the lens system may be measured directly. When the surface is convex, the distance may be measured with an ordinary steel centimetre scale. When the surface is concave, the device shown in Fig. 129 may be used. A hole about 0·2 cm. in diameter is drilled through

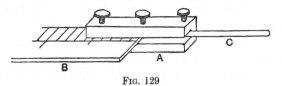

Fig. 129

a brass block A about 3 cm. in length and 1 cm. in width and thickness, and a slot, to admit a steel rule B, is cut to meet the hole; the block is secured to the rule by two set screws. A steel rod, C, with rounded ends, fits into the hole and is secured by a set screw, care being taken that the end of the rod is in contact with the end of the scale. The length of the rod is measured by calipers. Since the axis of the rod nearly coincides with the edge of the rule, the scale reading of the rule, with the correction for the rod, gives the required distance.

For measuring the distance between two surfaces, the device shown in Fig. 130 is useful. A rod A with a rounded or pointed end is held by a small 3-jaw chuck B. A rod C is soldered into the hollow shank of the chuck. A milled ring D enables a good grip to be obtained. The length of A projecting from B is varied until the instrument fit between the surfaces. The length is then measured by calipers.

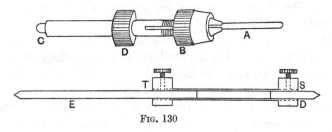

FIG. 130

For greater distances, a tube ST (Fig. 130) with adjustable pointed rods D, E may be used. The ends of the tube are soldered into blocks; the rods are secured by set screws passing through the blocks. A tube 20 cm. long, provided with three rods 3, 19, 19 cm. long, will measure distances from 22 to 53 cm.

207. The optical bench. The bench (Fig. 131) is a stout board to which are fixed a straight steel rod, a metal strip and a scale divided into millimetres, the three being parallel to the length of the bench. The benches at the Cavendish Laboratory have graduated scales 30, 50, 100 or 200 cm. in length. In the longer benches, the board may be braced and kept from twisting by attaching it to a light steel girder; the girder is hidden within the woodwork in Fig. 131. A 30 cm. bench is shown in Fig. 22.

The base of each carriage is a casting having three feet, in two of which a V-groove is cut. These two feet rest on the steel rod fixed to the bench and the third foot rests on the metal strip. There are thus five points of contact; the one remaining degree of freedom allows the carriage to slide along its track.

FIG. 131

If a metal scale be available, the third foot may slide on the scale, keeping clear of the graduations, and the metal strip is then unnecessary. An engineer's steel scale is preferable to a boxwood scale, for permanency and accuracy.

Each base, in addition to an index, carries a vertical steel rod, which passes through the base and is secured by a nut. If the rod be removed (Fig. 22), any object may be attached to the carriage by a bolt and nut. The steel guide rod on the bench is so arranged that the upper face of the carriage is horizontal whether the carriage stand on the bench or on a table.

Each carriage is provided with a cross-rod which can be clamped to the vertical rod. Other fittings can be fixed to the vertical rod or to the cross-rod. When attached to the cross-rod, the fitting has two angular adjustments, for (1) the cross-rod can be turned about the vertical rod, and (2) the fitting can be turned about the cross-rod. In this way the normal to a plane mirror attached to the fitting can be adjusted to any direction.

A small "vice" is used for holding a needle, a scale or other object. A block of metal of square section forms one jaw; the other jaw is a metal plate provided with a clamping-screw.

A second screw, seen on the left-hand carriage in Fig. 131, can be adjusted so that the jaws of the vice are parallel when clamping the object. The block may, conveniently, have two holes at right angles so that the vice can be fixed to a cross-rod in two positions, as in Fig. 131.

Each needle used on the bench is soldered into a plate pierced by a hole considerably larger in diameter than the clamping-screw of the vice, and the screw passes through this hole. Thus the plate, when held between the jaws of the vice, is capable not only of rotation but also of a limited translation in any direction in its own plane. If this plane be perpendicular to the length of the bench, the tip of the needle can be moved at right angles to the length of the bench, and this motion will not cause any displacement of the tip parallel to the length of the bench. To secure this condition, the cross-rod is set perpendicular to the length of the bench, and the vice is adjusted on the rod so that the plane of the needle plate is vertical. This adjustment facilitates the parallax test, which cannot be effectively applied unless the tip of the locating needle can be brought into the *same* position as the image of the tip of an object needle.

A plane mirror is shown on the right-hand carriage in Fig. 131. The mirror should be of plate glass of good quality and should be tested by a telescope. It should be mounted in a block or frame in such a way that it is not strained. The block or frame carries a boss by which it can be clamped to a cross-rod or a vertical rod.

A scale ruled in millimetres on plate glass is mounted in a frame fixed to a boss by which it can be clamped to a cross-rod or a vertical rod.

A lens system for use on the bench is shown in Fig. 131. The lenses are let into two frames rising from a base provided with a clamping boss. To obtain *complete* adjustability, the boss should be fixed to the under side of the base; the system can then be clamped to a cross-rod. The position of the boss in the Figure is, however, sufficient for most purposes.

A simple lens may be mounted in a block with a fitting similar either to that used for the plane mirror or to that used for the glass scale, so that it can be clamped to a cross-rod.

Each boss shown in Fig. 131 is pierced by two **V**-shaped holes at right angles. This allows the fittings to be rigidly clamped in two positions on rods from $\frac{1}{4}$ to $\frac{1}{2}$ inch in diameter.

A microscope, such as that of Fig. 23, is frequently used. It may be held in a suitable clamp, e.g. Pye's "Ideal Clamp," which is attached to a cross-rod or to a vertical rod. In some experiments the axis of the microscope is parallel to the length of the bench, in others it is at right angles to that direction.

A small platform, about 10 cm. by 10 cm., fitted with a boss, is convenient for supporting any object which cannot be held in a vice.

208. Adjustment of thin converging lens. Two needles whose tips are P, Q are mounted on two carriages, as in Fig. 131, and a third needle whose tip is R is fixed to a third carriage or is held in a *fixed* position in a stand near the bench. The P-carriage is adjusted on the bench, and the vice and the needle are adjusted so that P can be made to coincide with R. Then Q is set to R in the same manner. The line PQ is then parallel to the bench.

The lens L (Fig. 132) is mounted on a carriage and is placed between P and Q. Keeping the axis of the lens as nearly parallel to the bench as can be done by eye, the observer adjusts the clamp X on the cross-rod YZ and the clamp Y on the vertical rod (seen in section) so that it is possible, by sliding the carriages on the bench, to make the *tip* P coincide with the real image of the *tip* Q. Since the lens is thin, the nodal points are very close to the straight line PQ. The P-carriage is now moved until the image of P formed by reflexion at the B-face (as in Experiment 31) be near to P itself. It will be impossible to obtain coincidence unless the axis of the lens coincide with

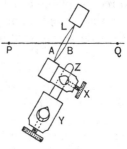

Fig. 132

PQ. The necessary small adjustment of L is made by turning X about YZ and YZ about the vertical rod.

209. Adjustment of plane mirror. Two needles are adjusted, as in § 208, so that the straight line joining their tips P, Q, is parallel to the bench. The plane mirror is mounted on a carriage and is placed as near the needle Q as the carriages allow. The mirror is then adjusted so that P, Q and their images in the mirror are in a straight line. The method is not very satisfactory as the points to be observed are at different distances from the eye.

A better method is to use a thin converging lens, adjusted, as in § 208, by aid of the needles P, Q. The plane mirror is then adjusted so that, when the *tip* P is placed in the focal plane of the lens, and the rays fall on the mirror, they return along their own paths and form an image of the *tip* P coincident with P. The image of the P-needle is inverted. (See §§ 139, 140.) The normal to the mirror is then parallel to the bench.

210. Adjustment of lens system. The system is mounted on a carriage with its axis as nearly parallel to the bench as it can be made by eye. Two needles, tips P, Q, are on two other carriages. First, Q is put as far from the system as the bench allows, and P is adjusted relative to its carriage so that P can be made to coincide with the real image of Q by sliding the P-

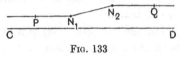

FIG. 133

carriage on the bench. Let N_1, N_2 be the nodal points of the system. Since N_2Q is large, N_2Q is nearly parallel to CD, the direction of the bench, and hence P, at the image of Q, is nearly on the straight line through N_1 parallel to CD. Next, the P-carriage is moved to a distance, and Q is adjusted relative to its carriage so that it can be made to coincide with the image of P. One or two repetitions of this process will ensure that PN_1, N_2Q are parallel to CD, as in Fig. 133. By comparison of P and Q with the tip of a third needle, as in § 208, or by interchanging the P- and Q-carriages on the bench, we can test whether N_1P, N_2Q are portions of a single straight line. Any small error is corrected by a slight adjustment of the system. Slight subsequent adjustments of P and Q may be necessary.

If the distance between the nodal points be small, as it may be although the system appear "thick," the method will, obviously, fail as far as the *direction* of the axis is concerned. In this case, we may employ the method of reflexion described in § 208, using one of the lenses of the system for the purpose.

211. Use of distance rod. The distance d cm. between two objects X, Y is often required. When X is mounted on one carriage and Y on another, the bench readings of the carriages are taken. Let the difference of readings be D_1 cm. One of the carriages is then moved until a rod of known length l cm. touch both X and Y, and the bench readings are again taken. Let the difference be D_2 cm. Then $d - l = D_1 - D_2$, or

$$d = D_1 - D_2 + l.$$

212. Experiments with system giving virtual image of distant object. When the system does not give a real image of a distant object, we cannot place needles at the foci. In this case, a long-focus microscope or a suitable telescope may be used to locate the foci. The distant object is viewed through the system by means of the microscope, which is adjusted so that the object is focused, without parallax, on the cross-wire. Lycopodium is then placed on the nearer surface—say the surface B—of the lens system, and the stand of the microscope is moved until the lycopodium be in focus. The distance through which the microscope is moved is equal to the distance of the focus from B. The method is repeated for the other focus. The microscope is used in like manner to locate the image of an object which is placed at a known distance from the A-face, and therefore at a known distance from the corresponding focus, since the distance of that focus from A is known. The measurements are facilitated if the microscope be mounted on a carriage on an optical bench.

The magnitude of the image can be found if the microscope have a micrometer eye-piece.

EXPERIMENT 37. **Determination of cardinal points by Newton's method.**

213. Method. The lens system AB (Fig. 134) is mounted on a carriage on the optical bench, and its axis is set as nearly as possible parallel to the length of the bench either by eye or by § 210. The bench reading of this carriage should be recorded. Two needles, whose tips are P, Q, are mounted on carriages and are required for marking the positions of the foci and of two conjugate points. The tips P, Q are adjusted relative to their carriages by the method of § 210.

The plane mirror M can be turned (1) about a vertical axis, represented diagrammatically by the rod K, and (2) about a horizontal axis, represented by the joint L, and thus the normal to the mirror can take any required direction. This double adjustment is necessary in many experiments; the student should take care to understand it.

To set the needle Q in one of the focal planes of the system, the plane mirror M is placed on the A-side of the system, and the *direction of the mirror* is adjusted so that the tip of the image of the needle coincides very nearly with the tip itself, the coincidence being tested by a motion of the eye from side to side.

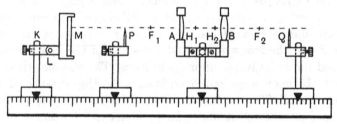

FIG. 134

Exact coincidence is obtained by a *slight* adjustment of the needle in the vice (§ 207). In this case, the tip Q is at F_2 in the focal plane. The needle P is then set at F_1 in the other focal plane in a similar manner, the mirror being on the B side. The bench readings of the carriages are recorded. The distances $F_1A = U$ and $F_2B = V$ from each focus to the nearest surface of the lens system are found by aid of a distance rod (§ 211). Since the medium at each end of the system is air, the two focal lengths are equal, but this does *not* imply that $U = V$.

The two carriages bearing P and Q are then adjusted so that P and Q mark two conjugate points. If necessary, a slight adjustment of one of the needles in its vice is made. The distances $PF_1 = p$ and $QF_2 = q$ are then found from the bench readings. Several positions for P are taken and the values of f, the focal length, are calculated from the corresponding values of p and q by Newton's formula, § 190, equation (11),

$$pq = f^2. \quad \dots\dots\dots\dots\dots\dots\dots(1)$$

The mean value of f is then found.

The principal points H_1, H_2 can now be located, for

$$AH_1 = f - U, \quad BH_2 = f - V. \quad \dots\dots\dots\dots(2)$$

The distance F_1F_2 is equal to $2f + t$, where t is the distance between the principal planes of the system. When the distance AB is known, $2f + t$ is given by

$$2f + t = F_1F_2 = U + V + AB. \quad \dots\dots\dots(3)$$

Since $pq = f^2$, $(p + q)^2 = 4f^2 + (p - q)^2$, and thus $(p + q)^2$ has its smallest value when $p = q$. In this case, $(p + q)^2 = 4f^2$, and thus, when PQ is greater than F_1F_2, the least distance between P and Q is $4f + t$. Since the distance is a minimum when $p = q$, it will be practically independent of any small difference between p and q.* Hence, if a value of f be first found approximately from a pair of values of p and q and if PF_1 be then made equal to that value, the distance from P to the conjugate point Q may be taken as being very nearly equal to $4f + t$.

The distances F_1F_2 and PQ can be deduced from the readings of the carriages bearing the needles, by aid of a rod of known length.

The minimum value of PQ gives $4f + t$, while the distance F_1F_2 gives $2f + t$. From these results the value of t and that of f are found; the latter may be compared with the mean value obtained from the formula $pq = f^2$.

It may be found that the distance F_1F_2 is less than $2f$. This implies that t is negative and that the principal points are inverted, as in Fig. 121.

A figure should be drawn to scale showing the positions of the foci and of the principal points relative to the system.

Instead of using a plane mirror to find U and V, we may use a distant object, turning first A and then B towards the object.

* The solution $p + q = -2f$, leading to $p = -f$, $q = -f$, corresponds to the principal points H_1, H_2. If P move along the axis, the distance PQ is "stationary" as P passes through H_1. But H_1H_2 is not necessarily the *shortest* distance between image and object, for in certain cases that distance may be zero, though not stationary. See § 247.

When the object is at "infinity," its image in each case coincides with a focus of the system.

When, however, the distance of the object is finite, the distance of its image from the focus must be taken into account. When $PF_1 = d$, $F_2Q = q = f^2/d$. Hence, if V' be the distance from B to the image of the distant object,

$$V = V' - q = V' - f^2/d. \qquad \dots\dots\dots\dots(4)$$

Similarly, if U' be the distance from A to the image of the distant object, when B is turned towards the object,

$$U = U' - f^2/d. \qquad \dots\dots\dots\dots\dots\dots(5)$$

In the small term f^2/d we may, in place of f, use f', the value of f obtained on the assumption that the object is at an infinite distance; since the term is small, an approximate value of d will suffice.

In the following method only *one* focus is directly located by aid of a distant object. The bench is directed so that the image of the distant object is at F_2. The needle on the A-side of the system is then put (1) at P, and (2) at R, where $PF_1 = p$, $RF_1 = r$, and the images Q and S of P and R are found. Let $QF_2 = q$, $SF_2 = s$. Then, by (1), $p = f^2/q$, $r = f^2/s$, and hence $f^2 (1/q - 1/s) = p - r$, or

$$f^2 = sq\, (p - r)/(s - q). \qquad \dots\dots\dots\dots(6)$$

We can, of course, find $p - r$ without knowing the position of F_1. When f has been found by (6), the distance PF_1 can be found (thus locating F_1), since

$$p = f^2/q = s\, (p - r)/(s - q). \qquad \dots\dots\dots\dots(7)$$

If necessary, the correction indicated by (4) must be applied to the position of the image of the distant object.

214. Practical example.
Mr C. W. Kearsey used system having $AB = 15\cdot19$ cm. The foci were found by plane mirror;

$$U = AF_1 = 7\cdot23 \text{ cm.,} \qquad V = BF_2 = 8\cdot66 \text{ cm.}$$

Minimum distance method. From above,

$$2f + t = F_1F_2 = 7\cdot23 + 15\cdot19 + 8\cdot66 = 31\cdot08 \text{ cm.}$$

Minimum distance between image and object $= 4f + t = 67\cdot67$ cm. Hence

$$f = \tfrac{1}{2}(67\cdot67 - 31\cdot08) = 18\cdot30 \text{ cm.,} \qquad t = -5\cdot51 \text{ cm.}$$

Newton's method. Distances p, q of object and image from F_1, F_2 were measured.

$$\begin{array}{ccc}
p & q & f = (pq)^{\frac{1}{2}} \\
24\cdot46 \text{ cm.} & 13\cdot65 \text{ cm.} & 18\cdot27 \text{ cm.} \\
18\cdot31 & 18\cdot35 & 18\cdot33 \\
13\cdot50 & 24\cdot92 & 18\cdot34
\end{array}$$

Mean $f = 18\cdot31$ cm. Hence

$$AH_1 = f - U = 11\cdot08, \qquad BH_2 = f - V = 9\cdot65 \text{ cm.}$$

Since $AB = 15\cdot19$, $t = H_1H_2 = 15\cdot19 - 11\cdot08 - 9\cdot65 = -5\cdot54$ cm.

The principal points are therefore inverted, as in Fig. 121. The sequence of points is $F_1AH_2H_1BF_2$.

EXPERIMENT 38. **Determination of cardinal points by Gauss's method.**

215. Method. The lens system is placed so that its axis passes through a *distant* object. A pin is then *carefully* adjusted so that its tip coincides with the image, and the distance of the tip from the nearest surface of the system is measured; the device shown in Fig. 129 may be used when the surface is concave. The tip of the pin is in the focal plane and thus, if the tip be nearly on the axis, the measured distance is very nearly equal to the distance of the corresponding focus from the surface. The system is then turned end for end, and the distance of the other focus from the nearest surface of the system is found in a similar manner. The adjustment of the pin is complete when a motion of the eye causes no apparent motion of the image relative to the pin. If a large lens be used, a diaphragm with an aperture not more than 3 cm. in diameter should be fitted to it in order to reduce the effects of spherical aberration. Two or three settings of the pin should be made for each position of the lens.

When a distant object is not available, a plane mirror is used, as in § 213, and the tip of the pin is made self-conjugate.

Let the axis cut the extreme surfaces of the system in A and B (Fig. 135), let F_2 be the focus when the rays from the distant point fall first on the surface A and let F_1 be the focus when they fall first on the surface B. Let $AF_1 = U$, $BF_2 = V$.

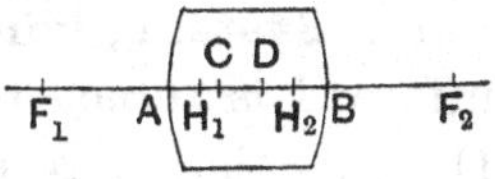

Since we know the distance of A from F_1, we can find f by Newton's formula (§ 190) if we find the distance between F_2 and C the image

FIG. 135

of A. This measurement is easily made by means of a microscope of low power, such as that of Fig. 23, attached to a
carriage on an optical bench. The axis of the microscope is
parallel to the length of the bench. Lycopodium is placed on
the surfaces A, B, and the axis of the system is made to coincide
with that of the microscope. The surface B is turned towards
the microscope, which is focused first on the grains at B and
then on the images (formed by the system) of the grains at A.
The displacement of the microscope is equal to the distance
between B and C, the image of A. Let the distance BC be b,
and let b be reckoned positive when C and A are on the same
side of B. The system is now turned end for end, and the microscope is focused first on A and then on D, the image of B, and
the distance AD, or a, is reckoned positive when D and B are
on the same side of A. Each setting of the microscope is repeated
two or three times.

Since A and C are conjugate, Newton's formula gives

$$f^2 = AF_1.CF_2 = U(V + b). \ \ \dots\dots\dots\dots(1)$$
Similarly $\qquad f^2 = BF_2.DF_1 = V(U + a). \ \ \dots\dots\dots\dots(2)$

The mean value of f derived from these equations is taken as
the focal length of the system. The focal length is positive if the
image of a distant object be inverted.

When the thickness AB has been measured, the positions of
the principal points relative to the system can be found. For, if
H_1 be the principal point corresponding to F_1, we have $F_1H_1 = f$,
and thus, when f is positive,

$$AH_1 = f - U, \qquad BH_2 = f - V.$$

The results may be tested by finding the distances of two
conjugate points, P and Q, from A and B respectively, the points
being identified by the tips of pins. Putting $PA + AH_1 = u$ and
$QB + BH_2 = v$, the focal length is calculated by the formula

$$1/f = 1/u + 1/v.$$

216. System giving virtual image of distant object. The
surface A is turned towards the microscope, which is focused
(1) on F_1 the image of a distant object on the B-side of the
system, (2) on A, (3) on D, the image of B. The point F_1 will
be on the B-side of the system. The image at F_1 will always

be *erect* when f is negative, but inverted when f is positive. The system is then turned end for end, and the microscope is focused (1) on F_2, (2) on B, (3) on C, the image of A. The bench readings of the microscope give AF_1, AD, BF_2, BC. If D lie between A and F_1, and C between B and

FIG. 136

F_2, as in Fig. 136, we have $DF_1 = AF_1 - AD$, $CF_2 = BF_2 - BC$.

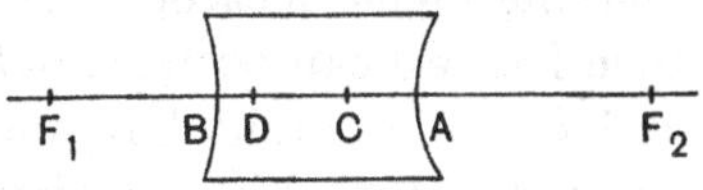

Since B and D and also A and C are conjugate, we have

$$f^2 = BF_2 . DF_1, \qquad f^2 = AF_1 . CF_2. \qquad \ldots\ldots\ldots\ldots(3)$$

To avoid confusion, the student should mark off to scale, as in Fig. 136, the positions of A, B, F_1, F_2, C, D before calculating f. The principal point H_1 lies at the distance $|f|$ from F_1, where $|f|$ is the numerical value; when f is positive, H_1 lies between F_1 and the corresponding distant object, but, when f is negative, F_1 lies between H_1 and that object. Similarly H_2 lies at the distance $|f|$ from F_2.

217. Practical example. For a convexo-concave lens, A was on convex face; $AB = 1\cdot425$ cm. The image of a distant object was inverted; hence the focal length is positive. Mean values were

$$U = 9\cdot657, \qquad V = 7\cdot383, \qquad a = 1\cdot113, \qquad b = 0\cdot863 \text{ cm.}$$

By (1), (2), $f = \{U(V + b)\}^{\frac{1}{2}} = 8\cdot924$, $f = \{V(U + a)\}^{\frac{1}{2}} = 8\cdot917$ cm. Mean $f = 8\cdot920$ cm. Hence

$$AH_1 = 8\cdot920 - 9\cdot657 = -0\cdot737, \qquad BH_2 = 8\cdot920 - 7\cdot383 = 1\cdot537 \text{ cm.}$$

$$AH_2 = \text{thickness} - 1\cdot537 = -0\cdot112 \text{ cm.}$$

Thus both the principal points are outside the lens and between A and F_1. The sequence of points is $F_1 H_1 H_2 A B F_2$.

EXPERIMENT 39. **Determination of focal length by the method of magnification.**

218. Introduction. The focal length may be deduced from a series of determinations of the ratio of the size of the image to that of the object. By (5a), § 188,

$$\frac{h_1}{h_2} = -\frac{u_1 - f_1}{f_1} = 1 - \frac{u_1}{f_1}, \qquad \frac{h_2}{h_1} = -\frac{u_2 - f_2}{f_2} = 1 - \frac{u_2}{f_2}. \ldots(1)$$

Here u_1 and u_2, the distances of object and image from the corresponding principal planes, are measured in opposite directions, and these distances cannot be found unless the positions of the principal points H_1 and H_2 be known. If x_1 and x_2 be the bench readings of the carriages bearing the object and the image locator respectively, and if x_2 and u_2 increase together, we can write $u_1 = k_1 - x_1$ and $u_2 = x_2 - k_2$, where k_1 and k_2 are constant but unknown; actually, k_1 is the bench reading of the object-carriage corresponding to H_1, and similarly for H_2. The equations of magnification now become

$$\frac{h_1}{h_2} = 1 - \frac{k_1}{f_1} + \frac{x_1}{f_1}, \qquad \frac{h_2}{h_1} = 1 + \frac{k_2}{f_2} - \frac{x_2}{f_2}. \qquad\ldots\ldots\ldots(2)$$

Hence, if we plot one curve with x_1 as abscissa and h_1/h_2 as ordinate and a second curve with x_2 as abscissa and h_2/h_1 as ordinate, each will be a straight line. From the slopes of these lines f_1 and f_2 can be determined, for, when h_1/h_2 increases by unity, x_1 increases by f_1, and, when h_2/h_1 increases by unity, x_2 diminishes by f_2. Due account must be taken of whether h_1/h_2 is positive or negative. The media at the two ends of the system are identical and thus, by § 189, f_1 and f_2 are equal. Hence, if the two lines be drawn on the *same* diagram, they will be parallel, if x_1 and x_2 be measured in opposite directions on the diagram.

219. Method. When the optical system gives a real image of a real object for a range of positions of the object, the measurements involved in (2) are easily made upon an optical bench. The bench is provided with three carriages R, S, and T. The carriage S bears the optical system and remains in a fixed position, intermediate between R and T; the bench reading of S should be recorded. The A-face of the system is towards R.

The carriage R bears the object, which may be a scale on glass or celluloid divided to millimetres, with *fine* dividing lines. A celluloid scale should be compared directly with a scale ruled on metal or glass, since celluloid scales are liable to considerable changes in length.

The carriage T bears a scale on glass divided to millimetres. The scales are placed at right angles to the length of the bench and are brought to the same level as the axis of the system. The *divided* faces of the scales are turned toward the system. If the object be a glass scale, a sheet of ground glass, with a lamp behind it, may be set up as a background.

The carriage R is placed at a distance from the system, and the carriage T is adjusted so that the divided face of the glass scale coincides with the plane of the image, in which case a motion of the eye from side to side causes no apparent motion of the image relative to the glass scale. The bench readings of the carriages R and T are then recorded.

The readings on the glass scale of the images of two lines on the object are taken, and these readings as well as the distance between the two lines are recorded. The distance between the lines is denoted by h_1 and that between their images by h_2, the latter being positive when the image is erect. The observations are repeated for each of a series of positions of R, till R comes close to S. A record is made as to whether the image be erect or inverted. If more convenient, two lines on the glass scale may be chosen and their readings on the image of the object scale may be recorded.

If the system do not give a real image for any position of the (real) object, an auxiliary lens L may be placed between the object scale and the system, so that the image formed by L acts as a virtual object for the system. The size of this image is h_1, and the size of its image formed by the system is h_2. If L be suitable, we can place the system so that this final image is real.

In some cases, we need not use L, if we can observe the virtual image of the object scale with a micrometer microscope.

220. Positions of principal points. When f has been determined, the principal points are easily found. From (1) we see that, when $f_1 = f_2 = f$,

$$u_1 = f(1 - h_1/h_2), \qquad u_2 = f(1 - h_2/h_1).$$

Hence, when h_1/h_2 has been found for any position of R and the corresponding position of T, the values of u_1 and u_2 can be calculated. If, now, we measure the distance of the object from the A-face of the system, and the distance of the glass scale from the B-face, these measurements, together with the corresponding values of u_1 and u_2, will determine the positions of H_1, H_2 relative to the system.

If we take $h_1/h_2 = -1$, then $u_1 = 2f$, $u_2 = 2f$. The bench readings of R and T corresponding to $h_1/h_2 = -1$ and $h_2/h_1 = -1$ are found from the graphs.

221. Practical example. Mr E. H. Taylor used a system for which $AB = 15\cdot19$ cm. This lens was also used in §§ 214, 233. The table gives the bench readings x_1, x_2, of the carriages R, T, bearing the object scale and the glass scale.

x_1 cm.	x_2 cm.	Readings for h_1 cm.	h_1 cm.	h_2 cm.	h_1/h_2	h_2/h_1
35·00	119·71	9·07 – 3·78	5·29	– 2·0	– 2·645	– 0·378
50·00	122·81	9·27 – 4·69	4·58	– 2·5	– 1·832	– 0·546
65·00	130·94	7·64 – 5·12	2·52	– 2·5	– 1·008	– 0·992
73·00	144·60	7·86 – 5·56	2·30	– 4·0	– 0·575	– 1·739
77·00	164·08	7·44 – 5·67	1·77	– 5·0	– 0·354	– 2·825

The graph for x_1 and h_1/h_2 gave $f_1 = 18\cdot350$ cm. and that for x_2 and h_2/h_1 gave $f_2 = 18\cdot325$. Mean $f = 18\cdot34$ cm.

The points for which $u_1 = 2f$, $u_2 = 2f$ are denoted by G_1, G_2. The graph showed that the bench reading of R when $h_1/h_2 = -1$ was 65·22 cm.; the object was then at G_1. When the object was 15 cm. from A, the bench reading was 75·80. Thus $U = AF_1 = (75\cdot80+15) - (65\cdot22+f) = 90\cdot80 - 83\cdot56 = 7\cdot24$ cm. The graph showed that the bench reading of T when glass scale was at G_2 was 131·20 cm. When this scale was 15 cm. from B, the bench reading was 119·21 cm.

Thus $V = BF_2 = (131\cdot20 - f) - (119\cdot21 - 15) = 8\cdot65$ cm.

Hence $AH_1 = f - U = 11\cdot10$ cm., $BH_2 = f - V = 9\cdot69$ cm.,

and $H_1H_2 = AB - 11\cdot10 - 9\cdot69 = -5\cdot60$ cm.

The sequence of the points is $F_1AH_2H_1BF_2$.

EXPERIMENT 40. Determination of focal length by micrometer method.

222. Method. A microscope fitted with a micrometer eyepiece may be employed to determine the focal length of an optical system, by a method due to T. H. Blakesley.

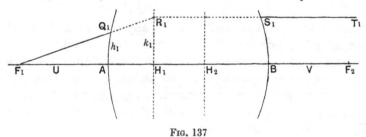

FIG. 137

Let F_1, F_2 (Fig. 137) be the foci, H_1, H_2 the principal points and A, B the points in which the axis intersects the extreme

surfaces of the system. Let $AF_1 = U$ and $BF_2 = V$. Let Q_1 be
a point on the surface near A, and let the *straight line* F_1Q_1 cut
the first principal plane in R_1. Let $R_1S_1T_1$ be drawn parallel to
the axis and let it cut the B-surface in S_1. Since the ray F_1Q_1
passes through the focus F_1, it gives rise to the emergent ray
S_1T_1, and since the ray F_1Q_1 passes through Q_1, the image of
Q_1 lies somewhere on the straight line $R_1S_1T_1$. Hence R_1H_1 is
equal in magnitude to the image of Q_1A.

Let $Q_1A = h_1$, $R_1H_1 = k_1$. By the triangles $R_1H_1F_1$, Q_1AF_1,

$$f_1 = F_1H_1 = F_1A \cdot R_1H_1/Q_1A = U \cdot k_1/h_1. \qquad \ldots\ldots\ldots(1)$$

Similarly, if Q_2 be a point on the B-surface at a distance h_2
from B, and if k_2 be the magnitude of the image of Q_2B,

$$f_2 = F_2H_2 = V \cdot k_2/h_2. \qquad \ldots\ldots\ldots\ldots\ldots\ldots(2)$$

To mark the distance Q_1A on the A-surface, a piece of tinfoil
may be used. From a strip of foil with parallel sides, a piece
about 1 cm. in length is cut off; if the surface of the lens be
breathed upon, the foil will adhere to it. The strip should be cut
with a *sharp* knife upon a card. The foil is placed centrally on
the face A. The width of the strip should be such that its image
covers about two-thirds of the micrometer scale. In place of the
foil, two grains of lycopodium may be used.

The eye-piece of the microscope is now adjusted so that the
micrometer scale is in focus. The foil on the A-surface is then
viewed *directly* by the microscope, which is adjusted so that the
image of the foil is focused, without parallax, on the micrometer;
the micrometer lines must be parallel to the edges of the image
of the strip. The scale readings of the edges of the strip are now
taken. The process is repeated two or three times, the lens being
slightly moved so as to furnish independent readings. Let the
mean number of scale divisions covered by the strip be m_1.

The lens system is now turned round, and the strip of foil is
viewed by the microscope *through* the system; let the mean
number of divisions covered by it be n_1.

A strip of foil is now placed on the B-surface, and the measurements are repeated. When seen directly it covers m_2 divisions,
and when seen through the system it covers n_2 divisions.

The distances, $AF_1 = U$ and $BF_2 = V$, of the foci from the
surfaces of the system are found as in § 213, and AB is measured.

The values of f_1 and f_2 are calculated by the equations

$$f_1 = \frac{k_1}{h_1} U = \frac{n_1}{m_1} U, \qquad f_2 = \frac{k_2}{h_2} V = \frac{n_2}{m_2} V. \quad \dots\dots\dots(3)$$

Since the system is in air, f_1 is theoretically equal to f_2; the mean of f_1, f_2 is taken as the value of f. From f, U, V and AB the distances AH_1, BH_2 and H_1H_2 are found, each with its proper sign.

223. Practical example. G. F. C. Searle used system of two lenses. The A-surface was convex, B-surface concave; $AB = 1\cdot21$ cm.

When foil was at A, it covered $m_1 = 29\cdot63$ micrometer divisions when seen directly, and $n_1 = 27\cdot17$ divisions when seen through system. When foil was at B, $m_2 = 29\cdot63$, $n_2 = 35\cdot20$. Also $AF_1 = U = 11\cdot90$, $BF_2 = V = 9\cdot21$ cm. By (3)

$f_1 = n_1U/m_1 = 10\cdot91$, $f_2 = n_2V/m_2 = 10\cdot94$.   Mean $f = 10\cdot92$ cm.

Hence $U - f = 0\cdot98$, $f - V = 1\cdot71$; $f - V - AB = 0\cdot50$ cm.

Both H_1 and H_2 are outside system; $AH_1 = \cdot98$, $AH_2 = \cdot50$, $H_1H_2 = \cdot48$ cm. The sequence of points is $F_1H_1H_2ABF_2$.

EXPERIMENT 41. **Direct determination of positions of principal points.**

224. Method. Let H_1K_1, H_2K_2 (Fig. 138) be the principal planes of the lens AB. Then any ray which is directed to K_1 before incidence on the A-face is directed from K_2 after emergence from the B-face, and $K_1H_1 = K_2H_2$. This equality can be used to give the actual positions of the principal planes.

The apparatus is arranged on an optical bench. A strip of tinfoil T is mounted on a plate of glass G and is illuminated by

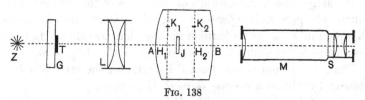

FIG. 138

a sodium flame Z placed at a safe distance from G. An auxiliary lens L is placed so that TL is about twice the focal length of L; this lens may be similar to that used in § 105. Let the image of T formed by L be denoted by J. Beyond L are placed the lens AB and a microscope M, with a micrometer scale S in the eye-piece. A pair of fine parallel lines ruled (§ 331) on celluloid may be used in place of the foil.

The scale S is focused, and the microscope is moved along the bench until, when AB is absent, the image of J be seen in focus on S. Let it cover n divisions of S. The lens AB is now placed on the bench and M is moved along the bench to bring J again into focus. If the image do not cover n divisions, AB and M are moved until the image do cover n divisions. In practice, this double adjustment is easily made. The image J is now in the plane $H_1 K_1$, and the image of J formed by AB is in $H_2 K_2$, and M is focused on H_2.

The two bench readings of M, when focused (1) on J (observed when AB is absent), and (2) on the image of J formed by AB, give the distance $H_1 H_2$ and show whether H_1 or H_2 is the nearer to T.

The microscope is now focused on lycopodium on the B-face; the bench readings for B and for H_1 and H_2 give BH_1 and BH_2. Thus, when AB is known, $H_1 A$ can be found.

A set of observations may be made with AB reversed, so that B is towards T. In this case, the plane $H_2 K_2$ is made to coincide with J. We now measure AH_1 and AH_2, and BH_2 is deduced from AB and AH_2.

If the principal plane $H_1 K_1$ be outside the system *on the A-side*, the real object T can be placed in that plane, and the lens L is not required when A is towards T. The lens L will, however, be necessary when B is towards T, unless H_2 be outside the system on the B-side.

225. Practical example. Mr C. F. Sharman used lens with A-face convex, B-face concave; $AB = 1\cdot 27$ cm. The bench readings increased in the direction T to M.

(i), A-face towards T		(ii), B-face towards T	
Bench reading of M	Distances	Bench reading of M	Distances
J and H_1 31·13 cm.	$BH_1 = 2\cdot 36$ cm.	A 30·10 cm.	
H_2 31·65	$BH_2 = 1\cdot 84$	J and H_2 30·67	$AH_2 = 0\cdot 57$ cm.
B 33·49		H_1 31·16	$AH_1 = 1\cdot 06$

By table (ii), sequence of points is $BAH_2 H_1$.

By table (i), $\quad AH_1 = BH_1 - AB = 2\cdot 36 - 1\cdot 27 = 1\cdot 09$ cm.,

and $\qquad\qquad AH_2 = BH_2 - AB = 0\cdot 57$ cm.

Mean values, $\quad AH_1 = 1\cdot 075, \quad AH_2 = 0\cdot 570$ cm.; hence $H_1 H_2 = 0\cdot 505$ cm.

EXPERIMENT 42. **Determination of cardinal points by nodal point method.**

226. Introduction. We have seen in § 191 that the distance of the first focus from the first nodal point is equal to the second focal length, and that the distance of the second focus from the second nodal point is equal to the first focal length. Thus, if we can determine the positions of the foci and of the nodal points, we shall be able to find the focal lengths of the system.

The positions of the nodal points N_1, N_2 can be easily determined by the aid of the special property of these points, viz. that any ray starting in the first medium along a straight line which passes through N_1 gives rise to a ray which emerges into the second medium along a straight line which passes through N_2 and is *parallel* to the incident ray.

Let the system be placed on a revolving table, and let the vertical axis of revolution intersect the axis of the system, which

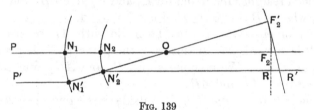

FIG. 139

is horizontal, in some point O (Fig. 139). Let ON_2N_1 be the axis and F_2 the second focus of the system, for some standard position of the table. Then rays from a very distant point P on the axis ON_1 will come to a focus at F_2. If the table be now turned round through a small angle, the points N_1, N_2 and F_2 will describe arcs of circles about O and will come to N_1', N_2' and F_2'. Then a ray $P'N_1'$, parallel to PN_1, will emerge along $N_2'R$, cutting the tangents at F_2 and F_2' to the circle F_2F_2' in R and R'. By the property of the nodal points, $N_2'R$ is parallel to $P'N_1'$ and therefore to POF_2. Since the ray $P'N_1'$ is parallel to PN_1, it comes from the distant point P and therefore, since R' is in the focal plane through F_2', an image of P is formed at R'. The angle F_2OF_2' being small, the point R' may be considered to coincide with the point R. Thus, as the table is turned from the standard position, the image of P will remain in the focal plane through F_2, but, in the absence of special adjustment of the lens system on the table, the nodal point N_2 will move away from the line POF_2 and the image of P will move in the same direction and through the same distance.

If N_2, the nodal point of emergence, be made to coincide with O, N_2 will not move away from the straight line POF_2 when the table is rotated, and hence the image of P will remain at rest at F_2. When, therefore, the lens system has been so adjusted on the rotating table that the image of a distant point does not move when the table is turned through a small angle, the axis of rotation passes through the nodal point of emergence. If we now measure the distance of the second focus F_2 from the axis of rotation, the first focal length f_1 is found at once.

If the system be such that it gives a *real* image of a distant object, a ground glass screen may be placed in the focal plane to receive the image. In cases where a distant object is not available, a plane mirror may be set up so that the rays from the tip of a needle placed at F_2 fall normally on the mirror after passing through the system. It will be easily seen that, if the image of the tip continue to coincide with the tip itself in spite of the rotation of the table, the axis of rotation passes through N_2, the nodal point of emergence.

227. Apparatus. A simple form of apparatus for this method is shown in Fig. 140. A heavy foot carries a table, which revolves

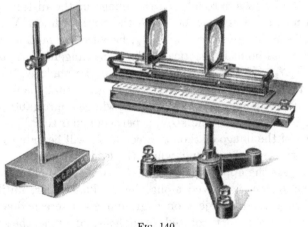

FIG. 140

about a vertical axis; the fitting should be good, since any "play" renders it difficult to make the optical adjustments with accuracy. The lens system is mounted on a base, which slides against a graduated bar, the position of the system relative to the scale being determined by the reading of the index mark. The *same* side of the base remains in contact with the scale throughout the experiment, and the optical axis of the system should be parallel to this side. The graduated bar can be moved perpendicularly to itself and can be clamped to the table in any position.

In the lens system shown in Fig. 140, the lens holders are mounted so that the distance between the lenses can be varied. The focal lengths of the lenses are best *unequal*.

228. Adjustment of graduated bar. The first step is to make the optical axis of the system intersect the axis of revolution of the table by adjusting the graduated bar. From Fig. 141 we see that, even if the axis of revolution, which is represented

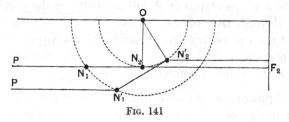

Fig. 141

by O, do not pass through N_2, the image of the distant point P will not move appreciably from the straight line $PN_1N_2F_2$, if ON_2 be perpendicular to PN_1N_2, because the motion of the second nodal point is practically along the tangent at N_2 to the circle N_2N_2'. If ON_2 be not large and if the angle N_2ON_2' be small, the distance N_2N_2' will not be so great that the focusing of the distant point on the screen at F_2 will be appreciably spoilt by the motion. Thus, when ON_2 is perpendicular to PN_1N_2, the position of the image of P on a screen at F_2 will be independent of the (small) angle through which the table is turned.

To adjust the apparatus so that the two axes shall intersect, the lens system is moved along the graduated bar, until the image of a distant object on a ground glass screen does not move when the table is rotated through a small angle, the screen

being adjusted for each new position of the lens system so that
the image is sharply focused upon it. The screen is left in
position. The table is then turned through approximately 180°,
and the lens system is moved along the bar until the image of
the distant object is again focused on the screen. Since the
two focal lengths are equal, it will be found that the image of
the distant object is again unaffected by a small rotation of the
table. The needle used in § 230 may be used in place of the
ground glass to locate the image and is often more convenient.

The two positions F_2, F_1 (Fig. 142) of the image of the distant
object P cannot coincide unless the optical
axis intersect the axis of revolution. Thus,
if x be the shortest distance between the
optical axis $N_1 N_2$ and the axis of revolution, which is represented by O, the

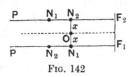

Fig. 142

two positions of the image will be separated by a distance $2x$.
Hence, if it be found that the two positions of the image of
P are separated by a distance $2x$, the axes will intersect each
other if the graduated bar be moved perpendicularly to itself
through a distance x in the proper direction.

When a ray falls on a mirror, the angle between the incident
and reflected rays is twice the angle of incidence. Hence, when
a plane mirror (§ 231) is used, the bar is moved through $\frac{1}{2}x$.

229. Determination of distance between nodal points.
The system is adjusted so that one nodal point is on the axis
of revolution, and the reading of the index mark on the graduated
bar on the table is taken. This is repeated two or three times,
and the mean reading of the index is found. The table is then
turned through 180°, and similar observations are made for the
other nodal point. The difference between the two mean readings
of the index gives the distance between the nodal points. By
§ 191, this distance is equal to that between the principal points.

From the two positions of the system relative to the table
corresponding to a stationary image, we can decide whether the
distance between the nodal points be positive or negative. Thus,
suppose that the position for a stationary image of a distant
object is found with the A-end of the system turned towards
the object. The table is now turned so that the B-end faces the

object. If, to obtain a stationary image, it be necessary to move the system *towards* the object, the distance between the nodal points is *positive* as in Fig. 124, § 191.

230. Determination of focal length. The A-end of the system is turned towards the distant object, and the index mark of the system is brought to its mean reading for the nodal point N_2. A sharp vertical needle is then adjusted so that there is no apparent motion of the needle relative to the image of a distant object when the eye is moved and the table is stationary. The distance BF_2 or V from the needle to the nearer face of the lens system is measured.

The lens system is now removed from the table, and the distance of the needle from the axis of the table is determined by a horizontal scale carried by the table top. The arrangement shown in Fig. 143 is convenient. A cranked rod can be secured by a set-screw in a socket in the casting which supports the table top, and this rod is fitted with a clip in which a scale can be held. When the table is not fitted with a rod, the scale may be held in any clip or holder which will rest upon the table top.

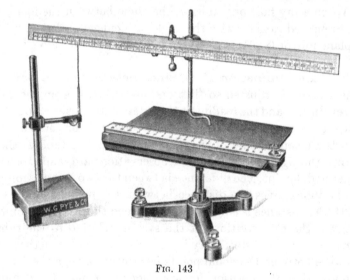

Fig. 143

The height of the scale is first adjusted so that its graduated face will meet the point of the needle when the table is turned, and then this face is made to intersect the axis of revolution. The last setting is easily made by means of a triangular drawing square; the projection on the plane of the table of the graduated face of the scale is compared with a line ruled on the table so as to pass through the point which remains at rest when the table is revolved.

The scale is now brought into contact with the needle and the reading of the needle on the scale is taken. The table is then turned so as to bring the other end of the scale to the needle and a second reading is taken. Since the nodal point N_2 was adjusted so as to lie on the axis of revolution, the difference between the two readings gives $2N_2F_2$ or $2f_1$.

The system is replaced on the table, which is turned so that the B-end of the system is towards the distant object. After the index mark has been set to the mean reading corresponding to the nodal point N_1, the focal length f_2 is determined. The distance AF_1 or U from the needle to the nearer surface of the system is measured.

The values obtained for the two focal lengths will be equal within the errors of observation (see § 189). The mean is taken as the focal length of the system.

The rod shown in Fig. 143 is unnecessary if measurements be made from the faces A and B, when the system is kept in a *fixed* position relative to the table. The distance r from the needle to A is found when A is towards the needle. The table is then turned through 180° and the distance s from the needle to B is found. The axis of the system must pass through the needle in each case. If p be the distance from the needle to the axis of the table,

$$p = \tfrac{1}{2}(AB + r + s). \quad \ldots\ldots\ldots\ldots(1)$$

231. Use of a plane mirror. When a distant object is not available, the system is placed between a *good* plane mirror and a vertical needle, and the tip of the needle is made to coincide with its own image. If the nodal point lie on the axis of revolution, the image will continue to coincide with the needle when the table is turned through small angles. The distance between the nodal points and the focal lengths are then found just as in §§ 229 and 230.

232. System giving virtual image of distant object. A virtual image cannot be received on a screen nor can a needle be made to coincide with it. We can, however, locate it by a telescope or by a microscope of low power. The microscope is mounted on a short optical bench so that it can move parallel to its own axis. If a distant object be not available, a collimator with a cross-wire may be used.

The lens system is adjusted on the table and the microscope is adjusted on its bench so that the image of the distant object (1) is in focus on the cross-wire or micrometer scale of the microscope, and (2) is stationary when the table is turned through small angles. The table is then turned through 180°, and the system is moved on the table until the image is again in focus. If the image be not in the same position as before relative to the cross-wire or scale, the graduated bar is adjusted as in § 228.

With the A-face towards the object, so that the image is at F_2, the double adjustment just described is made. When the image (1) is in focus, and (2) is stationary, the reading of the lens index on the table scale and the bench reading of the microscope are taken. The microscope is now focused on lycopodium on the B-face; the change of bench reading gives BF_2.

Without moving the system relative to the table, the latter is turned through 180°, and the microscope is focused on lycopodium on the A-face. The mean of the bench readings for B and A corresponds to a point at a distance $\frac{1}{2}AB$ from the axis in the direction of the microscope. Combining this mean reading with the reading for F_2, f_1 is found. If the image be erect, as seen by the naked eye, the focal length is negative.

The observations are repeated, starting with the B-face towards the object, and AF_1 and f_2 are found. The difference between the index readings of the lens on the table scale gives the distance between the nodal points.

233. Practical examples. Mr E. H. Taylor used lens system of §§ 214, 221; $AB = 15 \cdot 19$ cm. A plane mirror was used, as the most distant available object was only 65 metres from the system.

The system was adjusted so that image of needle was stationary, and index readings of lens carriage on table scale were taken.

Surface A facing mirror, mean reading 10·67 cm.

The table top was now turned through 180°. It was then necessary to move system *away* from mirror to obtain stationary image of needle. Hence t is negative and the nodal points are inverted. Then new index readings were taken.

Surface B facing mirror, mean reading 5·24 cm.

Hence distance between nodal points $= t = 5·24 - 10·67 = - 5·43$ cm.

With A towards mirror and index at 10·67 cm., needle was made to coincide with its own image. Distance of needle from axis of table was found by § 230. Scale readings were 58·46, 21·82 cm. Hence $f_1 = \frac{1}{2} (58·46 - 21·82) = 18·32$ cm.

Distance from surface B to needle $= BF_2 = 8·68$ cm.

With B towards mirror and index at 5·24, needle was made self-conjugate. Scale readings gave $f_2 = \frac{1}{2} (58·50 - 21·86) = 18·32$ cm.

Distance from surface A to needle $= AF_1 = 7·36$ cm.

Mean $f = \frac{1}{2} (f_1 + f_2) = 18·32$ cm.

Hence $\quad AN_1 = f - AF_1 = 10·96, \quad BN_2 = f - BF_2 = 9·64$ cm.

The sequence of points is $F_1 A N_2 N_1 B F_2$.

The thickness AB was calculated. Thus

$$AB = AN_1 + N_1 N_2 + N_2 B = 10·96 - 5·43 + 9·64 = 15·17 \text{ cm.}$$

Direct measurement gave $AB = 15·19$ cm.

Mr J. A. Pattern used system formed of two diverging lenses; $AB = 6·02$ cm. This system was used in § 238. Image of distant object was virtual. Image was erect; f is therefore negative.

When (1) the A-face, (2) the B-face was towards object, readings of lens index on table scale for stationary image were (1) 56·97, (2) 55·85 cm. Hence distance between nodal points is 1·12 cm. The scale readings increased in direction B to A, and hence direction N_1 to N_2 is the same as that from A to B.

When A-face was towards object and image (1) was stationary, so that N_2 was on the axis, and (2) was in focus, the bench reading of microscope was 9·69 cm. When lycopodium on B-face was in focus, reading was 16·17 cm. Hence $BF_2 = 6·48$ cm. Without moving system relative to table, the latter was turned through 180°. The bench reading for lycopodium on A-face was 19·19 cm. The mean for A and B is 17·68 cm. Since $AB = 6·02$ cm., the distance of F_2 from the axis is $17·68 - 9·69 - \frac{1}{2} AB = 4·98$ cm. Hence $f_1 = - 4·98$ cm.

When N_1 was on the axis, similar observations gave $AF_1 = 8·40$ cm., $f_2 = - 4·97$ cm.

The mean value is $f = - 4·975$ cm. The sequence of points is $F_1 B N_2 N_1 A F_2$.

EXPERIMENT 43. Measurement of focal lengths by goniometer.

234. Introduction. In Fig. 144, let $F_1 P$ and $F_2 Q$ be the two focal planes of the system, $H_1 K_1$ and $H_2 K_2$ the two principal or unit planes and N_1 and N_2 the two nodal points. The lenses forming the system are *not* shown in the diagram. Let $F_1 H_1 = f_1$ and $F_2 H_2 = f_2$; then f_1 and f_2 are the two focal lengths of the system. Since N_1, N_2 are the nodal points, an incident ray QN_2 gives rise to an emergent ray N_1R, the directions of QN_2 and N_1R being parallel; further, $F_1N_1 = f_2$ and $F_2N_2 = f_1$.

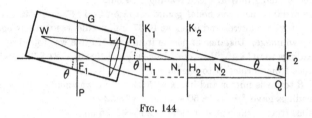

FIG. 144

When, as in the present experiment, the initial and final media are identical, N_1, N_2 coincide with H_1, H_2 and $f_1 = f_2$.

The point Q, in the focal plane F_2Q, gives rise to rays which emerge parallel to N_1R, and F_2 gives rise to rays which emerge parallel to the axis F_2N_2. If $QN_2F_2 = \theta$, the two beams of emergent rays are inclined to each other at the angle θ.

If we can measure the angle θ and the distance F_2Q, we can find f_1, for $f_1 = N_2F_2 = F_2Q/\theta$, for small angles. Denoting F_2Q by h, we have

$$f_1 = h/\theta. \quad ....(1)$$

The second focal length f_2 can be found in a similar manner.

The angle between the two beams of parallel rays may be measured in the manner indicated in Fig. 144. An arm G carries a lens L and a wire W. This wire is adjusted to be in the focal plane of the lens L and is at right angles to the plane of the diagram. When the arm is placed in the proper position, the lens L will bring the beam of parallel rays corresponding to Q to a focus at W. By moving the arm, the beam of parallel rays corresponding to F_2 can be brought to a focus at W. The angle through which the arm is turned is equal to the angle between the beams of parallel rays and is, therefore, equal to QN_2F_2 or θ.

If G be the arm of the goniometer described in § 71, the angle through which it is turned can easily be measured.

The method of auto-collimation may also be used. A plane mirror is mounted on the arm G and the arm is adjusted so that the beam of parallel rays corresponding to Q

falls normally on the mirror. The rays then return along their own paths and produce an image of Q coincident with Q itself. By turning the arm, the image of F_2 may be made to coincide with F_2; the angle turned through by the arm will be equal to QN_2F_2.

The mirror may be mounted on the table of a spectrometer or, more conveniently, on the arm of a goniometer. In the latter case, the mirror is so placed that the rays falling on it do not pass through the lens of the goniometer. The mirror forming the "index glass" of a sextant may also be used; the "index arm" of the sextant then serves as the arm G.

235. Experimental details. The distance F_2Q or h is measured by a scale placed with its *divided* face in the focal plane of the system. The adjustment is easily made if the scale be mounted on a carriage sliding on an optical bench as in Fig. 131. Care must be taken that the plane of the scale is perpendicular to the axis of the lens system. The divided part of the scale must be at the same level as the axis of the lens system, which is horizontal, and the division lines must be vertical.

A glass scale is convenient but any other scale may be used if the division lines be narrow and well defined.

The goniometer is placed so that, when the arm is in the central position, the line of collimation is parallel to the axis of the optical system and is as nearly as possible coincident with it. The centre of the goniometer lens should, therefore, be on the same level as the axis of the system. The goniometer should not be placed further from the lens system than is necessary. The arm must, of course, be free to turn without touching the system.

To obtain a suitable illumination of the glass scale, a sheet of white paper, well illuminated, may be placed behind the scale, or a sheet of ground glass with a lamp behind it may be used.

The carriage is moved along the bench until the divided face of the scale coincide with the focal plane of the system. The adjustment is complete when one of the lines on the scale is sharply focused, without parallax, upon the wire W (Fig. 144) of the goniometer, or when the image of W is focused upon one of the lines. In the latter case, an illuminated surface must be placed behind W.

The readings can now be taken. The goniometer arm is first adjusted so that, while it is nearly in the central position, the goniometer wire coincides with the image of a division line on the scale, and the index of the goniometer is read. The arm is then moved to the left so that the wire passes over the images of a convenient number of millimetre divisions of the glass scale, and the index is again read. This process is repeated with equal steps as long as the images of the division lines are clear and sharp. A similar set of readings with the same step is then made with the arm on the right of the central position.

The distance from the divided surface of the scale to the nearer surface of the lens system is then measured. The adjustable distance-piece shown in Fig. 130 is convenient. The distance can also be found from the bench readings, if a rod of known length be used (see § 211).

The lens system is then turned end for end, and the whole set of readings is repeated. The distance of the scale from the nearer surface of the system is again noted.

Let r_1 be the mean value of the ratio of the distance along the goniometer scale to the corresponding distance along the glass scale, when the latter is in the second focal plane, as in Fig. 144. In this case, the B-face of the lens system is towards the glass scale. Let the distance from the centre of the spherical pivot to the edge of the goniometer scale be l cm. Since 1 cm. on the glass scale corresponds to r_1 cm. on the goniometer scale, and since the angles QN_2F_2, RN_1F_1 are equal, we have

$$1/F_2N_2 = 1/f_1 = r_1/l.$$

Hence $$f_1 = l/r_1. \qquad \dots\dots\dots\dots\dots(2)$$

If r_2 correspond to f_2, $$f_2 = l/r_2. \qquad \dots\dots\dots\dots\dots(3)$$

The values found for f_1 and f_2 will be nearly equal. The mean may be taken as the focal length of the system.

The focal length is positive or negative according as the image of a distant object is inverted or erect.

The positions of the focal planes relative to the system being already known, the positions of the principal points can be found, and a diagram to scale may be constructed.

236. Application to microscope objective. The focal length of an objective can be measured if a micrometer eye-piece with a scale divided to tenths of a millimetre be available. The objective is placed between the goniometer and the eye-piece, the micrometer plate being turned towards the objective, and the micrometer is adjusted so that the image of the goniometer wire is sharply focused on the micrometer scale. The same end of the objective should face the micrometer as faces the specimen when the objective is used in a microscope. The goniometer arm is then moved step by step, and the corresponding readings of the goniometer scale and of the image of the wire on the micrometer scale are taken. From these readings the focal length is deduced.

The objective may then be turned end for end, and a similar set of measurements may be made. Since the objective is not "corrected" for this second position, the results are likely to be less satisfactory than those obtained in the first position.

The divided surface of the micrometer is protected by a plate of glass which is cemented to it, and thus the rays from a point on the divided surface of the micrometer pass through the plate of glass before they fall on the objective. But the image of the micrometer scale formed by refraction through the plate is of the same size as the scale itself, and thus the change of gonio-meter reading for 1 cm. on the micrometer scale is unaffected by the presence of the plate.

The distances U and V of the foci from the A- and B-faces may be measured by a microscope mounted on a sliding carriage as in § 215. The distance U (V) cannot be deduced from the distance of the outer surface of the micrometer plate from the A-face (B-face) of the objective unless we know the small dis-tance d by which the image of the micrometer scale (formed by refraction through the plate) lies behind the outer surface of the plate. The distance d can, however, be found by a moving microscope.

237. System giving virtual image of distant object. When the image of a distant object is virtual, we cannot place a glass or metal scale in such a position that rays from a point on the scale emerge from the system as a parallel beam, and the method of § 234 must be modified.

The goniometer is placed so that the rays from the wire W (Fig. 144) fall first upon the A-face of the system. A virtual image of W will be formed in the focal plane through F_2. This image is observed by a micrometer microscope calibrated by aid of a millimetre scale. Let m divisions of the micrometer scale correspond to 1 cm. The microscope M is mounted on an optical bench and is directed towards the B-face, and its carriage is adjusted so that W is seen in focus on the micrometer scale. The bench reading corresponds to F_2. The goniometer arm is then moved by equal steps, and the reading of the image on the micrometer scale is taken at each stage. Let n_1 micrometer divisions correspond to a change of z_1 cm. in the goniometer reading. The actual movement of the image is therefore n_1/m cm. If the length of the goniometer arm be l cm., the angle is z_1/l radians. The angle also equals $(n_1/m)/f_1$, since $F_2N_2 = f_1$. Hence

$$f_1 = n_1 l/(mz_1). \qquad \dots\dots\dots\dots\dots\dots(4)$$

The difference between the bench readings when M is focused (1) on F_2, (2) on lycopodium at B gives BF_2.

Similar observations are made when A is turned towards M.

238. Practical examples. Mr E. H. Taylor used system formed of a converging lens and a diverging lens. The point A was on the converging lens; $AB = 8\cdot20$ cm. Pivot to scale of goniometer $= l = 40$ cm.

When glass scale was in first focal plane, so that A-face was towards scale, readings were:

Readings of glass scale		Goniometer scale		Change of goniometer reading
0·5	3·0 cm.	6·74	10·51 cm.	3·77 cm.
1·0	3·5	7·42	11·30	3·88
1·5	4·0	8·23	12·07	3·84
2·0	4·5	9·00	12·82	3·82
2·5	5·0	9·75	13·61	3·86
3·0	5·5	10·51	14·38	3·87

Mean ratio of change of goniometer reading to interval on glass scale $= r_2 = 3\cdot840/2\cdot5 = 1\cdot536$. Since the distance from the first focal plane to the first nodal point is f_2, we have $f_2 = l/r_2 = 40/1\cdot536 = 26\cdot04$ cm. Distance A to glass scale $= AF_1 = U = 40\cdot10$ cm.

Lens system was turned end for end, and glass scale was set in second focal plane. Readings gave $f_1 = 26\cdot18$ cm., $BF_2 = V = 11\cdot90$ cm.

Mean focal length $= f = 26\cdot11$ cm.

The focal length is positive, since the image of a distant object was inverted.

Since medium at each end is air, principal points H_1, H_2 coincide with nodal points. Thus $AH_1 = U - f = 13\cdot99$; $BH_2 = f - V = 14\cdot21$ cm.

Since, as the numbers show, the sequence of points is $F_1H_1H_2ABF_2$, we find $H_1H_2 = AB + 13\cdot99 - 14\cdot21 = 7\cdot98$ cm.

System giving virtual image of distant object. G. F. C. Searle tested system of § 233. Image of distant object is erect, and hence focal length is negative. The sequence is F_1BAF_2 and $AB = 6\cdot02$ cm. One cm. corresponded to $m = 88\cdot0$ micrometer divisions in microscope.

With A towards goniometer, movement of $z_1 = 4\cdot0$ cm. caused image of goniometer wire W to move over $n_1 = 44\cdot38$ micrometer divisions. Since $l = 40\cdot00$ cm.; $f_1 = - n_1l/(mz_1) = - 44\cdot38 \times 40/(88\cdot0 \times 4) = - 5\cdot043$ cm.

With B towards goniometer, $n_2 = 44\cdot36$ divisions when $z_2 = 4\cdot0$ cm. Hence $f_2 = - n_2l/(mz_2) = - 5\cdot041$ cm. Mean $f = - 5\cdot04$ cm.

Bench readings of microscope gave $AF_1 = 8\cdot58$, $BF_2 = 6\cdot63$ cm. Thus N_1 is between A and F_1, and $AN_1 = AF_1 - |f| = 3\cdot54$ cm.; N_2 is between B and F_2, and $BN_2 = BF_2 - |f| = 1\cdot59$ cm. Since $AB = 6\cdot02$ cm., N_2 is between B and N_1, and $N_1N_2 = 6\cdot02 - (3\cdot54 + 1\cdot59) = \cdot89$ cm. The sequence is thus

$$F_1BN_2N_1AF_2.$$

EXPERIMENT 44. **Determination of the optical constants of a model eye.**

239. Introduction. By means of the eye, images of external objects are formed on the retina, which is in contact with a medium known as the vitreous humour having a refractive index of $1\cdot338$. The objects themselves are, however, situated in air of index $1\cdot000$. The object and image are in media of different indices and hence the two focal lengths of the eye are not equal and the nodal points do not coincide with the principal points. In the optical system described below, the initial and final media are air and water and thus the system may be called a model eye.

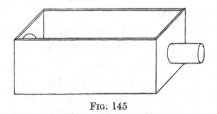

FIG. 145

A suitable system is shown in Fig. 145. A tank 18 cm. long, 9 cm. wide and 8 cm. deep, formed of brass plates, contains the water or other liquid. A window, 5 cm. in diameter, of *tested* plane glass is fitted into one end. Through the plate at the other end passes a brass tube 5·5 cm. in length and 3·7 cm. in external diameter; at each end is a double convex lens of about 10 cm. focal length. An optician should be instructed to "centre" and to "edge" the lenses to fit the tube. A ring is soldered into the tube 0·5 cm. from each end; the lenses rest against these rings and are secured with sealing wax.

240. Method of magnification. Let h_1 be the magnitude, measured at right angles to the axis of the system, of an object in a medium of index μ_1 at a distance u_1 from the corresponding principal point, and let h_2 be the magnitude of its image formed in a medium μ_2 at a distance u_2 from the corresponding principal point. Then, by (5a), § 188, if h_1 and h_2 have the same sign when the image is *erect*,

$$\frac{h_1}{h_2} = -\frac{u_1 - f_1}{f_1} = \frac{-p}{f_1}, \qquad \frac{h_2}{h_1} = -\frac{u_2 - f_2}{f_2} = \frac{-q}{f_2},$$

where $p = u_1 - f_1$, $q = u_2 - f_2$, are the distances of the object and its image from the corresponding foci. Hence we have

$$f_1 = -ph_2/h_1, \qquad f_2 = -qh_1/h_2. \quad \ldots\ldots\ldots\ldots(1)$$

Hence, if we can measure p and q and can determine the ratio h_1/h_2, we can find both f_1 and f_2.

By § 189, if the index of the second medium relative to the first be μ,

$$\mu = \mu_2/\mu_1 = f_2/f_1. \quad \ldots\ldots\ldots\ldots\ldots(2)$$

The fork (Fig. 146) forms an object for use in the water. Three sharp needles X, Y, Z are soldered into holes in a strip of brass. The tips of the needles may be made to be in one plane by allowing the tips to rest on a plane surface while the solder is still melted. The distance $XZ = d$, about 1·5 cm., is measured by a travelling microscope. The needles should be oiled after use to prevent rusting.

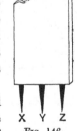

FIG. 146

The tank is filled with water. Two short optical benches S, T (Fig. 147) are placed parallel to the axis of the system. Carriages K, L slide on S, T (see Fig. 22). The extreme surfaces of the system are A and

B; A is in contact with air (μ_1) and B in contact with water (μ_2).

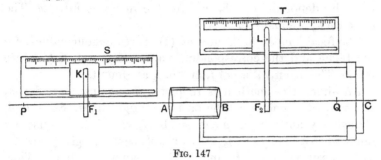

FIG. 147

The carriage K carries a glass scale, mounted as in Fig. 131; the graduated side faces A. The stem of the fork (Fig. 146) is clamped to an arm projecting from L; the bench along which L moves may be raised above the table. The tip Y of the central needle is adjusted to lie on the axis of the system inside the tank, and the line XZ is made horizontal and at right angles to the axis.

The glass scale is set in the focal plane through F_1 by aid of a plane mirror placed in the tank close to B, the adjustment of the carriage K on its bench being exact when the images of the division lines on the scale are in the same plane as the lines themselves. By turning the mirror through a small angle, the images may be made to coincide with the lines and then the parallax test is more easily applied. The bench reading of K is taken and $F_1 A = U$ is measured. Efficient illumination may be obtained by aid of a sloped glass plate (§ 100).

The fork is then set in the focal plane through F_2 by aid of a plane mirror placed in the air near A, the observations being made through the window C. Exact coincidence of Y with its image may be secured by turning the mirror. The bench reading of L is then taken and $F_2 B = V$ is measured by aid of a rod of known length (§ 211).

The fork is now moved away from B and the glass scale is moved away from A until the image of Y be focused without parallax on the scale, and the bench readings of the carriages are taken. The differences between these readings and those corresponding to the focal planes are q (for the fork) and p (for the scale). The distance between the images of X, Z is measured on the glass scale. If this be h_1 cm., the magnitude of the image must be denoted by $-h_1$ cm., for the image is *inverted*. The distance h_2 is equal to d or XZ.

The focal lengths are found by (1). Three or four values of q are taken and the corresponding values of p and of h_1/h_2 are found. The mean values of f_1 and of f_2 are found.

An alternative method is to plot h_1/h_2 against p, and h_2/h_1 against q; the points will lie on two straight lines. Since h_1/h_2 changes by unity when p changes by f_1, and h_2/h_1 changes by unity when q changes by f_2, the slopes of these lines give f_1 and f_2.

The positions of H_1, H_2 and of N_1, N_2 can now be found. Thus

$$AH_1 = F_1 H_1 - F_1 A = f_1 - U, \quad BH_2 = F_2 H_2 - F_2 B = f_2 - V, ...(3)$$

$$AN_1 = F_1 N_1 - F_1 A = f_2 - U, \quad BN_2 = F_2 N_2 - F_2 B = f_1 - V. ...(4)$$

The distance $N_1 N_2$ is equal to $H_1 H_2$. If $N_1 N_2 = H_1 H_2 = t$, and $AB = c$,

$$t = c - AH_1 - BH_2. \qquad ....(5)$$

If t be *negative*, H_2 lies between H_1 and F_1. A diagram is made showing the positions of the cardinal points relative to the system. The value of μ is deduced from (2).

241. Newton's method. The distances of object and image from the corresponding foci are measured and the product $f_1 f_2$ is found by (9), § 190. Thus $f_1 f_2 = pq$. From the values of p and of q, obtained as in the method of magnification, the mean value of $f_1 f_2$ may be found.

It should be noted that f_1 and f_2 cannot be found separately by Newton's method but only their product, and hence the positions of H_1, H_2 and N_1, N_2 cannot be determined without further data. If we assume a knowledge of μ, we obtain, by (2), $f_1^2 = pq/\mu$, $f_2^2 = \mu pq$.

242. Goniometer method. The fork is mounted on the carriage L as in § 240, and Y is made to lie on the axis of the system. A goniometer (Fig. 44) is placed to the left of A (Fig. 147), and L is then adjusted so that Y is focused without parallax on the wire of the goniometer; Y is then in the focal plane through F_2. The distance $V = BF_2$ is measured. The goniometer arm is now moved to make the wire coincide first with the image of X and then with the image of Z. The goniometer reading is taken in each case. If the difference of readings be r_1 cm., if the distance of the centre of the pivot from the edge of the scale be l cm., and if θ_2 be the corresponding angle, then $\theta_2 = r_1/l$.

Since $XZ = d$ cm., we have, by § 234,

$$f_1 = d/\theta_2 = dl/r_1. \quad \ldots\ldots\ldots\ldots\ldots\ldots(6)$$

We obtain f_1 and not f_2, since F_2N_2 is not f_2 but f_1.

In finding f_2, the carriage L is removed, and the goniometer, raised to the proper height, is placed near the tank with its scale parallel to the sides of the tank. A plane mirror is attached to a rod clamped to the goniometer arm, and the mirror is immersed in the tank, the plane of the mirror being approximately normal to the axis of the system. The mirror should be capable of adjustment about a horizontal axis parallel to its plane. The fork is placed in the air near A and is adjusted so that Y coincides with its own image formed by reflexion at the mirror, the light having passed twice through the lenses; Y is now in the focal plane through F_1. The distance $U = AF_1$ is measured. The goniometer arm is then moved to make (1) X coincide with

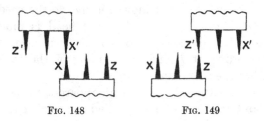

FIG. 148   FIG. 149

its image X', and (2) Z coincide with its image Z', as in Figs. 148 and 149. The rays which fall normally on the mirror start from X in case (1) and from Z in case (2). If the difference of goniometer readings be r_2 cm., and θ_1 be the corresponding angle, $\theta_1 = r_2/l$. Since the distance XZ is d cm. we have, by § 234,

$$f_2 = d/\theta_1 = dl/r_2. \quad \ldots\ldots\ldots\ldots\ldots\ldots(7)$$

From f_1, f_2, U and V the positions of H_1, H_2, N_1, N_2 are found as in § 240.

243. Revolving table method. The optical system is placed on a revolving table (§ 227) and the graduated bar of the table is adjusted, as in § 228, so that the optical axis passes through the axis of the table when one side of the tank is in contact with the bar; the *same* side is kept in contact with the bar throughout. This adjustment is made while the tank is empty. The tank is then filled with water. It may be necessary to load the base of the table to secure adequate stability.

The tank is turned so that A (Fig. 147) faces a distant object and the bench T is raised to a suitable height at such a distance from the table that the top can turn freely through a complete revolution. An arm attached to L carries a rod bearing a *single* needle. The needle is immersed in the water and the tip is adjusted to lie in the same horizontal plane as the optical axis.

The needle and the image of a distant object are viewed through the window C, and L is moved until the needle coincide with the image. If the table be turned through a small angle, the image will move relative to the (fixed) needle unless the nodal point N_2 lie on the axis of the table. The direction of motion of the image relative to the needle indicates the direction in which the tank must be moved along the graduated bar. When N_2 is on the axis of the table and the tip of the needle is at F_2, the image does not move relative to the needle either when the table is turned or when the head is moved from side to side. The distance $F_2 B = V$ is then measured, and the bench reading of L and the reading of the tank on the graduated bar are taken and *recorded*.

The rotation of the table causes an apparent motion of both the image and the needle. The rays which reach the eye from the image or from the tip of the needle suffer deviation at C, and this deviation varies as the tank is turned.

Without disturbing the bench T, the carriage L is removed. The tank is removed from the table and L is replaced on T at the *same* bench reading as before. A scale is then mounted on the table so that its graduated face intersects the axis of the table, and the table is turned to bring first one and then the other end of the scale into contact with the needle, as in § 230. Half the difference of scale readings gives $F_2 N_2$ or f_1.

If a very distant object be not available, a *fixed* plane mirror is placed in the air in front of A, and the needle and the tank are adjusted so that the image of the needle does not move relative to the (fixed) needle either when the table is turned or when the head is moved. The needle is observed through the window C.

We have now to find f_2. The needle is removed, and a *fixed* plane mirror is held in the tank with its reflecting face towards B. The mirror may be held by the arm carried by L or by the goniometer as in § 242. A needle is placed in the air in front of A, and the tank and needle are adjusted so that the image of the needle does not move from the needle either when the table is turned through a small angle or when the head is moved. The tip of the needle is now at F_1. Then $F_1 A = U$ is measured, and the index reading of the tank on the bar is taken. The tank is now removed, and the distance of the needle from the axis of rotation is determined by a scale mounted on the table. Half the difference of scale readings gives $F_1 N_1$ or f_2.

The difference between the two readings of the tank on the graduated bar gives t, the distance between the nodal points. The *sign* of t may be found by the rule given in § 229.

244. Practical example. In an experiment by Mr G. M. B. Dobson, $AB = c = 5.84$ cm. The distance between X, Z of fork was $d = 1.520$ cm.

Magnification method. The bench readings of K and L are denoted by K and L; $K = \infty$ indicates that *fork* was at F_2, and $L = \infty$ that glass scale was at F_1. The image was inverted.

$$p = 25.23 - K, \qquad q = L - 8.71, \qquad f_1 = -h_2 p/h_1 = -dp/h_1,$$
$$f_2 = -h_1 q/h_2 = -h_1 q/d, \qquad f_1 f_2 = pq.$$

K cm.	L cm.	Readings for image cm.	$-h_1$ cm.	p cm.	q cm.	f_1 cm.	f_2 cm.	$f_1 f_2$ cm.^{2}
∞	8.71		∞	∞	0			
14.90	16.20	8.50 − 6.33	2.17	10.33	7.49	7.24	10.69	77.37
16.05	17.20	8.39 − 6.53	1.86	9.18.	8.49	7.50	10.39	77.94
17.07	18.20	8.29 − 6.63	1.66	8.16	9.49	7.47	10.36	77.44
17.72	19.18	8.21 − 6.71	1.50	7.51	10.47	7.61	10.33	78.63
25.23	∞		0	0	∞			

Distance $AF_1 = U = 4.86$ cm.; $BF_2 = V = 4.70$ cm.

Goniometer method. Distance from pivot to scale $= l = 40.00$ cm.
Fork at F_2 in water. Goniometer readings for X, Z: 13.37, 5.35 cm.
Fork at F_1 in air. Goniometer readings for X, Z: 13.00, 7.01 cm.
Hence $r_1 = 13.37 - 5.35 = 8.02$ cm., $r_2 = 13.00 - 7.01 = 5.99$ cm.

Distance $AF_1 = U = 4 \cdot 92$ cm.; $BF_2 = V = 4 \cdot 78$ cm.

$$f_1 = \frac{dl}{r_1} = \frac{1 \cdot 520 \times 40 \cdot 00}{8 \cdot 02} = 7 \cdot 58 \, \text{cm.}, \quad f_2 = \frac{dl}{r_2} = \frac{1 \cdot 520 \times 40 \cdot 00}{5 \cdot 99} = 10 \cdot 15 \, \text{cm.}$$

Revolving table method. Distance $F_2 N_2 = f_1 = 7 \cdot 67$ cm. Scale reading of tank on table: $85 \cdot 63$ cm.

Distance $F_1 N_1 = f_2 = 10 \cdot 20$ cm. Scale reading of tank on table: $88 \cdot 03$ cm.

Distance between nodal points $= N_1 N_2 = t = - (88 \cdot 03 - 85 \cdot 63) = - 2 \cdot 40$ cm.

Distance $AF_1 = U = 4 \cdot 85$ cm.; $BF_2 = V = 4 \cdot 79$ cm.

Method	f_1	f_2	U	V
Magnification	7·46 cm.	10·44 cm.	4·86 cm.	4·70 cm.
Goniometer	7·58	10·15	4·92	4·78
Revolving table	7·67	10·20	4·85	4·79
Means	7·57	10·26	4·88	4·76

From the mean values we find

$$AH_1 = f_1 - U = 7 \cdot 57 - 4 \cdot 88 = 2 \cdot 69 \, \text{cm.}, \quad BH_2 = f_2 - V = 10 \cdot 26 - 4 \cdot 76 = 5 \cdot 50 \, \text{cm.}$$

Hence $H_1 H_2 = t = AB - AH_1 - BH_2 = 5 \cdot 84 - 2 \cdot 69 - 5 \cdot 50 = - 2 \cdot 35$ cm.

In theory $N_1 N_2 = H_1 H_2$. The revolving table method gave $N_1 N_2 = - 2 \cdot 40$ cm.

Refractive index of water relative to air $= \mu = f_2/f_1 = 1 \cdot 355$.

EXPERIMENT 45. Determination of radii of curvature of a lens.

245. Method. Let R, S (Fig. 150) be the centres of curvature of the faces A, B of the lens and let r, s be the radii of curvature; the radius of each surface is counted positive when the surface is convex. Let F_1, F_2 be the foci. Let K be conjugate to R, and L to S. Then the rays from K, after passing through the lens, proceed as if from R and are normal to the surface A. Some of the light is reflected at A and, since the rays are normal to A, the rays return along their own paths and converge in K. Hence, the tip of a pin placed at K will coincide with its image. Similarly,

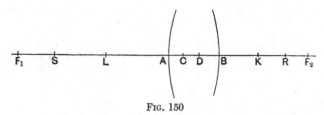

FIG. 150

the tip of a pin placed at L will coincide with its image formed by rays reflected at B. Let $AF_1 = U$, $BF_2 = V$, $BK = k$, and $AL = l$. By Newton's formula, § 190,

$$f^2 = KF_2 \times RF_1 = (V - k)(U + r). \quad \ldots\ldots\ldots(1)$$

Hence $r = f^2/(V - k) - U.$ ....(2)

Similarly, $s = f^2/(U - l) - V.$ ....(3)

The theory is applicable to *any* coaxal system of lenses and is not restricted to a lens formed of a single piece of glass.

Gauss's method (§ 215) is convenient for finding f^2, since U and V are required in that method. Let C (Fig. 150) be the image of A, and D that of B, let $AD = a$, $BC = b$; let a be positive when D, B are on the same side of A, and let b be positive when C, A are on the same side of B. Then $f^2 = U (V + b)$, and $f^2 = V (U + a)$. The mean of the two values of f^2 may be used in (2) and (3).

Practical details may be gathered from EXPERIMENTS 31, 38.

If we are dealing with a system of two or more lenses not cemented together, we have to decide which surface is acting as reflector when the pin coincides with an image of itself. Suppose the observer to be on the B-side of the lens system. If wet paper placed on A destroy the image, A was the reflector. If it do not destroy the image, an opaque screen is placed between the lens of which A is one surface and the next lens. If this destroy the image, the reflexion took place at the first surface beyond the screen. Proceeding in this way, the reflecting surface can be identified. But it is only when A acts as reflector that the theory applies. If B be proved to be the reflector, then B is concave.

If the calculation yield a negative value for the radius of A, the surface is concave, and an independent value of its radius may be determined optically by making a pin coincide with its own image, when A faces the pin.

246. Practical example. Mr G. M. B. Dobson used a system of two lenses cemented together; $AB = 1.07$ cm. In applying Gauss's method, the following values were found:

$$U = 30.89, \quad V = 29.82, \quad a = 0.70, \quad b = 0.66 \text{ cm.}$$

$$f^2 = U (V + b) = 941.5, \quad f^2 = V (U + a) = 942.0. \quad \text{Mean} f^2 = 941.8 \text{ cm.}^2.$$

The distances k, l were $k = BK = 7.46$, $l = AL = 62.29$ cm. By (2), (3),

$$r = \frac{f^2}{V - k} - U = 11.23 \text{ cm.}, \qquad s = \frac{f^2}{U - l} - V = - 59.81 \text{ cm.}$$

The negative value of s shows that B is concave. The radius, measured by making a pin coincide with its image by reflexion at B, was $s = - 59.64$ cm.

The radii as measured by a spherometer were $r = 11.52$, $s = - 59.65$ cm. The radius of A, measured by method of § 110, was $r = 11.35$ cm.

EXPERIMENT 46. **Determination of cardinal points of lens by Bravais points.**

247. Theory. The Bravais points are self-conjugate points on the axis of a lens system. We consider only the case where the media at the ends of the system are identical; then the nodal and principal points coincide.

Let F_1, F_2 (Fig. 151) be the foci; H_1, H_2 the principal points; let $F_1 H_1 = F_2 H_2 = f$, $H_1 H_2 = t$. Let K be a Bravais point and

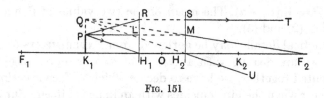

FIG. 151

let $F_1 K = x$. Since K is self-conjugate, Newton's formula (§ 190) gives $F_1 K . K F_2 = f^2$, or $x (2f + t - x) = f^2$. Hence

$$x = f + \tfrac{1}{2} t \pm \tfrac{1}{2} \{t (4f + t)\}^{\frac{1}{2}}, \quad \dots\dots\dots\dots(1)$$

and thus there are two Bravais points K_1, K_2. If O be the mid-point of $H_1 H_2$, $F_1 O = O F_2 = f + \tfrac{1}{2} t$, and thus K_1, K_2 lie on opposite sides of O and $O K_1 = O K_2$. If $K_1 K_2 = b$, then $b = \{t (4f + t)\}^{\frac{1}{2}}$, and

$$b^2 = t (4f + t). \quad \dots\dots\dots\dots\dots\dots(2)$$

There will be no Bravais points unless $t (4f + t)$ be positive, which is not always the case. Thus, a system of two *thin* lenses of focal lengths f_1, f_2, at distance a apart, has Bravais points only when $a^2 > 4 f_1 f_2$. If f_1, f_2 have opposite signs, this condition is satisfied for *all* values of a. If they have the same sign, there will be no Bravais points unless $a > 2 \sqrt{(f_1 f_2)}$.

The distance b is used with other measurements to give f and t.

We can measure $4f + t$ directly, for it is the minimum distance —say l—between an object and its real image. Then, by (2),

$$t = b^2/l. \quad \dots\dots\dots\dots\dots\dots(3)$$

When t has been found, we have, since $l = 4f + t$,

$$f = \tfrac{1}{4} (l - t). \quad \dots\dots\dots\dots\dots(4)$$

If the bench be too short to allow l to be measured, we may find the distance s between the foci; then $s = 2f + t$. By (2),

$$4f^2 = (2f + t)^2 - b^2 = s^2 - b^2. \quad\quad\quad\text{............(5)}$$

Then $$t = (2f + t) - 2f = s - (s^2 - b^2)^{\frac{1}{2}}. \quad\text{............(6)}$$

In Fig. 151, let PK_1 be an object perpendicular to the axis. Then a ray PL, parallel to the axis, will become the ray MF_2, their directions meeting at M on the principal plane through H_2. Similarly, a ray F_1PR, directed from the focus F_1, becomes the ray ST, parallel to the axis, their directions meeting at R on the principal plane through H_1. The directions of ST, MF_2 meet in Q, the image of P. When K_1 is a Bravais point, K_1PQ is a straight line.

A ray PH_1 gives rise to a ray H_2U, the two rays being parallel, since, in our case, the nodal points are at H_1, H_2. When K_1 is a Bravais point, QH_2U is a straight line. If $QK_1/PK_1 = m$,

$$m = QK_1/PK_1 = K_1H_2/K_1H_1 = (\tfrac{1}{2}b + \tfrac{1}{2}t)/(\tfrac{1}{2}b - \tfrac{1}{2}t),$$

or $$t = b\,(m - 1)/(m + 1). \quad\quad\text{.................(7)}$$

Then, by (2), $f = \tfrac{1}{4}(b^2/t - t) = bm/(m^2 - 1). \quad\text{............(8)}$

Hence, if we find m and b, we can calculate t and f.

If the rays ST, MF_2 be reversed, they will be directed to Q before entering the system, and to P on leaving the system, so that P is the image of Q, and K_1 is again self-conjugate. The magnification is now $1/m$ instead of m as in the previous case.

248. Simple method. The observations are made by aid of a microscope on an optical bench. The object is a patch of lycopodium or a strip of tin-foil on a small glass plate D (Fig. 152) which is fixed to a rod R and is illuminated by a sodium flame Z. The rod R is

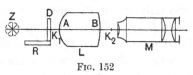

Fig. 152

attached to a carriage and is sufficiently long to allow D to be brought close to the lens L, whose faces are A, B. The microscope M must be such that an object, when seen in focus, is several centimetres from the objective. The M carriage is adjusted so that the object on D is seen in focus through M; the reading of the M carriage is then taken. The L carriage is now placed on the bench, the A-face of L being the nearer to D, and is adjusted so that D is again in focus; the M carriage is not moved in this adjustment. The Bravais point K_1 now lies on D, since the image of D coincides with D. The M carriage is now moved to bring lycopodium on the B-face into focus and the carriage reading of M is taken. The difference between the two readings of this carriage gives K_1B. The lens is now turned end for end and similar observations are made, the displacement of M now giving K_2A. The thickness AB is measured. Then

$$b = K_1K_2 = K_1B + K_2A - AB. \quad \ldots\ldots\ldots\ldots(9)$$

The microscope M is next placed so far from D that there are two positions of L for which D is in focus. As M is moved nearer to D, the distance between these two positions diminishes until they coincide. The reading of the M carriage then corresponds to the minimum distance. The difference between this reading and that found when M is focused directly on D gives l or $4f + t$. The observations may be made (1) with A, (2) with B nearer to D (see §§ 145, 213).

The value of t is found by (3), and then f is found by (4).

The distance of either K_1 or K_2 from the nearer principal plane is $\frac{1}{2}(b - t)$. Since K_1B, K_2A, AB are known, the positions of H_1, H_2 relative to the faces A, B are easily found.

This method fails if the Bravais points lie between A and B; then the auxiliary lens method (§ 250) must be used.

249. Micrometer method. A strip of tinfoil is placed on D. The microscope is focused on the foil and h, the number of divisions of the eye-piece micrometer covered by it, is found. Then L is put between M and D, A facing D. The lens (*not M*) is adjusted so that the foil is again in focus; it now covers k_1 divisions. The observations are repeated with B facing D, the foil seen through L covering k_2 divisions. Theoretically $k_1 = k_2$. Taking the mean, $m = \frac{1}{2}(k_1 + k_2)/h = k/h$. When the distance $K_1K_2 = b$ has been found, t and f are calculated by (7) and (8). Instead of the strip of tinfoil, two grains of lycopodium may be used.

If the Bravais points lie between A and B, an auxiliary lens (§ 250) is used.

250. Auxiliary lens method. A converging lens G (Fig. 153), placed in a *fixed* position at a suitable distance from D, forms a real image of D at E. The microscope M is focused

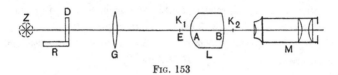

FIG. 153

on this image at E, and the carriage reading of M is taken. The lens L is now put on the bench, with the A-face towards G, and L (*not* M) is adjusted so that D is in focus in M. There are two positions of L in which this is secured, viz. when (1) K_1, and (2) K_2 coincides with E. The distance between these positions is given by the readings of the L carriage and equals K_1K_2 or b. When K_2 coincides with E, the M carriage is readjusted so that lycopodium on B is in focus; the displacement of M gives BK_2. Similar readings are taken with B towards G; a second value of b is thus found as well as AK_1. The minimum distance $l = 4f + t$, between E and its image, is found as in § 248.

If an auxiliary lens similar to that of § 105 be used at G, it should be placed at about twice its focal length from D. A well-corrected magnifying lens will give good results, if the proper side face D and if DG be small compared with GE.

The definition will be poor unless the lenses G and L be nearly coaxal.

The micrometer method (§ 249) can be employed if the image E be used in place of the object D.

251. Practical example. G. F. C. Searle used a lens for which $AB = 1·20$ cm.; A-face was convex, B-face concave. The bench readings increased from D to M.

Simple method. Mean reading of M carriage when D was in focus, 55·90 cm. The M carriage was then set to this reading. The lens was set so that, with A facing D, K_1 coincided with D. Microscope was then moved to bring lycopodium on B into focus; new mean reading 60·29 cm. Hence

$$BK_1 = 60·29 - 55·90 = 4·39 \text{ cm}.$$

The lens was turned round so that B faced D. Similar measurements gave $AK_2 = 1·60$. By (9), $b = BK_1 + AK_2 - AB = 4·79$ cm.

Mean readings of M for minimum distance between image and object for (1) A, (2) B facing D, were 100·61, 100·59; mean 100·60 cm. Hence

$$l = 4f + t = 100·60 - 55·90 = 44·70 \text{ cm.}$$

By (3) and (4), $t = b^2/l = 0·513$ cm., $f = \frac{1}{4}(l - t) = 11·05$ cm.

The positions of H_1, H_2 relative to AB can now be found. We have

$$K_1A = K_1B - AB = 3·19 \text{ cm.;} \qquad K_1H_1 = \tfrac{1}{2}(b - t) = 2·14 \text{ cm.;}$$
$$K_1H_2 = \tfrac{1}{2}(b + t) = 2·65 \text{ cm.;} \qquad K_2B = K_2A - AB = 0·40 \text{ cm.}$$

The sequence is $F_1K_1H_1H_2ABK_2F_2$.

Micrometer method. Seen directly, strip covered $h = 44·5$ divisions. When lens was interposed with A facing D, strip covered $k_1 = 55·1$ divisions, when K_1 coincided with D. On reversing lens, $k_2 = 55·0$. Thus $k = \tfrac{1}{2}(k_1 + k_2) = 55·05$. Hence $m = k/h = 1·237$. As above, $b = 4·79$ cm. By (7), (8),

$$t = b(m - 1)/(m + 1) = ·507 \text{ cm.;} \qquad f = bm/(m^2 - 1) = 11·18 \text{ cm.}$$

Auxiliary lens method. The M carriage was set to 110·77 cm., the mean found when E, the image of D by lens G, was in focus. Then L was placed between G and M, with A facing G. The mean reading of L carriage, when K_2 coincided with E, was 94·45 cm.

The L carriage was then adjusted so that K_1 coincided with E; mean reading 99·24 cm.

Then $b = K_1K_2 = 99·24 - 94·45 = 4·79$ cm.

The L carriage was set at 94·45 cm., and M was moved to bring lycopodium on B into focus; mean reading of M was 110·27 cm.

Hence $BK_2 = 110·77 - 110·27 = 0·50$ cm.

Similar measurements with B facing G gave $b = 4·80$ cm., $AK_1 = 3·26$ cm.

EXPERIMENT 47. Anharmonic property of lens system.

252. Anharmonic ratios. Let A_1, B_1, C_1, D_1 (Fig. 154) be four points on a straight line, and O any point not on A_1D_1. Let $OZ = p$ be the perpendicular from O on A_1D_1. Then

$$p.A_1B_1 = OA_1.OB_1 \sin A_1OB_1,$$
$$p.C_1D_1 = OC_1.OD_1 \sin C_1OD_1,$$
$$p.A_1D_1 = OA_1.OD_1 \sin A_1OD_1,$$
$$p.B_1C_1 = OB_1.OC_1 \sin B_1OC_1.$$

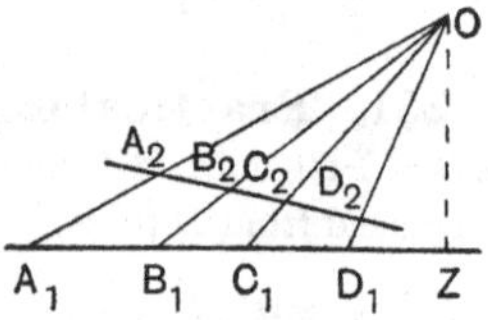

FIG. 154

Hence $$\frac{A_1B_1.C_1D_1}{A_1D_1.B_1C_1} = \frac{\sin A_1OB_1.\sin C_1OD_1}{\sin A_1OD_1.\sin B_1OC_1}. \quad \text{........(1)}$$

The quantity on the left in (1) is the anharmonic ratio of the "range" $A_1B_1C_1D_1$ and is denoted by $[A_1B_1C_1D_1]$. The quantity on the right is the anharmonic ratio of the "pencil" OA_1, OB_1, OC_1, OD_1, and is denoted by $O[A_1B_1C_1D_1]$. Thus

$$[A_1B_1C_1D_1] = O[A_1B_1C_1D_1]. \dots\dots\dots(2)$$

If the pencil cut any other straight line in the range $A_2B_2C_2D_2$, $[A_2B_2C_2D_2] = [A_1B_1C_1D_1]$, for each equals $O[A_1B_1C_1D_1]$.

Let N_1, N_2, H_1, H_2, F_1, F_2 (Fig. 155) be the nodal, principal and focal points of a lens system. Let R_1 be a point in the

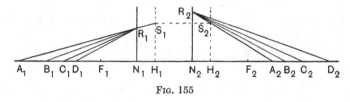

FIG. 155

nodal plane through N_1, and let R_2 be its image, which lies in the nodal plane through N_2. Let $R_1N_1 = y_1$, $R_2N_2 = y_2$.

Let A_1 be a point on the axis, and A_2 its image. Since R_2 is the image of R_1, the ray A_1R_1 becomes R_2A_2. Let A_1R_1, R_2A_2 cut the principal planes in S_1, S_2; then $S_1H_1 = S_2H_2$. Let $A_1H_1 = u$, $A_2H_2 = v$. Since, by § 191, $N_1H_1 = f_1 - f_2$,

$$S_1H_1 = y_1u/(u - f_1 + f_2).$$

If Δ be the deviation, $\Delta = S_1H_1(1/u + 1/v)$. But, by (5), § 188, $1/v = (1 - f_1/u)/f_2$, and hence, on reduction, $\Delta = y_1/f_2$. Similarly, $\Delta = y_2/f_1$. Hence,

$$\Delta = y_1/f_2 = y_2/f_1. \dots\dots\dots\dots(3)$$

The deviation is thus independent of the position of A_1 on the axis. Hence, if A_1, B_1, C_1, D_1' be object points on the axis, and if A_2, B_2, C_2, D_2 be their images, the angles $A_1R_1B_1$, $B_1R_1C_1$, $C_1R_1D_1$ are respectively equal to $A_2R_2B_2$, $B_2R_2C_2$, $C_2R_2D_2$, and thus

$$R_1[A_1B_1C_1D_1] = R_2[A_2B_2C_2D_2]. \dots\dots\dots(4)$$

Hence, by (2), the anharmonic ratio of the range $A_1B_1C_1D_1$ equals that of the range $A_2B_2C_2D_2$, or

$$[A_1B_1C_1D_1] = [A_2B_2C_2D_2]. \dots\dots\dots(5)$$

Thus, if any four points A_1, B_1, C_1, D_1 be taken on a straight line and any lens system be placed so that its axis coincides with A_1D_1, the anharmonic ratio of the range of image points equals that of the range of object points. The result does not depend upon the focal lengths f_1, f_2.

The result (5) is easily established by Newton's formula. For, if $A_1 F_1 = a_1, \ldots A_2 F_2 = a_2, \ldots$, we have $a_2 = f_1 f_2 / a_1, \ldots$. Hence

$$[A_2 B_2 C_2 D_2] = \frac{A_2 B_2 . C_2 D_2}{A_2 D_2 . B_2 C_2} = \frac{(b_2 - a_2)(d_2 - c_2)}{(d_2 - a_2)(c_2 - b_2)} = \frac{(1/b_1 - 1/a_1)(1/d_1 - 1/c_1)}{(1/d_1 - 1/a_1)(1/c_1 - 1/b_1)}$$

$$= \frac{(a_1 - b_1)(c_1 - d_1)}{(a_1 - d_1)(b_1 - c_1)} = [A_1 B_1 C_1 D_1].$$

The product $f_1 f_2$ can be found from the distances $A_1 B_1$, $B_1 C_1$, $A_2 B_2$, $B_2 C_2$. Let $A_1 F_1 = a_1$, $A_2 F_2 = a_2$, $B_1 F_1 = a_1 - r_1$, $B_2 F_2 = a_2 + r_2$, $C_1 F_1 = a_1 - s_1$, $C_2 F_2 = a_2 + s_2$. By Newton's formula,

$$a_1 a_2 = f_1 f_2, \quad (a_1 - r_1)(a_2 + r_2) = f_1 f_2, \quad (a_1 - s_1)(a_2 + s_2) = f_1 f_2.$$

Subtracting the first equation from the other two, we have

$$a_1 r_2 - a_2 r_1 = r_1 r_2, \qquad a_1 s_2 - a_2 s_1 = s_1 s_2.$$

Hence $\qquad a_1 = \dfrac{r_1 s_1 (s_2 - r_2)}{r_1 s_2 - r_2 s_1}, \qquad a_2 = \dfrac{r_2 s_2 (s_1 - r_1)}{r_1 s_2 - r_2 s_1}. \qquad \ldots\ldots\ldots(6)$

Equations (6) determine the positions of the foci relative to A_1, A_2. Since $a_1 a_2 = f_1 f_2$,

$$f_1 f_2 = \frac{r_1 s_1 (s_1 - r_1) r_2 s_2 (s_2 - r_2)}{(r_1 s_2 - r_2 s_1)^2}$$

$$= \frac{A_1 B_1 . A_1 C_1 . B_1 C_1 . A_2 B_2 . A_2 C_2 . B_2 C_2}{(A_1 B_1 . A_2 C_2 - A_1 C_1 . A_2 B_2)^2}. \qquad \ldots\ldots(7)$$

In the ordinary case in which the system is surrounded by air, $f_1 = f_2$; then (7) gives the square of *the* focal length.

If *four* object points be used, as is necessary for testing (5), each set of three will give a value of $f_1 f_2$, and thus we obtain four values for $f_1 f_2$, of which we can take the mean.

253. Experimental details. A carriage P on an optical bench bears the object, which may be a needle point, a divided scale or some lycopodium on a glass plate, as in § 248. The lens system, whose end faces are L_1, L_2, is supported by a carriage G in a *fixed* position. A carriage Q bears the image locator—a needle point, a microscope or a micrometer eye-piece. For each of four positions of P, corresponding to A_1, B_1, C_1, D_1, the bench reading of Q is found. The four positions of P should be such that A_2B_2, B_2C_2, C_2D_2 are large enough for accurate measurement. If P_A, P_B ... Q_A, Q_B ... be corresponding bench readings, and if the zero of the bench be to the left in Fig. 155,

$$A_1B_1 = P_B - P_A, \qquad A_2B_2 = Q_B - Q_A,$$

and so on. Then $[A_1B_1C_1D_1]$, $[A_2B_2C_2D_2]$ are calculated.

The product f_1f_2 is found from $A_1B_1C_1$ and $A_2B_2C_2$, as in (7). Three other values of f_1f_2 may be found by using $A_1B_1D_1$, $A_1C_1D_1$, $B_1C_1D_1$ instead of $A_1B_1C_1$. If a_1, a_2 be calculated from (6), and if we measure the distances of A_1, A_2 from the corresponding faces L_1, L_2, the distances L_1F_1, L_2F_2 can be found. Thus, if $L_1F_1 = U$, $L_2F_2 = V$, we have $U = L_1A_1 - a_1$, $V = L_2A_2 - a_2$. When there is air at each end of the system, $f_1 = f_2 = f$, and then the positions of the principal points can be found.

254. Practical example. G. F. C. Searle and J. A. Crowther used a photographic lens system and located image by micrometer eye-piece. The sequence of the object points was $A_1B_1C_1D_1$. They found

$$A_1B_1 = 35, \qquad B_1C_1 = 10, \qquad C_1D_1 = 10, \qquad A_1D_1 = 55 \text{ cm.},$$
$$A_2B_2 = 5{\cdot}86, \qquad B_2C_2 = 4{\cdot}53, \qquad C_2D_2 = 10{\cdot}02, \qquad A_2D_2 = 20{\cdot}41 \text{ cm.}$$
$$[A_1B_1C_1D_1] = A_1B_1 . C_1D_1/(A_1D_1 . B_1C_1) = 0{\cdot}6364; \quad [A_2B_2C_2D_2] = 0{\cdot}6351.$$

Since $f_1f_2 = f^2$, the values of f given by the four sets $A_1B_1C_1$, $A_1B_1D_1$, $A_1C_1D_1$, $B_1C_1D_1$ were 20·85, 20·88, 20·90, 20·94 cm. Mean 20·89 cm. System gave inverted image of distant object; hence f is positive.

EXPERIMENT 48. **Study of telescopic system.**

255. Apparatus. A telescopic system may be constructed of two lenses A, B (Fig. 156) fitted into holders R, S clamped to a board by screws passing through slots in the bases of R, S. The bases slide between guides which keep the lenses coaxal. If the lenses be thin, the distance AB will be approximately equal to the algebraical sum of the focal lengths; the final adjustment is made optically. A telescope with cross-wires is first focused on a *distant* object P (*at least* 100 metres) so that a motion of the eye causes no motion of the image relative to the wires. The system is now interposed between the telescope

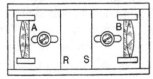

FIG. 156

and P, and R or S is adjusted so that P is again in focus. The holder is then secured by its screw.

The axis of the system may be set parallel to the optical bench in the following manner. The normal to a plane mirror M is set parallel to the bench by § 209. A converging lens L is placed in front of M, and a pin is adjusted so that the image of its tip T, formed by L and M, coincides with T itself. The system is then interposed between M and L, and is adjusted in direction so that T is again self-conjugate.

If the system be achromatic, an observing telescope of considerable power should be used, since an error in AB is more easily detected with a high than with a low power. In the system used at the Cavendish Laboratory, the lenses are telescope objectives, the convex sides facing outwards, as in Fig. 156, so that the spherical aberration may be small when *distant* objects are viewed through the system.

Instead of a telescope, the goniometer of § 71 may be used. The goniometer is placed on one side of the system and a plane mirror on the other, and AB is adjusted so that the goniometer wire is self-conjugate. Chromatic errors will be avoided if an auto-collimating device (§ 71) be used and the wire be illuminated by sodium light. The goniometer also may be used as a collimator. The wire, illuminated by sodium light, is viewed through the system by the telescope.

256. Measurement of transverse magnification. The system is mounted on a carriage, with its axis parallel to the bench, and two scales, C, D (Fig. 157) are mounted on two other carriages. The *divided* face of each scale is towards the system.

The scale D is divided on glass and is mounted as in Fig. 131. The scales are set at right angles to the axis of the system, which is placed between them. With C in any given position, D is

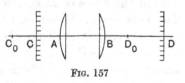

FIG. 157

adjusted so that the images of the lines on C are in focus on the divided face of D. A little lycopodium placed on the latter face will allow the parallax test to be more easily applied. Let h_2 be the distance, measured on D, between the images of two lines on C, the distance between the lines being h_1. If m be the transverse magnification, $m = h_2/h_1$; we count m positive when the image is erect (§ 201). Determinations of h_2/h_1 are made for several positions of C. The bench readings of C and D are recorded in each case, for use in § 257. A progressive change in m, as C is moved in one direction, indicates imperfect adjustment of the distance AB.

The scale C may be of white celluloid with fine dividing lines. Astigmatism and curvature of the field prevent accurate observations when a considerable length of C is used, and thus h_1 should not exceed 2 or 3 cm. If a glass scale, backed by a sodium flame, be used for C, chromatic error is avoided. (See § 219.)

In place of the scale D, it may be convenient to use a microscope with a micrometer eye-piece. The number of micrometer divisions, n_2, covered by the image of a definite distance on C is found; the number, n_1, covered by the image of the same distance, when C is viewed directly, is also found. If both images be erect, $m = n_2/n_1$; if one be erect and the other inverted, $m = -n_2/n_1$. The bench reading for the microscope is taken in each case.

257. Measurement of longitudinal magnification. Let C_0, D_0 (Fig. 157) be two conjugate positions of the scales, C_0 being as far from A as the bench allows. Let C, D be any other pair of conjugate positions, AC being less than AC_0. Then, by § 203, BD will be greater than BD_0, as in Fig. 157. Let $C_0C = l_1$, $D_0D = l_2$. Then l_1 is the difference of bench readings of C in the two positions, and similarly for l_2. If e be the longitudinal magnification, or the "elongation,"

$$e = l_2/l_1. \quad \ldots\ldots\ldots\ldots\ldots\ldots\ldots(1)$$

The ratio l_2/l_1 is found for each of a series of values of l_1. Since the system is in air, $\mu_2 = \mu_1$; thus, by § 204,

$$e = l_2/l_1 = m^2. \quad \ldots\ldots\ldots\ldots\ldots(2)$$

258. Measurement of magnifying power. The magnifying power, M, of the system is the ratio θ_2/θ_1, where θ_1, θ_2 are the angles of Fig. 128; to find M we compare these angles. A beam of rays parallel to the axis gives rise to an emergent beam parallel to the axis, and a beam of rays parallel to RP'

(Fig. 128) gives rise to an emergent beam parallel to SQ'. If these emergent beams be received on a goniometer and the arm be adjusted to bring each beam in turn to a focus on the wire, the angle turned through by the arm equals the angle between the beams.

If two *distant* objects X, Y be available, the goniometer arm is turned to bring the image first of X and then of Y to the wire. The angle between the two positions of the arm equals that subtended by XY at the observer's station; this is θ_1. The system is then interposed, and the angle θ_2 between the apparent directions of X and Y is measured. We count θ_2/θ_1 *positive* when in each case the arm has to be moved in the *same* direction to pass from X to Y.

When *distant* objects are not available, two goniometers with arms G_1, G_2 may be used, as in Fig. 158. The wire of G_2 is set

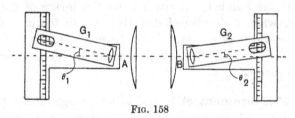

Fig. 158

on the image of the wire of G_1 for a number of positions of G_1. The angles θ_1, θ_2 are measured by G_1, G_2 respectively. The goniometers are placed as near the system as the fittings allow.

In the appliance shown in Fig. 159, a lantern projection lens L and a glass scale S, divided to mm., are fixed to a board so that S is in the focal plane of L. The same end of L faces S as would face the slide in the lantern. The appliance is substituted for the goniometer G_1. The image of the scale S will then be in focus on the wire of G_2. Readings of G_2 are taken for a number of division lines on S, and the change of angle

Fig. 159

of G_2 for one cm. on S is found. We may call this θ_2. The appliance is then put directly in front of G_2 and the measurements are repeated. The change of angle for one cm. on S is now θ_1.

259. Practical example. Observations made by Miss Slater.

Transverse and longitudinal magnifications. Distance h_1 was a whole number of mm. on scale C (Fig. 157); h_2 was measured on glass scale D. Columns C, D give bench readings of C, D. Image was inverted. "Stopping down" lens A improved definition.

C cm.	D cm.	h_1 cm.	h_2 cm.	h_2/h_1	l_1 cm.	l_2 cm.	l_2/l_1
77·7	131·3	1·20	− 1·79	− 1·492	0	0·0	—
79·7	135·6	0·80	− 1·20	− 1·500	2	4·3	2·150
81·7	140·2	0·70	− 1·05	− 1·500	4	8·9	2·225
82·7	142·2	0·80	− 1·20	− 1·500	5	10·9	2·180
83·7	145·1	0·60	− 0·90	− 1·500	6	13·8	2·300
84·7	147·1	0·90	− 1·32	− 1·467	7	15·8	2·257

Mean value of h_2/h_1 is − 1·493. Hence $m = h_2/h_1 = -$ 1·493.

Mean value of l_2/l_1 is 2·222. Hence $m = - (l_2/l_1)^{\frac{1}{2}} = -$ 1·491. The sign is negative since the image is inverted.

Magnifying power. Goniometers G_1, G_2 were used; pivot to scale, 40 cm. for each. Since angles are small, displacements of goniometer arms along scales are $40\theta_1$ and $40\theta_2$ cm. First and second columns give pairs of readings of G_1, third and fourth columns give corresponding readings for G_2.

Readings of G_1		Readings of G_2		$40\theta_1$	$40\theta_2$
8·0	10·5 cm.	11·30	9·64 cm.	2·5	− 1·66
8·5	11·0	10·99	9·33	2·5	− 1·66
9·0	11·5	10·64	8·99	2·5	− 1·65
9·5	12·0	10·31	8·66	2·5	− 1·65

Hence $M = \theta_2/\theta_1 = -$ 1·655/2·5 $= -$ ·6620; hence $m = 1/M = -$ 1·511.

EXPERIMENT 49. **Determination of effective aperture of stop of photographic lens*.**

260. Notation for stops. The stops of a photographic lens are usually marked $f/8$, $f/16$, etc. The symbol $f/8$ means that the effective diameter of the stop is one-eighth of the focal length of the lens system. This effective diameter is not, as a rule, the actual diameter of the hole in the diaphragm used to regulate the light entering the camera, but is, in the case of a camera adjusted for landscape photography, the diameter of that incident beam of rays parallel to the axis, which, in its passage through the lens system, exactly fills the opening in the actual

* See *Proc. Cambridge Phil. Soc.* Vol. XVIII, p. 195, for an account of the effect of the distance of the object on the effective aperture.

diaphragm. The photographic speed of the lens, for landscapes, is proportional to the square of the diameter of the incident beam.

The "f" notation is not the only system in use. The stops on some cameras are graduated in "U.S." numbers (Uniform System, not United States). The stop denoted by $f/4$ is taken as the unit, and the numbers are so chosen that the "U.S." number of a stop is proportional to the exposure required. The systems correspond as follows:

"f" system	$f/4$	$f/8$	$f/16$	$f/32$	$f/64$
Uniform System	1	4	16	64	256

If the "f" number be f/n, we have $f/n = f/(\frac{1}{4} n \times 4)$, and hence the "U.S." number is $(n/4)^2$. Thus, if f/n correspond to the "U.S." number 8, we have $(n/4)^2 = 8$, or $n^2 = 128$, or $n = 11\cdot31$; this stop is often marked $f/11$.

The number n in f/n expresses the ratio of f to the effective diameter of the stop for landscapes. It is convenient to call the reciprocal number $1/n$ the "aperture ratio."

261. Stop in front of lens. When the stop S (Fig. 160) is placed in front of the lens system, its effective diameter for

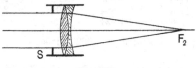

Fig. 160

landscape work equals the diameter of S. If the diameter of S be a cm., and the focal length of the lens be f cm., the aperture ratio is a/f; if the "f" number be f/n, we have $1/n = a/f$, and

$$n = f/a. \quad \dots\dots\dots\dots\dots\dots\dots\dots\dots(1)$$

262. Stop between components of lens system. A stop does more than regulate the amount of light passing through the lens. The position of a stop of given aperture influences the five "defects" of a lens, viz. Spherical aberration, Coma, Astigmatism, Curvature of image surfaces, Distortion. In the case of lenses consisting of two separated components, it is advantageous to place the stop *between* the components, as in Fig. 161.

The effective diameter of the stop will now not be equal to its actual diameter. The effective diameter can be calculated from the actual diameter when the position of the stop relative to the front component and also the optical constants of that component are known, but we may arrange the measurements so that there is no need to take the lens system to pieces.

Let S (Fig. 161) be the stop and T_1 the image of the real object S formed by the front lens L. This image will usually be virtual, as in Fig. 161. Then, if the rays be reversed, S is the (real) image

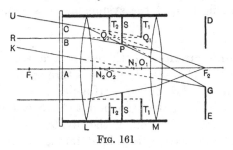

FIG. 161

of T_1 by the same lens. Hence a ray RQ_1, which before incidence on L is parallel to the axis AF_2 and is directed to Q_1 on the edge of T_1, will, after traversing L, pass (actually) through P on the edge of the stop itself. Hence the distance of RQ_1 from the axis AF_2 is equal to the radius ($\frac{1}{2}a$) of the effective aperture.

If L be converging, and if S lie between L and the focal plane of L, the distance of the image T_1 from L will be greater than the distance of S from L, and consequently the diameter of the aperture in the image will be greater than in the actual stop.

In Fig. 161, T_2 is the image of the stop S formed by the lens M. Each of the rays RB and UC in its path through the system passes actually, or in direction, through (1) Q_1, (2) P, (3) the edge Q_2 of T_2. It will be seen that Q_2 is the image of Q_1 by the complete system.

263. First method. The lens system is *firmly* supported with its axis horizontal. A microscope is mounted on a carriage, moving on a short optical bench, so that its axis is at right angles to the direction of motion of the carriage and is parallel to F_1F_2, the axis of the system; the two axes are at the same height above the table. The microscope is then focused through L

upon one end of the horizontal diameter of the stop S, and the carriage is adjusted so that the image of the edge is brought to the cross-wire of the microscope or to a selected line of the micrometer scale. The microscope is adjusted in the direction of its own axis so that there is no parallax between the image and the cross-wire. The carriage is then moved along the bench to bring the image of the other end of the diameter to the cross-wire. The displacement of the microscope equals a, the effective diameter of the stop.

The focal length, f, of the system is measured, and $n = f/a$ is calculated.

264. Second method. If there be a luminous *point* at the focus F_2 (Fig.161), the rays, after passing through the system, form a beam parallel to the axis AF_2. If a glass scale ABC be placed with its *divided* face against the mount of the lens, the diameter of the beam can be read off, if the divided face have a matt surface, which so scatters any light falling upon it that the illuminated part is easily located. The diameter of the bright patch equals the effective diameter of the stop. A matt surface may be obtained by aid of paraffin wax. The scale is *gently* heated over a flame until warm enough to melt the wax. A *thin* layer of wax may be obtained by wiping off most of the melted wax with a piece of paper with a straight edge. A matt surface may also be obtained by dabbing putty against the glass.

To obtain a beam from a small area in the focal plane, we allow light from a *flame* to pass through a small opening in a metal plate placed in that plane. The flame should be near enough to the plate to ensure that the stop is *filled* with light. Since the rays do not now proceed from a *point*, the emergent beam will not be made up only of rays which are parallel to the axis, and consequently the diameter of the bright patch on ABC will no longer be equal to the effective diameter of the stop.

Let G be a point, in the plane of the diagram, on the edge of the circular opening in the focal plane, let N_2 be the nodal point corresponding to F_2, and let N_1 be the other nodal point. Then, if a ray from G be directed towards N_2 before it strikes M, it will, on emergence from L, proceed in the direction N_1K, the directions GN_2 and N_1K being parallel, by the property of the nodal points. Since G is in the focal plane, *all* the rays due to G which emerge from the lens L are parallel to N_1K. The extreme ray is the one which, in its course, passes through the point P (in the plane of the diagram) on the edge of the stop. But Q_1 is the image of P formed by L, and thus the extreme ray, on emerging from L, must be directed from Q_1. Hence, Q_1CU, drawn parallel to GN_2, is the extreme emergent ray. If it strike the matt surface ABC in C, AC is the radius of the bright patch*.

Since CQ_1 is parallel to N_2G, the triangles CBQ_1, GF_2N_2 are similar, and hence $CB/BQ_1 = GF_2/F_2N_2$. If d be the diameter of the opening, $F_2G = \frac{1}{2}d$. If $BC = \frac{1}{2}b$ and $BQ_1 = x$, we have $b/x = d/f$, or $b = xd/f$. If c be the diameter of the bright patch, $c = 2(AB + BC) = a + b = a + xd/f$. Hence

$$a = c - xd/f. \qquad\qquad\dots\dots\dots\dots\dots\dots(2)$$

To find BQ_1 or x, a plate of glass of good quality is fixed against the mount of the lens in place of the glass scale, and lycopodium is placed on the face nearer to L. A microscope on a sliding carriage is arranged so that its axis and the direction of motion of the carriage are parallel to the axis of the system, and the microscope is focused through the plate ABC first on the edge of the stop and then on the lycopodium. The displacement of the carriage equals BQ_1 or x.

The diameter d of the hole is measured by a scale or by a travelling microscope.

265. Third method. The determination of BQ_1 may be avoided by using *two* plates with apertures of diameters d_1 and

---

* It has been assumed that the effective diameter of the incident beam is limited by S and not by the mountings of the lenses. If the aperture in S be so large that the ray which passes through P in its course to F_2 passes close to the lens mountings, it may happen that, unless F_2G be small, the ray from G which is directed to Q_2 is caught by the lens mountings and does not penetrate the system. In that case, the edge of the luminous patch is not at C but at some point nearer to A, and then the diameter of the patch depends not on S but on the diameter of one of the lens mountings. If such a system were used for landscape photography, the illumination would fall off towards the edges of the plate. This defect is avoided by keeping the stop of greatest aperture sufficiently small.

d_2. Let c_1, c_2 be the diameters of the corresponding patches. By (2), $(c_1 - a)/(c_2 - a) = d_1/d_2$. Thus

$$a (d_1 - d_2) = c_2 d_1 - c_1 d_2 = c_1 (d_1 - d_2) - d_1 (c_1 - c_2),$$

and $$a = c_1 - d_1 (c_1 - c_2)/(d_1 - d_2). \quad \ldots\ldots\ldots\ldots(3)$$

266. A system for laboratory work. A lantern projection lens is convenient. The cylindrical tube forming the mount is easily attached to a base. The stop may be inserted in the tube between the two components of the system. Fig. 162 shows the arrangement used at the Cavendish Laboratory. The lens system AB, of about 8 inches focal length, is mounted at one end of a board. At the other end is fixed an upright C with a circular opening. The metal plate D is held by a spring against that face of the upright which is furthest from the lens. The lens is adjusted,

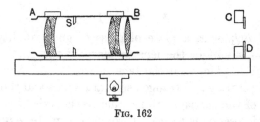

FIG. 162

by aid of a plane mirror or a *distant* object, so that its focal plane coincides with the face of C against which D rests. The stop S should have a bevelled edge.

If the focal length be measured by the goniometer method (§ 234), a glass scale is substituted for D, its divided face being in contact with C.

A projection lens so mounted is very useful. If a cross-wire be fitted to the metal plate D, the arrangement acts as a collimator. The spherical aberration will be small if the lens system be mounted on the board so that the end which faces the lantern slide, when the lens is used for projection, faces D. If the lens be mounted the wrong way round, it will be much less satisfactory.

267. Practical example. A projection lens was tested.

First method. Readings of microscope carriage gave $a = 3\cdot18$ cm.

Second method. Diameter of hole in plate $= d_1 = 0\cdot80$ cm. Diameter of patch $= c_1 = 3\cdot31$ cm. Thus $c_1 > a$. Readings of microscope carriage for B, Q_1, gave $x = 3\cdot42$ cm. Nodal point method (§ 226) gave $f = 21\cdot59$ cm. By (2),

$$a = c_1 - xd_1/f = 3\cdot18 \text{ cm.}$$

Third method. Diameters of holes, $d_1 = 0\cdot80$, $d_2 = 0\cdot42$ cm., of patches $c_1 = 3\cdot31$, $c_2 = 3\cdot25$ cm. By (3),

$$a = c_1 - d_1 (c_1 - c_2)/(d_1 - d_2) = 3\cdot18 \text{ cm.}$$

Hence $n = f/a = 21\cdot59/3\cdot18 = 6\cdot79$. The marking on "$f$" system is $f/6\cdot79$; the aperture ratio is $1/6\cdot79$ or $0\cdot147$.

CHAPTER XI

ASTIGMATISM AND FOCAL LINES

268. Non-spherical wave front. When a wave front is spherical, the rays meet accurately in a point and the beam is stigmatic. To deal with non-spherical wave fronts, we first consider the surface represented by

$$x - y^2/2b - z^2/2c = 0. \ldots\ldots\ldots\ldots\ldots(1)$$

This passes through the origin O. The sections of the surface by the planes $y = 0$ and $z = 0$ are the parabolas $x = z^2/2c$ and $x = y^2/2b$. The radii of curvature of these curves at O are c and b. We denote their centres of curvature by C, B.

It will be seen that the value of x is determined when y and z are known. If Q be a fixed point at a finite distance from the surface and if QP be normal to the surface at P, the distance from Q to a point P' on the surface *very near* P is independent of the position of P'.

Let Q be on the axis Ox and let $OQ = u$. If (x, y, z) be a point on the surface and r be its distance from Q, we have

$$r^2 = (u - x)^2 + y^2 + z^2.$$

Now x is of the second order of small quantities when y, z are small, and therefore x^2 is of the fourth order. Neglecting x^2 and substituting for x from (1), we find

$$r^2 = u^2 + y^2 (1 - u/b) + z^2 (1 - u/c).$$

If QP be normal to the surface, *both* the conditions

$$\partial r/\partial y = 0, \quad \partial r/\partial z = 0,$$

or $\quad y (1 - u/b) = 0, \ \ldots\ldots(2) \qquad z (1 - u/c) = 0, \ \ldots\ldots(3)$

must be satisfied by the coordinates of P.

If $u = b$, both (2) and (3) are satisfied for any small value of y, if $z = 0$. If $u = c$, both (2) and (3) are satisfied for any small value of z, if $y = 0$. But (2), (3) are not both satisfied for any value of u if *both* y and z differ from zero, unless $b = c$. Hence, if P do not lie in one of the planes $y = 0$, $z = 0$, the normal at P does not meet Ox.

The planes $y = 0$, $z = 0$ are called the principal sections of the surface. The points B, C are called the centres of principal curvature, or the focal points, of the surface, and b, c are called the principal radii of curvature at O.

If near a point P on a surface, the equation to the surface, referred to the normal at P as axis of x and to two suitably chosen axes through P in the tangent plane, be of the form (1), P is called an "ordinary" point. If each point of a surface be "ordinary," the surface is called an "ordinary" surface.

The normals to a wave front remain unchanged as the wave advances in a uniform isotropic medium; hence, if at any instant any point of the wave front be ordinary, the corresponding point at any later time will also be ordinary.

269. Lines of curvature. A line of curvature on a surface is a curve such that, if P, P' be consecutive points on the curve, the normals to the surface at P, P' intersect. If the surface be ordinary, the curves in which the two principal sections cut the surface near P are elements of lines of curvature. Thus there are two, and only two, lines of curvature through every point of the surface and these intersect at right angles. If, however, the surface be symmetrical about the normal PN, every plane containing PN cuts the surface in a line of curvature.

270. Focal lines. The wave front due to a luminous point is spherical, and so is "ordinary." It can be shown that it remains ordinary, though generally not spherical, after any number of reflexions or refractions at ordinary surfaces.

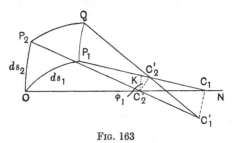

FIG. 163

Let O (Fig. 163) be a point on the wave front, and let ON, the normal at O, be taken as the chief ray. Let OP_1, OP_2 be the lines of curvature through O, and let C_1, C_2 be the focal points on ON, i.e. the centres of curvature corresponding to the principal sections. Thus OP_1, OP_2 intersect at right angles. Let $OC_1 = \rho_1$, $OC_2 = \rho_2$, $OP_1 = ds_1$, $OP_2 = ds_2$.

If OP_2 be small enough, all the rays which pass through OP_2 meet in C_2. If P_1Q be the line of curvature through P_1 conjugate to P_1O, all the rays which pass through P_1Q will cut P_1C_1 in C_2'; the line C_2C_2' lies in the principal section C_1OP_1. But P_1C_2' will differ from OC_2 by a small quantity of the *first* order, unless $d\rho_2/ds_1 = 0$. If C_2K be perpendicular to P_1C_1, we have ultimately $KC_2' = (d\rho_2/ds_1)\,ds_1$. Since $C_2C_1 = \rho_1 - \rho_2$, $KC_2 = (\rho_1 - \rho_2)\,ds_1/\rho_1$. Hence, if $KC_2C_2' = \phi_1$,

$$\tan \phi_1 = KC_2'/KC_2 = \rho_1\,(d\rho_2/ds_1)/(\rho_1 - \rho_2).$$

Thus ϕ_1 is zero only when $d\rho_2/ds_1 = 0$. A similar result holds for C_1C_1', where C_1' is the focal point conjugate to C_2 on P_2C_2.

All the rays from very narrow bands of the wave front lying about (1) OP_1, (2) OP_2, will intersect (1) C_2C_2', (2) C_1C_1'. The lines C_1C_1', C_2C_2' are called the focal lines corresponding to O. The rays from the remainder of OP_1QP_2 will pass *near* each of the lines C_2C_2', C_1C_1'. If the part of the wave front actually employed be small enough, the aberrations of the rays from the two lines may not be serious. We then have an astigmatic beam, each ray passing, at any rate very nearly, through each of two focal lines. The term "focal line" is due to Coddington.

271. Circle of least confusion. Let an opaque screen with a small circular opening of radius r, whose centre is O (Fig. 164), be placed in the path of a beam of rays so that the plane of the

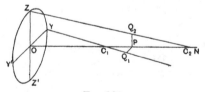

Fig. 164

opening is perpendicular to the ray ON which passes through O. We take ON as the chief ray of the beam passing through the opening. Let Y, Y', and Z, Z' be the points at which the edge is cut by those lines of curvature of the wave front which pass through O. Let C_1, C_2 on ON be the focal points of the wave front.

Draw C_1Y, C_2Z. Through P, a point on ON between C_1 and C_2, draw a plane perpendicular to ON, cutting C_1Y, C_2Z in Q_1, Q_2. Let $OC_1 = v_1$, $OC_2 = v_2$, $OP = w$. Then $PQ_1 = (w - v_1)\,r/v_1$, and $PQ_2 = (v_2 - w)\,r/v_2$. Thus $PQ_1 = PQ_2$, if $(w - v_1)/v_1 = (v_2 - w)/v_2$, or if

$$1/w = \tfrac{1}{2}\,(1/v_1 + 1/v_2). \quad\text{..................(4)}$$

If a screen be placed in the plane PQ_1Q_2, the breadth of the illuminated patch will be the same along PQ_1 as along PQ_2, when w is given by (4). Since the patch is oval in form, it cannot be far from circular; it may, in fact, be shown to be circular. A circle with its centre at P and passing through Q_1 and Q_2, where $PQ_1 = PQ_2$, is called the circle of least confusion.

The position of the circle of least confusion depends upon the position of the circular opening $YZY'Z'$ and not merely upon the positions of the focal points C_1, C_2. If $C_1P = p$, $C_1C_2 = h$, then $w = v_1 + p$, $v_2 = v_1 + h$, and (4) can be transformed into

$$1/p - 1/v_1 = 2/h, \quad\text{....................(5)}$$

which shows how p or C_1P depends upon v_1 or OC_1.

EXPERIMENT 50. Focal lines formed by concave mirror.

272. Introduction. Let O (Fig. 165) be the centre of curvature of a concave spherical mirror, and P a luminous point.

Let two rays PR, PS in the plane of the diagram meet the mirror in R, S. Let the reflected rays RQ_1Q_2, SQ_1Q_2' intersect in Q_1, and let them meet PO in Q_2, Q_2'. If RS be small compared with OR, the rays reflected at the points between R and S will pass very nearly through Q_1, and all of them will *cut* OP between Q_2 and Q_2'.

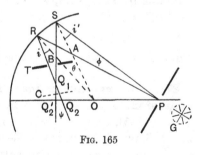

FIG. 165

If RS revolve about OP, so as to trace out part of a zone on the sphere, Q_1 describes an arc of a circle which has its centre on OP and its plane perpendicular to OP. Thus all the rays reflected from the zone will actually or very nearly intersect the circle through Q_1 and will also *cut* OP between Q_2 and Q_2'.

The points Q_1, Q_2 are the primary and secondary focal *points* corresponding to the chief ray PR. The circular arc through Q_1 is the primary focal line, and $Q_2 Q_2'$ is the secondary focal line. The primary line is perpendicular to the rays, but the secondary line is not perpendicular to them, unless $RQ_2 P = \frac{1}{2}\pi$.

Let $OR = r$, $PR = u$, $Q_1 R = v_1$, $Q_2 R = v_2$. By the law of reflexion, the angles PRO, $Q_1 RO$ at R are equal, and similarly for S. Let $PRO = Q_1 RO = i$, $PSO = Q_1 SO = i'$; let $ROS = \theta$, $RPS = \phi$, $RQ_1 S = \psi$, and let PR, OS intersect in A and $Q_1 S$, OR in B. Then $\phi + i' = SAR = \theta + i$, and $\psi + i = RBS = \theta + i'$. Hence

$$\phi + \psi = 2\theta. \qquad \dots\dots\dots\dots\dots\dots\dots(1)$$

The perpendiculars from S on PR and from R on $Q_1 S$ are each equal to $\cos i . RS$ when θ is infinitesimal, and thus $\phi = \cos i . RS/u$, $\psi = \cos i . RS/v_1$. Also $\theta = RS/r$. Hence, by (1),

$$\cos i . RS \,(1/u + 1/v_1) = 2RS/r,$$

and thus
$$\frac{1}{u} + \frac{1}{v_1} = \frac{2}{r \cos i}. \qquad \dots\dots\dots\dots\dots(2)$$

To find Q_2, we note that the sum of the areas of the triangles $Q_2 RO$, PRO equals the area of PRQ_2. Hence

$$\tfrac{1}{2}rv_2 \sin i + \tfrac{1}{2}ru \sin i = \tfrac{1}{2}uv_2 \sin 2i.$$

Dividing by $\frac{1}{2}ruv_2 \sin i$, we have

$$\frac{1}{u} + \frac{1}{v_2} = \frac{2 \cos i}{r}. \qquad \dots\dots\dots\dots\dots\dots(3)$$

Equations (2) and (3) determine the positions of the focal lines.

273. Observation of focal lines and of circle of least confusion. A metal plate pierced by a small hole is placed at P (Fig. 165), and rays from a flame G pass through the hole and fall upon the mirror. The centre of the aperture of the mirror and its centre of curvature O, should be on the same level as P. If a screen of ground glass be now placed near the mirror and then be drawn back along $RQ_1 Q_2$, the primary line will appear as a bright vertical line on the ground glass at Q_1, and the secondary line as a horizontal line at Q_2.

If an electric lamp be used at G, a piece of ground glass, of fine grain, slightly oiled to make it nearly transparent, is placed between G and P. The glass will scatter the light sufficiently to ensure that the mirror is filled with light.

The focal lines may also be examined by aid of a lens of *small* focal length, e.g. 5 cm., without using the ground glass screen. The unaided eye, wherever placed along RQ_1, perceives only an ill-defined patch of brightness against the background of the mirror.

The position of the circle of least confusion depends, as shown in § 271, not only upon the positions of the two focal points on the chief ray, but also upon the position of the circular opening through which the beam passes. A plate T (Fig. 165), with a circular hole, about 1 cm. in diameter, is placed about midway between the mirror and Q_1; the plane of T is perpendicular to the reflected beam. The primary focal line will now be more sharply defined. If the ground glass be held in *contact* with T, the luminous patch should show that the opening is *filled* with light. If the screen be placed in the proper position C between the two focal lines, the patch is very nearly circular. The distances TC, TQ_1, TQ_2 are measured, and the value of $1/TQ_1 + 1/TQ_2$ is compared with that of $2/TC$. (See § 271.)

274. Experimental test of formulae. For accurate work, the mirror should give only a single reflexion. A mirror of speculum metal or of glass silvered on the *front* gives bright focal lines, but the concave face of a lens may be used as the mirror if the other face be smeared with vaseline and lampblack or be

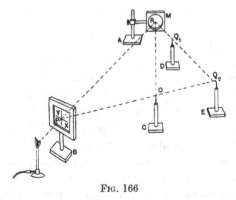

FIG. 166

coated with black varnish to stop reflexion at that face. The mirror M (Fig. 166) is mounted in a frame carried by a cross-rod attached to the stand A. The mirror turns about the cross-rod and can be clamped in any position. A small mark R is made on the mirror near its centre.

The point of intersection of horizontal and vertical lines PX, PY drawn on a piece of ground glass, illuminated (but not destroyed) by a flame from behind, serves as the point P of Fig. 165. The glass is carried on the stand B, and the ground side faces the mirror. The stand C carrying a pin is adjusted so that the tip O is at the centre of curvature of M (see § 100); then O coincides with its own image. The stand D carries a vertical pin, and E a horizontal pin.

The first step is to adjust the mirror and the C-pin so that O, when coincident with its image, is at the same height above the table as the mark R. The point P is then set to the same height.

The tip of the D-pin is now made to coincide with the primary focal line Q_1, where an "image" of the line PY is formed. The adjustment is complete when, on moving the eye from side to side, the line and the pin appear to move across the mirror together. If the distance of the eye from the mirror be sufficiently greater than RQ_2, the horizontal secondary line will also be seen. The height of the D-pin is adjusted so that, when the eye sees both focal lines crossing R, the tip is in the line of sight. The E-pin is adjusted so that the tip lies at Q_2 on the horizontal focal line, where an "image" of PX is formed. The adjustment is tested by moving the eye up and down. The E-pin is adjusted horizontally so that, when the eye sees both lines crossing R, the tip is in the line of sight. With perfect adjustment, RQ_1Q_2, POQ_2 are straight lines.

If a luminous point travel along PX, the focal points will describe curves whose tangents are horizontal at Q_1 and Q_2. If the point travel along PY, the focal points will describe curves having vertical tangents at Q_1 and Q_2. The primary focal lines due to successive luminous points on PY will overlap at Q_1, and an eye on RQ_1 will see a sharp image of PY. The *secondary* focal lines (near Q_2) due to successive luminous points on PY will *not* overlap, but will be practically parallel to each other, and thus the rays from points on PY will be scattered over a vertical strip whose horizontal width subtends at Q_1 the same angle as the horizontal diameter of the mirror. If the rays be received on a ground glass screen, the illumination at Q_2 due to points on PY will thus be very feeble, and there will be no suggestion of an "image." Similar remarks apply to PX. Thus an "image" of PY is formed at Q_1 and an "image" of PX at Q_2.

The distance RP may be measured by an adjustable distance-piece with *pointed* ends (Fig. 130). For RO, RQ_1, RQ_2, OP, OQ_2 the appliance of Fig. 129 is convenient. The distance-piece or scale should not be held in the hand, but supports should be arranged so that it can rest upon them. Then a full- or half-size diagram is made. First RO is marked down and then by circles about R with radii RP, RQ_2 and by circles about O with radii OP, OQ_2 the points P and Q_2 are located.

The angles ORP, ORQ_2 can be measured on the diagram by a protractor; they are theoretically equal and their mean is taken as i. We may also find $\cos i$ from the equations

$$\cos i = (u^2 + r^2 - OP^2)/2ur, \qquad \cos i = (v_2{}^2 + r^2 - OQ_2{}^2)/2v_2 r.$$

275. Practical example. Measurements by Mr W. Gray gave $r = 21{\cdot}49$, $u = 33{\cdot}41$, $v_1 = 12{\cdot}40$, $v_2 = 21{\cdot}60$, $OP = 20{\cdot}09$, $OQ_2 = 12{\cdot}82$ cm. Triangles PRO, ORQ_2 gave $\cos i = {\cdot}8179$, $\cos i = {\cdot}8230$. Mean value of $\cos i$ is ${\cdot}8204$. Hence $i = 34° 53'$.

For (i) primary, (ii) secondary line,

    (i) $1/u + 1/v_1 = {\cdot}1106$ cm.$^{-1}$, $2/(r \cos i) = {\cdot}1134$ cm.$^{-1}$,

    (ii) $1/u + 1/v_2 = {\cdot}07623$ cm.$^{-1}$, $(2 \cos i)/r = {\cdot}07635$ cm.$^{-1}$.

EXPERIMENT 51. Focal lines by refraction at plane surface.

276. Introduction. The object of this experiment is to find the positions of the two focal lines seen when a luminous point placed in a tank of water is viewed by rays which strike the surface obliquely.

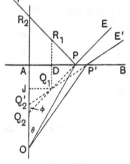

Let AB (Fig. 167) be the surface of the water, and O the luminous point below AB, the axis OA being normal to the surface. Let OP, the chief incident ray, and OP', a neighbouring ray, each in the plane of the diagram, meet the surface in P, P'. Let the directions of the refracted rays PE, $P'E'$ intersect in

FIG. 167

Q_1, and cut the axis in Q_2, $Q_2{}'$. Then PE is the chief refracted ray.

If APP' turn through a small angle about the axis OA, PP' will sweep out a small portion S of an annular area on the surface. By symmetry, the principal sections at P of the refracted wave front are (1) the plane of the diagram, and (2) a plane through the chief refracted ray $Q_2 Q_1 PE$ perpendicular to the plane of the diagram. The points Q_1, Q_2 are the primary and secondary focal points on the chief ray.

The incident rays in the plane of the diagram which lie between OP and OP' give rise to refracted rays whose directions pass actually or very nearly through Q_1. As APP' revolves about OA, Q_1 describes a small arc of a horizontal circle having its centre at J on OA. Thus the directions of all the rays emerging from S pass actually or very nearly through this small arc, which is the primary focal line. Since any ray from O after refraction is directed from a point on the axis, the directions of all the rays emerging from S cut OA between Q_2 and Q_2', and thus the secondary focal line is a small length of OA near Q_2.

We will now find the positions of the focal lines. Let $OP = u$, $PQ_1 = v_1$, $PQ_2 = v_2$. Let $AOP = \theta$, $AQ_2 P = \phi$; then θ and ϕ are also the angles of incidence and refraction at P. If μ be the index of water, $\sin \phi = \mu \sin \theta$. But $\sin \phi = AP/v_2$, $\sin \theta = AP/u$. Hence

$$v_2 = u/\mu. \quad\quad\quad\quad\quad\quad (1)$$

If $AOP' = \theta + \alpha$, $AQ_2'P' = \phi + \beta$, then

$$\sin (\phi + \beta) = \mu \sin (\theta + \alpha),$$

or $\sin \phi \cos \beta + \cos \phi \sin \beta = \mu (\sin \theta \cos \alpha + \cos \theta \sin \alpha)$.

Since α and β are very small, we may write

$$\sin \phi + \beta \cos \phi = \mu \sin \theta + \mu \alpha \cos \theta.$$

But $\sin \phi = \mu \sin \theta$, and hence

$$\beta \cos \phi = \mu \alpha \cos \theta. \quad\quad\quad\quad\quad (2)$$

The perpendicular from P on $Q_1 P'$ is ultimately $PP' \cos \phi$ and is also equal to $v_1 \beta$. Hence $PP' = v_1 \beta/\cos \phi$. Similarly $PP' = u\alpha/\cos \theta$. Thus

$$v_1 \beta/\cos \phi = u\alpha/\cos \theta. \quad\quad\quad\quad\quad (3)$$

Dividing (3) by (2), we have $v_1/\cos^2 \phi = u/(\mu \cos^2 \theta)$, or

$$v_1 = u \cos^2 \phi/(\mu \cos^2 \theta). \quad\quad\quad\quad (4)$$

277. Experimental details. At the bottom of a tank containing water is placed a piece of unglazed white porcelain T (Fig. 168) (a menu tablet) ruled with cross-lines OX, OY. This leans against the side of the tank and is held in position by a weight L. The cross-line OX is horizontal. To mark the point P, a needle is *firmly* supported in a clip so that its tip nearly, but not quite, touches the surface of the water. If there be actual contact, the surface is disturbed by capillary action. The point P

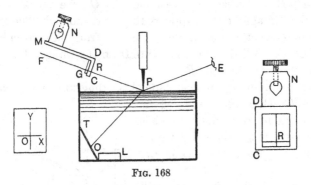

Fig. 168

should lie in a plane through O perpendicular to the horizontal line OX. If the tank be suitable, a bridge-piece carrying the needle clip should be arranged so that it can be clamped to the top of the tank in any required position. The tank stands on a *steady* table out of reach of draughts; the observations cannot be made satisfactorily if the surface be disturbed.

A second pair of cross-lines ruled on ground glass is placed at R; the ground side must face P. The plate of ground glass G (Fig. 168), about 2 cm. square, is attached to the end CD of a metal strip MDC bent at right angles at D. The end view shows an opening in CD to allow the cross-lines to be seen. By a clamp N the strip is fixed to a horizontal rod carried by a stand. The point R in which the cross-lines intersect should be near the lower edge C of CD, so that R can be brought close to the water.

An eye at E (Fig. 167) sees the cross-lines at O by rays refracted at points of the surface very near P, and will judge O to be in the direction EQ_1Q_2. If a ray EP fall on the surface, it will be reflected along PF, and, if R be placed on PF, R will appear to the eye at E to be in the same direction EP as O. The surface forms by reflexion a perfect image of R at a point vertically below R and as far beneath the surface as R is above it. If R be at R_1, its image by reflexion will coincide with Q_1; when it is at R_2, vertically above O, its image will coincide with Q_2. For reasons similar to those given in § 274, an "image" of the horizontal line OX will be seen at Q_1 and an "image" of OY will be seen at Q_2.

When R is to be set to correspond with Q_2, the eye is placed so that the "image" of OX appears to pass through P. The adjustment is correct if, on moving the eye from side to side, the "image" of OY does not separate from the image of the non-horizontal line through R.

The emergent rays in the plane of the diagram touch a caustic curve HK (Fig. 169). Let PE be the emergent ray which touches the caustic at Q_1, the point whose image by *reflexion* at AB is R_1. Then, to an eye at E'', the "image" of the line OX will appear to be above the image of R_1. If the eye be moved from E'' through E to E', the "image" of OX will first move down into coincidence with the image of R_1, and then will move up again as E' is approached.

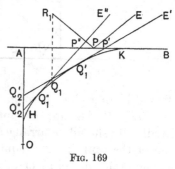

Fig. 169

If the arc $HQ_1 = s$, and if $AP = x$, it can be shown that $ds/dx = 3\,(\mu^2 - 1)\sin\theta/(\mu\cos^2\theta)$, where $AOP = \theta$ (Fig. 167). This varies from zero when $\theta = 0$, to 3 when θ has the critical value $\sin^{-1}(1/\mu)$. For water $\mu = 4/3$.

θ	$0°$	$10°$	$20°$	$30°$	$40°$	$45°$	$48°\,35'$
ds/dx	0	·313	·678	1·167	1·917	2·475	3·000

In § 278, $\theta = 41°\,57'$; thus, if the eye be placed so that the line of sight misses P by 1 mm. measured horizontally, the apparent position of the cross-line OX will be 2 mm. from Q_1 measured along the arc.

When R is to be set to correspond with Q_1, the eye is placed so that the "image" of OY appears to pass through P. The adjustment is correct if, on moving the eye up and down, the "image" of OX move down to the image of the horizontal line through R, but do not pass beyond it.

The illumination of the cross-lines at O and of the ground glass at R should be adjusted so that the images seen (1) by refraction, and (2) by reflexion, are nearly equal in intensity.

The distances are measured by the instrument in Fig. 170. On the rod C slide adjustable pieces carrying stout bent needles

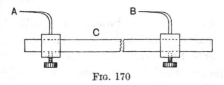

FIG. 170

A, B. To measure the depth, the tip of A is placed at O on the porcelain plate, and B is adjusted to touch the surface of the water. To avoid changes of level of the surface, the instrument should be kept in the water, immersed to the proper depth, while the optical adjustments are made.

The angle θ is found from the triangle AOP (Fig. 167), in which the sides AO, OP and the right angle OAP are known.

278. Practical example. Observations by S. E. Brown and G. F. C. Searle gave $AO = 11 \cdot 75$, $OP = u = 15 \cdot 80$ cm.

Hence $v_2 = u/\mu = \frac{3}{4}u = 11 \cdot 85$ cm. Observed $PR_2 = PQ_2 = 11 \cdot 78$ cm.

From triangle AOP, $\cos \theta = 11 \cdot 75/15 \cdot 80 = \cdot 7437$. Hence $\theta = 41° \ 57'$. Then $\sin \phi = \mu \sin \theta = \frac{4}{3} \sin 41° \ 57' = \cdot 8913$. Hence $\phi = 63° \ 2'$.

Then $v_1 = u \cos^2 \phi/(\mu \cos^2 \theta) = 4 \cdot 406$ cm. Observed $PR_1 = PQ_1 = 4 \cdot 45$ cm.

EXPERIMENT 52. **Measurement of powers of thin astigmatic lens.**

279. Introduction. The simplest form of astigmatic lens has one face cylindrical and the other plane or spherical. Such lenses are used in spectacles for correcting astigmatic defects of the eye.

Let the radius of the spherical surface be r and that of the cylindrical surface s, each counted positive when the corresponding surface is convex. Let O be the centre of the spherical surface and YZ the axis of the cylindrical surface, i.e. the axis of the circular cylinder of which that surface is a part. Let two planes be drawn through O. Let the first plane be perpendicular to YZ; Fig. 171 shows this section. Let the second plane contain

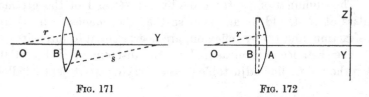

FIG. 171   FIG. 172

the axis YZ; this section is shown in Fig. 172. These sections are the "principal sections" of the lens. The straight line through O which intersects YZ at right angles in Y is the axis of the lens. Let it meet the spherical face in A and the cylindrical face in B.

If rays from a luminous point P on the axis of the lens meet the lens, the wave front before incidence will be spherical. After the passage through the lens, the wave front will no longer be spherical, and thus the rays will not come to a focus in a single point on the axis. The emergent wave front will, however, have two principal sections and, when the aperture is small, all the emergent rays lying in each of these sections will meet the axis in the focal point corresponding to that section. By symmetry, these sections must coincide with the principal sections of the lens. The rays which leave P in either of these planes remain in that plane and meet the axis in the corresponding focal point.

We have, then, only to consider the rays in the planes of Figs. 171, 172. For rays in the plane of Fig. 171, the lens acts as a double-convex lens with radii r and s. If the lens be thin, the power, $1/f_1$, in this plane is given by

$$1/f_1 = (\mu - 1)(1/r + 1/s), \quad \dots\dots\dots\dots(1)$$

where μ is the index.

For rays in the plane of Fig. 172, the lens acts as a plano-convex lens. If $1/f_2$ be the power in this plane,

$$1/f_2 = (\mu - 1)/r. \quad \dots\dots\dots\dots\dots(2)$$

If $r = \infty$, so that one face is cylindrical and the other plane,

$$1/f_1 = (\mu - 1)/s, \qquad\qquad 1/f_2 = 0.$$

Let u be the distance of the luminous point P from the thin lens, and v_1, v_2 the distances of the points Q_1, Q_2 in which the rays in the principal sections of Figs. 171, 172 meet the axis of the lens. Then

$$1/u + 1/v_1 = 1/f_1, \quad \dots\dots\dots\dots\dots(3)$$

$$1/u + 1/v_2 = 1/f_2. \quad \dots\dots\dots\dots\dots(4)$$

The radius of curvature of either principal section of the emergent front has a maximum or a minimum value where the axis of the lens meets the section. Hence, by § 270, the focal lines will be perpendicular to that axis. The focal line through Q_1 is parallel to YZ, the axis of the cylinder, and the focal line through Q_2 is perpendicular to YZ. Since the focal lines pass through Q_1 and Q_2, their distances from the lens obey (3) and (4).

For a lens of finite thickness c, the two powers are given by (24), § 196. Thus

$$\frac{1}{f_1} = (\mu - 1) \left\{ \frac{1}{r} + \frac{1}{s} - \frac{(\mu - 1)\,c}{\mu r s} \right\}, \quad \frac{1}{f_2} = \frac{\mu - 1}{r},$$

but the principal or unit planes are not identical in the two cases.

We have supposed one face of the lens to be spherical and the other cylindrical, but the method of measuring the powers of the lens is independent of this supposition. For instance, the lens might have both faces cylindrical, with unequal radii, the axes of the cylinders being inclined at any angle.

280. Circle of least confusion. We suppose that the lens is mounted in a frame having a circular opening with its centre on the axis of the lens. If the distance of the circle of least confusion from the lens be w, we have, by § 271,

$$1/w = \tfrac{1}{2}\,(1/v_1 + 1/v_2) = \tfrac{1}{2}\,(1/f_1 - 1/u + 1/f_2 - 1/u).$$

If we denote $\tfrac{1}{2}\,(1/f_1 + 1/f_2)$ by $1/f_3$, the circle of least confusion obeys the formula

$$1/u + 1/w = 1/f_3. \quad \dots\dots\dots\dots\dots(5)$$

281. Experimental details. Light from a bright flame L (Fig. 173) passes through a hole P, not more than 0·2 cm. in diameter, in the metal plate C. The lens AB is mounted in D, a frame with a circular opening. After passing through the lens, the light falls on a screen H of *finely*-ground glass mounted in the frame K; the ground side of the glass faces P. The plate C and the frames D, K are arranged on an optical bench. The fittings C, D, K may be clamped to cross-rods in the same way as the

plane mirror in Fig. 131. The axis of the lens should be parallel
to the bench and should pass through P. We will suppose that

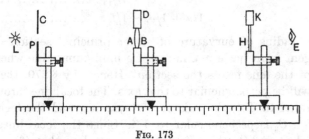

the A-face of the lens is towards P. If the lens be thin, we write,
as a rough compromise, $u = AP + \frac{1}{2}c$, $v_1 = BH_1 + \frac{1}{2}c$ and
$v_2 = BH_2 + \frac{1}{2}c$, where c is the thickness of the lens, and H_1, H_2
are the two positions of the screen.

On examining the screen H by the eye at E, a luminous patch
is seen. By adjusting H, either focal line can be focused. If
the axis of the cylindrical surface of the lens be vertical, one
line will be vertical and one horizontal; the vertical line will
be the nearer to the lens. A magnifying lens is useful in examining
the focal lines; their edges will be sharp when the lines are in
focus. The settings can be more accurately made if *fine* cross-
wires be stretched across the opening P, their directions being
accurately parallel and perpendicular to the axis of the cylindrical
surface. Unless these wires be carefully adjusted, they are a
hindrance rather than a help. After a little practice, the observer can find, with surprising accuracy, the position of the
screen for which the luminous patch is most nearly circular.
A table of reciprocals is used in the calculations.

282. Practical example. Mr G. M. B. Dobson used lens 0·45 cm.
thick. Rough correction for thickness made by measuring u, v_1, v_2, w from
centre of lens; plane of aperture assumed to pass through this point. Readings
were made for u, v_1, v_2, w; only their reciprocals, in *dioptres* (D), are here
recorded. Thus, when $u = 24\cdot78$ cm., $1/u = 4\cdot036\ D$.

$1/u$	$1/v_1$	$1/v_2$	$1/w$	$1/f_1$	$1/f_2$	$\frac{1}{2}(1/f_1 + 1/f_2)$	$1/f_3$
4·036	5·015	2·036	3·544	9·051	6·072	7·562	7·580
3·358	5·701	2·701	4·202	9·059	6·059	7·559	7·560
2·875	6·188	3·208	4·704	9·063	6·083	7·573	7·579
2·514	6·540	3·569	5·063	9·054	6·083	7·568	7·577

The mean value of $1/f_3$, viz. 7·574 D, found by observations on the circle of least confusion, is close to that deduced from the powers of the lens for the two focal lines, viz. 7·566 D.

EXPERIMENT 53. **Study of astigmatism due to pair of plano–cylindrical lenses.**

283. Introduction. Plano-cylindrical lenses are used by opticians in testing patients' eyes for astigmatism. The lenses supplied for opticians' use in "trial cases" are suitable for the present experiment. These are "edged" to a circular form, and two marks are engraved on the plane face of each lens to show the direction of the axis of the cylindrical face. The lenses used for the experiment have their cylindrical faces convex, and have *equal* powers. If the powers be unequal, the theory is less simple.

Let AB, CD (Fig. 174) be two *equal* thin plano-cylindrical lenses placed with their plane faces in contact, and let the lines AB, CD, which are drawn on the plane faces to show the directions of the axes of the cylinders, intersect in O at an angle θ. In the experiment, θ is varied. The straight line through O normal to both faces of each lens is called the axis of the system; the axis of each *lens*, i.e. the line perpendicular to both faces of the lens, coincides with it.

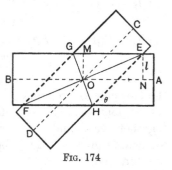

FIG. 174

A luminous point P is placed on the axis of the system. Since this axis is a line of symmetry for each surface of each lens, it is also a line of symmetry for the wave front on its emergence from the system. The ray passing along the axis may be taken as the chief ray; on account of the symmetry, the focal lines will cut the axis at right angles in the two focal points. The focal lines will be in the principal sections of the emergent wave front.

When the two lenses are exactly similar, the principal sections of the emergent wave front must, by symmetry, pass through EF, GH, the diagonals (perpendicular to each other) of the rhombus enclosed by the edges of the lenses.

284. Focal lengths of system. To find the "focal points" (§ 268) of the emergent beam, we find the focal lengths of the system for rays which pass through it in the planes of the two principal sections. We suppose that the edges of the lenses are sharp so that the thickness vanishes at the edges. For each lens, let l be its half-width, measured perpendicular to the axis of its cylindrical surface, and let $\frac{1}{2}c$ be its thickness along its axis. The diagonals bisect the angles between AB and CD. Hence $EON = GOM = \frac{1}{2}\theta$, and thus

$$OE = h_1 = l/\sin\tfrac{1}{2}\theta, \qquad OG = h_2 = l/\cos\tfrac{1}{2}\theta. \qquad\ldots\ldots\ldots(1)$$

The focal lengths we seek are the same as those of two double-convex lenses with spherical faces, each of small thickness c, and with sharp edges of diameters $2h_1$ and $2h_2$ respectively.

By (27a), § 132,

$$1/f_1 = 2(\mu - 1)c/h_1{}^2 = 2(\mu - 1)(c/l^2)\sin^2\tfrac{1}{2}\theta$$
$$= (\mu - 1)(c/l^2)(1 - \cos\theta), \qquad\ldots\ldots\ldots(2)$$
$$1/f_2 = 2(\mu - 1)c/h_2{}^2 = 2(\mu - 1)(c/l^2)\cos^2\tfrac{1}{2}\theta$$
$$= (\mu - 1)(c/l^2)(1 + \cos\theta). \qquad\ldots\ldots\ldots(3)$$

When the lenses are "crossed," so that $\theta = \frac{1}{2}\pi$, and $\cos\theta = 0$, the two powers are equal, and the system acts as a thin double-convex lens with spherical faces. If the lenses were now fused together, we should obtain a form of lens sometimes employed in reading glasses. Let f_0 be the focal length when the lenses are "crossed." Then $1/f_0 = (\mu - 1)c/l^2$, and thus

$$1/f_1 = (1 - \cos\theta)/f_0, \qquad 1/f_2 = (1 + \cos\theta)/f_0. \qquad\ldots(4)$$

In the experiment, one lens is fixed and the other can be rotated. Since the two powers change most rapidly with changes of θ when the lenses are "crossed," an accurate reading of the divided circle (§ 286) is more easily obtained for that position than for the "parallel" position. We therefore measure the angle through which the rotatable lens has been turned from its position when the lenses are crossed; let this angle be ϕ. Then $\phi = \frac{1}{2}\pi - \theta$, and hence, in terms of ϕ,

$$1/f_1 = (1 - \sin\phi)/f_0, \qquad 1/f_2 = (1 + \sin\phi)/f_0, \qquad\ldots(5)$$

and thus $\qquad\qquad 1/f_1 + 1/f_2 = 2/f_0. \qquad\ldots\ldots\ldots\ldots\ldots\ldots(6)$

If the distances v_1, v_2 of the focal lines from the lens be measured when the distance of the luminous point from the lens is u, the powers $1/f_1$, $1/f_2$, are found by

$$1/f_1 = 1/u + 1/v_1, \qquad 1/f_2 = 1/u + 1/v_2. \qquad \ldots\ldots(7)$$

The focal line corresponding to v_1 is perpendicular to the plane containing the axis and OE.

285. Circle of least confusion. Let a diaphragm with a circular opening be placed in contact with the system. If w be the distance of the circle of least confusion from the system, we have, by § 271, $1/w = \frac{1}{2}(1/v_1 + 1/v_2)$. Hence, by (7) and (6),

$$1/w = \frac{1}{2}(1/f_1 + 1/f_2 - 2/u) = 1/f_0 - 1/u,$$

or $\qquad\qquad\qquad 1/w + 1/u = 1/f_0. \qquad \ldots\ldots\ldots\ldots\ldots\ldots\ldots(8)$

The position of the circle of least confusion is thus independent of the angle between the axes of the cylinders. It is at Q_0, the point which is conjugate to P when the lenses are crossed. Let the diameter of the opening in the diaphragm be D, and that of the circle of least confusion d. Then, by the method of § 271, we can show that

$$d = (Dw/f_0).|\sin\phi|. \qquad \ldots\ldots\ldots\ldots\ldots(9)$$

The diameter of the circle changes from zero to the maximum Dw/f_0, as ϕ increases from zero to $\frac{1}{2}\pi$. When, as in practice, the hole in the plate is of finite size, d will not be zero when $\phi = 0$. If it then be d_0, we have

$$d = d_0 + (Dw/f_0).|\sin\phi|. \qquad \ldots\ldots\ldots\ldots(10)$$

286. Practical details. The arrangement for holding the lenses is shown in Fig. 175. A hole is bored through a block of wood W, and one lens A is let into a suitable recess. The other lens C is let into a recess turned in the end of a metal tube T which fits an enlarged part of the hole. The tube carries a divided circle R; readings are taken by the two indicators I_1, I_2. The arrangement is attached to the fittings of an optical bench by the clamp S.

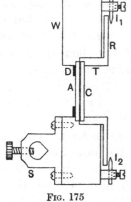

FIG. 175

The whole aperture of a plano-cylindrical lens as supplied for "trial cases" is not available, since

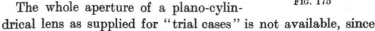

one face is "ground" except a central strip with its sides parallel to the axis of the cylinder. A diaphragm D should therefore be fitted in contact with one lens.

The measurements are similar to those of EXPERIMENT 52; the apparatus is arranged as in Fig. 173, the appliance of Fig. 175 taking the place of the frame D.

The first step is to find the reading of the circle when the cylinders are "crossed." The circle of least confusion, when the axes are inclined at any angle, is easily found. The divided circle is then turned to make this circle as small as possible, care being taken to make an accurate adjustment of the ground glass screen. Since, by (10), $d - d_0$ is proportional to $|\sin \phi|$, the diameter of the circle changes rapidly as ϕ is changed in either direction from zero, and hence the reading of the divided circle when the cylinders are "crossed" can easily be found. This reading, which corresponds to the zero of ϕ, is then taken and recorded.

The distance u of the pin-hole P from the lenses is now measured and also the distance w of the circle of least confusion when the cylinders are "crossed." Then $1/f_0$ is found by (8).

If the divided circle be rotated, the luminous patch on the ground glass remains circular, though its diameter increases; thus we see that w, the distance from the lenses of the circle of least confusion, is independent of the angle between the axes of the cylinders, as was shown in § 285.

The divided circle is now set to $5°$, $10°$, ... from its zero position, and the distances v_1 and v_2 of the two focal lines from the lenses are found. The line corresponding to v_2 moves towards the lenses and the other away from them. Each line turns about the axis of the system through $\frac{1}{2}\phi$ when the rotatable lens turns through ϕ. The least value of f_2 is $\frac{1}{2}f_0$, and the greatest value of f_1 is infinity, these values occurring when $\phi = \frac{1}{2}\pi$. It is obvious that we need consider only those values of ϕ for which $0 \lessgtr \phi \lessgtr \frac{1}{2}\pi$. It will thus be possible to obtain values for f_2 for *all* values of ϕ within this range. As regards f_1, the focal line will move away from the lenses until $f_1 = u$, which occurs when $\sin \phi = 1 - f_0/u$; the focal line is then at infinity. If ϕ be further increased, f_1 becomes greater than u, and the focal line is *virtual* and moves in from infinity until, when $\phi = \frac{1}{2}\pi$, it coincides with the object P, since f_1 is infinite when

$\phi = \frac{1}{2}\pi$. Thus, only a limited range of ϕ is available in the case of this focal line, unless an auxiliary converging lens be employed to locate it.

The observations are repeated with the angles $-5°$, $-10°$,..., and the mean values obtained from the two sets of observations are found.

The powers $1/f_1$ and $1/f_2$ are found by (7) for each value of ϕ, and are compared with the powers given by (5).

Cross-wires stretched across the opening P in the metal plate C (Fig. 173) are useless unless they be carried in a fitting which allows them to be rotated in the plane of the plate. Their directions will need a fresh adjustment for each new setting of the divided circle R (Fig. 175). For each focal line, a double adjustment is required. The screen H (Fig. 173) is adjusted on the bench and the fitting holding the cross-wires is rotated until the focal line "image" of the appropriate wire be sharply defined.

287. Practical example. Each lens had one power zero and the other nominally two dioptres (2 D). For $u = 107$ cm., $w = 32{\cdot}70$ cm. Hence

$$1/f_0 = 1/u + 1/w = {\cdot}00935 + {\cdot}03058 \text{ cm.}^{-1} = 3{\cdot}993 \; D.$$

With $u = 107$ cm., readings for v_1, v_2 were taken at intervals of $5°$ for both $\phi = n°$ and $\phi = -n°$. A few mean values are given.

ϕ	v_1 cm.	v_2 cm.	$1/f_1$ D	$1/f_2$ D	$\frac{1}{2}(1/f_1+1/f_2)$ D	$(1-\sin\phi)/f_0$ D	$(1+\sin\phi)/f_0$ D
$0°$	32·70	32·70	3·99	3·99	3·99	3·99	3·99
10	42·37	26·35	3·29	4·73	4·01	3·30	4·69
20	57·90	22·55	2·66	5·37	4·02	2·63	5·36
30	87·35	19·95	2·08	5·95	4·01	2·00	5·99

The last two columns give values of $1/f_1$, $1/f_2$ as calculated by (5). By (6), $\frac{1}{2}(1/f_1 + 1/f_2) = 1/f_0$. There is fair agreement between observed and calculated values. For greater accuracy it would be necessary to allow for the thickness of the system.

EXPERIMENT 54. **Focal lines due to sloped lens.**

288. Introduction. If a luminous point P be placed on the axis of a lens with spherical faces and of small aperture, the emergent rays meet in a point Q on the axis. If P move perpendicular to the axis through an infinitesimal distance to P', the emergent rays meet in a *point* Q' very near Q, where $Q'Q$ is also perpendicular to the axis. When $P'P$ subtends a finite angle at the lens, this result is no longer even approximately true. We will therefore examine the action of a lens upon an incident beam when the chief ray of the beam makes any finite angle with the axis. For simplicity, we shall consider only a thin lens.

If the chief ray of the incident beam meet the first surface of a very thin lens at its vertex A, it meets the second surface very near the other vertex B. Whatever the thickness of the lens, the plane containing the chief ray and the axis of the lens is a plane of symmetry for each face and is, therefore, one of the principal sections of the emergent wave front. The other principal section is perpendicular to this plane and contains the emergent chief ray. Since the lens is very thin, we may treat this section as containing also the *incident* chief ray. We have, thus, only to find the action of the lens on rays which pass through it in these two planes.

289. Method of least time. We first find the optical distance between two points on a ray, when the ray passes obliquely through a parallel plate of glass. Let the plane of Fig. 176 contain the normal AB to the faces of the plate and also the incident ray LA. Let $LACD$ be the path of the ray; since the plate has parallel faces, CD is parallel to LA. Let $AB = c$ be the thickness of the plate. Let θ, θ' be the angles of incidence and refraction at A, and μ the index of the glass. Let CM, DN be perpendiculars upon the straight line LAN. Then $MAC = \theta - \theta'$. The time of passage of light from A to C is $\mu.AC/V$, where V is the velocity of light in air. In this time, light would travel a distance $\mu.AC$ in air; $\mu.AC$ is known as the "optical distance" from A to C. Let the optical distance

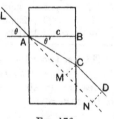

FIG. 176

from L to D be d. Then, adding and subtracting AM, and noting that $CD = MN$ and that $AM = AC\cos(\theta - \theta')$, we have

$$d = LA + AM + MN + \mu.AC - AM = LN + AC\{\mu - \cos(\theta - \theta')\}$$
$$= LN + AC\{\mu - \cos\theta\cos\theta' - \sin\theta\sin\theta'\}.$$

Now $\sin \theta = \mu \sin \theta'$, and hence

$$\mu - \sin \theta \sin \theta' = \mu - \mu \sin^2\theta' = \mu \cos^2\theta'.$$

But $AC \cos \theta' = AB = c$, and hence

$$d = LN + c\,(\mu \cos \theta' - \cos \theta). \quad \dots\dots\dots\dots(1)$$

Let the A- and B-faces of the lens have radii a, b, counted positive when convex, and let these faces meet in a sharp edge of radius h. Let OX (Fig. 177) be the axis of the lens. When the thickness $AB (= c)$ is very small compared with a or b, we have by (23), (27), § 132,

$$c = \tfrac{1}{2}h^2\,(1/a + 1/b) = \tfrac{1}{2}h^2/\{f\,(\mu - 1)\}, \quad \dots\dots\dots(2)$$

where f is the focal length.

Let POQ_1, a ray passing through the centre of the lens, be taken as the chief ray. Let PK_1Q_1 be the path of a ray in the plane containing OP and the axis OX, and let this ray meet PO in Q_1. Let $OP = u$,

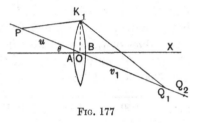

Fig. 177

$OQ_1 = v_1$, and let $1/f_1$ be the power of the lens in this plane. Then

$$PK_1 = \{u^2 + h^2 - 2uh \sin \theta\}^{\frac{1}{2}} = u\,\{1 + (- 2\sin \theta . h/u + h^2/u^2)\}^{\frac{1}{2}}$$
$$= u\{1 + \tfrac{1}{2}(- 2\sin\theta . h/u + h^2/u^2) - \tfrac{1}{8}(- 2\sin\theta . h/u + h^2/u^2)^2 - \dots\}.$$

Since $1 - \sin^2\theta = \cos^2\theta$, we have, as far as terms in h^2,

$$PK_1 = u - h \sin \theta + \tfrac{1}{2}\cos^2\theta . h^2/u.$$

Similarly,

$$Q_1K_1 = \{v_1{}^2 + h^2 + 2v_1 h \sin \theta\}^{\frac{1}{2}} = v_1 + h \sin \theta + \tfrac{1}{2}\cos^2\theta . h^2/v_1.$$

By (1) and (2), the optical distance from P to Q_1 along the chief ray is $u + v_1 + c\,(\mu \cos \theta' - \cos \theta)$, or

$$u + v_1 + \tfrac{1}{2}h^2\,(\mu \cos \theta' - \cos \theta)/\{f\,(\mu - 1)\}.$$

Equating this to $PK_1 + Q_1K_1$, we have

$$\cos^2\theta(1/u + 1/v_1) = \cos^2\theta\,(1/f_1) = (\mu\cos \theta' - \cos\theta)/\{f(\mu - 1)\}. \;\dots(3)$$

Let the plane through PO perpendicular to the plane of the diagram cut the edge of the lens in K_2; then OK_2 is perpendicular to OP. Let the rays which pass through the lens in this plane meet OP in Q_2, where $OQ_2 = v_2$, and let $1/f_2$ be the power of the lens in this plane. Then

$$PK_2 = \{u^2 + h^2\}^{\frac{1}{2}} = u + \tfrac{1}{2}h^2/u, \qquad Q_2K_2 = v_2 + \tfrac{1}{2}h^2/v_2.$$

The optical distance from P to Q_2 along the chief ray is

$$u + v_2 + \tfrac{1}{2}h^2 (\mu \cos \theta' - \cos \theta)/\{f(\mu - 1)\}.$$

Equating this to $PK_2 + Q_2K_2$, we have

$$1/u + 1/v_2 = 1/f_2 = (\mu \cos \theta' - \cos \theta)/\{f(\mu - 1)\}. \quad ...(4)$$

By (3) and (4),

$$\frac{\cos^2\theta}{f_1} = \frac{1}{f_2} = \frac{\mu \cos \theta' - \cos \theta}{(\mu - 1)f} = \frac{\omega}{f}, \quad............(5)$$

where $\omega = (\mu \cos \theta' - \cos \theta)/(\mu - 1)$. The following table shows how ω depends upon μ and θ. Throughout, ω is greater than 1. The values of $1/\omega$ are *printed in italics*. For $\mu = 1\cdot5$, the values of ω for 70°, 80°, 90°, are 1·654, 1·916, 2·236, and those of $1/\omega$ are ·604, ·522, ·447 respectively.

θ $\backslash$ μ	1·50	1·52	1·54	1·56	1·58	1·60
0°	1·000	1·000	1·000	1·000	1·000	1·000
	1·000	*1·000*	*1·000*	*1·000*	*1·000*	*1·000*
10°	1·010	1·010	1·010	1·010	1·010	1·010
	·990	*·990*	*·990*	*·990*	*·990*	*·990*
20°	1·041	1·041	1·040	1·040	1·040	1·039
	·960	*·961*	*·961*	*·962*	*·962*	*·962*
30°	1·096	1·095	1·094	1·092	1·091	1·090
	·912	*·913*	*·914*	*·916*	*·917*	*·918*
40°	1·179	1·176	1·173	1·170	1·168	1·165
	·848	*·850*	*·852*	*·854*	*·856*	*·858*
50°	1·294	1·288	1·284	1·279	1·274	1·270
	·773	*·776*	*·779*	*·782*	*·785*	*·788*
60°	1·449	1·441	1·432	1·424	1·416	1·409
	·690	*·694*	*·698*	*·702*	*·706*	*·710*

The table shows that small changes in μ have little effect on the value of ω for a given angle. Spectacle lenses have an index between 1·52 and 1·53, and for them it may suffice to take the value of ω or of $1/\omega$ for $\mu = 1\cdot52$.

290. Practical details. The lens L (Fig. 178) is fixed to the plate C, which slides on the *square* shaft E and can be secured by a screw in the collar D. The ends of E are cylindrical and the shaft turns stiffly in the cradle F. The cradle is connected to the divided circle H by the tube J which fits and turns about the vertical shaft N rising from the base M. The base can slide on the optical bench. The circle is read by the pointers K. The plane of the plate C is parallel to the axis of N and is vertical when in use.

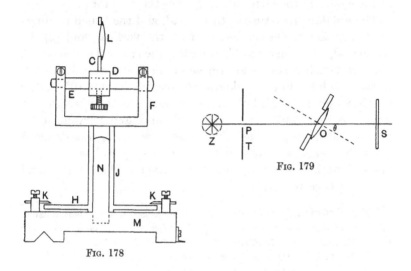

FIG. 179

FIG. 178

The measurements are made on a bench. A plate T (Fig. 179) carries fine vertical and horizontal cross-wires intersecting in P and illuminated by the flame Z. The directions of the wires must be accurately adjusted if well-defined focal lines are to be obtained. The centre of the lens is at O, and S is the observing apparatus, either a fine ground glass screen or an eye-piece.

To adjust the collar D on the shaft E (Fig. 178), the axis of the lens is set parallel to the bench, and the distance of some fitting on the bench from the face A is measured. The cradle is then turned through $180°$ about N, and the distance of the same fitting from the other face B is measured. If these two distances be made equal by adjusting D, the central plane of the lens contains the axis of N. An image of the cross-wires is now formed on S, with (1) the A-face, and (2) the B-face, towards S. The two images can be made to have the same position by turning the plate C and the shaft E about the axis of E. The lens is now in adjustment.

The plate T is adjusted, by the method of § 142, or otherwise, so that the line PO is parallel to the bench.

The lens holder is now placed so that PO is about $2f$, and the cradle is turned so that both wires can be simultaneously focused on S. The reading of S is taken and also the reading of the circle H; the latter reading gives the zero for θ.

The cradle is now turned to $\theta = 10°$, and the bench readings of S are taken when the image of (1) the vertical, and (2) the horizontal, wire is in focus. The cradle is then turned to $\theta = -10°$, and the bench readings are repeated. For each wire, the mean of the bench readings is taken as the reading for $\theta = 10°$. Other readings, for $\theta = 20°$, $30°$, ..., are taken. As θ increases, the images become more and more ill-defined and very accurate results will be impossible. The distances u, v_1, v_2, are deduced from the bench readings of the carriages by aid of a rod of known length. The vertical and horizontal focal lines correspond to v_1 and v_2 respectively.

291. Practical example. G. F. C. Searle used spectacle lens, 0·22 cm. thick, of nominal power 2·75 D. By § 171 (*First Method*), index was found 1·529. Throughout $u = 75·62$ cm., $1/u = ·01322$ cm.$^{-1} = 1·322$ D. For $\theta = 0$, $v = 71·80$ cm., $1/v = 1·393$ D. Hence power is

$$1/f = 1/u + 1/v = 2·715 \text{ dioptres.}$$

Angle θ was varied by steps of 10°, and v_1, v_2 were measured; a few values are given. Powers $1/f_1$, $1/f_2$ were found from $1/u$ and $1/v_1$, $1/v_2$. By (5), f_1/f_2 equals $\cos^2 \theta$. Sixth column gives ω/f, calculated for $\mu = 1·53$ by Table of § 289. By (5), $1/f_2$ equals ω/f.

θ	v_1 cm.	v_2 cm.	$1/f_1$ D	$1/f_2$ D	ω/f D	f_1/f_2	$\cos^2 \theta$
0°	71·80	71·80	2·715	2·715	2·715	1·000	1·000
20	53·34	66·44	3·197	2·827	2·825	·884	·883
40	25·26	53·66	5·281	3·186	3·189	·603	·587
50	14·91	46·14	8·029	3·489	3·491	·435	·413

EXPERIMENT 55. **Curvature of the field.**

292. Introduction. For any given position of the luminous point, we can find, by § 289, the positions of the two focal lines formed by a thin lens. But in practice we often desire to form a picture, as sharp as possible, of an assemblage of points, and not merely of a single point. We may, for instance, desire to obtain a sharp image of every point of a landscape upon a flat photographic plate, or to make a photographic copy of a flat diagram. It is thus convenient to have curves which exhibit the performance of a lens in this respect.

293. Image surfaces. Let OC (Fig. 180) be the axis of a lens. Let C be a luminous point, D its real image. Let planes through C, D, perpendicular to OC, cut the plane of the figure in CH, DK. Let P, in the horizontal plane of the diagram, be a luminous point on CH. We take as the chief ray from P the

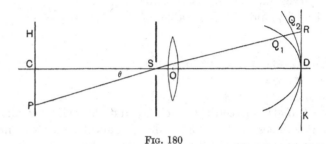

Fig. 180

ray which in its course passes through the centre S of a small stop. This stop may be outside the lens on either side, as in Fig. 180, or it may be between the components of a lens system, as in photographic lenses. Let the chief ray meet DK in R. The focal lines corresponding to P will cut the chief ray in Q_1 and Q_2. For a thin converging lens, Q_1, Q_2 will lie between R and the lens, and the vertical focal line through Q_1 will be further from R than the horizontal focal line through Q_2. If P move along CH, the points Q_1, Q_2 will describe curves DQ_1, DQ_2, and corresponding to the whole plane CH there will be curved surfaces which are formed by the revolution of DQ_1, DQ_2 about OC and are called the primary and secondary image surfaces. The "defect" of the system of lens and stop which causes the image surfaces to deviate from the plane DK is called "curvature of the field."

When the lens is *thin* and the stop is in *contact* with it, so that the chief ray passes through O, the centre of the lens, POR is a straight line. Putting $OP = u$, $OQ_1 = v_1$, $OQ_2 = v_2$, we can find, by (3) and (4) of § 289, the positions of Q_1, Q_2 for each position of P on CH. Fig. 181 shows the curves in three cases. For the curves which pass through G, where $OG = 2f$, the straight path of the object point cuts the axis at C to the left of O, where $OC = 2f$. For the curves which pass through the focus F, OC is infinite. The other two curves are formed when $OC = f$, so that P moves in the focal plane. The curves have been drawn for $\mu = 1\cdot5$, by aid of the table of § 289.

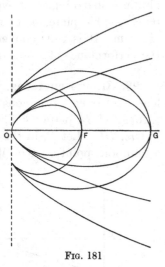

294. Curvature of image surfaces at vertex. In some instruments, such as telescopes, the

Fig. 181

greatest angle between the chief ray and the axis is small, and then it suffices to find the radii of curvature of the image surfaces at the vertex D (Fig. 180). In finding these radii for a single thin lens, we use expansions in powers of θ (§ 289) and neglect θ^3 and higher powers in comparison with unity. To this order, $\sin \theta = \theta$, $\sin \theta' = \theta'$, and $\sin \theta' = \sin \theta/\mu$ becomes $\theta' = \theta/\mu$. Further,

$$\cos \theta = 1 - \tfrac{1}{2}\theta^2, \qquad \cos \theta' = 1 - \tfrac{1}{2}\theta'^2 = 1 - \tfrac{1}{2}\theta^2/\mu^2.$$

Hence $\quad \omega = (\mu \cos \theta' - \cos \theta)/(\mu - 1) = 1 + \tfrac{1}{2}\theta^2/\mu. \quad \text{......(1)}$

Since, as far as θ^2, $1/\cos^2\theta = (1 - \tfrac{1}{2}\theta^2)^{-2} = 1 + \theta^2$, we have

$$\omega/\cos^2\theta = (1 + \tfrac{1}{2}\theta^2/\mu)(1 + \theta^2) = 1 + \tfrac{1}{2}\theta^2(2\mu + 1)/\mu. \quad...(2)$$

Hence, if we put V for v_1 or v_2, each of the equations (3) and (4) of § 289 can be written in the form

$$1/u + 1/V = (1 + n\theta^2)/f,$$

where $n = (2\mu+1)/(2\mu)$ when $V = v_1$, and $n = 1/(2\mu)$ when $V = v_2$.

Let P be on the line CH (Fig. 180), where $OC = h$. Then

$$1/u = \cos \theta/h = (1 - \tfrac{1}{2}\theta^2)/h.$$

Since $1/V = (1 + n\theta^2)/f - 1/u$, we have

$$\frac{1}{V} = \frac{1}{f} - \frac{1}{h} + \theta^2\left(\frac{n}{f} + \frac{1}{2h}\right) = \frac{h-f}{hf}\left\{1 + \theta^2 \cdot \frac{2hn + f}{2(h-f)}\right\}.$$

Hence $$V = \frac{hf}{h-f}\left\{1 - \theta^2 \cdot \frac{2hn+f}{2(h-f)}\right\}. \quad \ldots\ldots\ldots\ldots(3)$$

When $\theta = 0$, $V = hf/(h-f) = OD$, the points Q_1, Q_2 coinciding with D.

If T be a point on the curve given by (3), and TN be the perpendicular from T upon the axis, we have

$$TN = V \sin\theta = hf\theta/(h-f),$$

$$DN = OD - V\cos\theta = \frac{hf}{h-f}\left[1 - \left\{1 - \theta^2 \cdot \frac{2hn+f}{2(h-f)}\right\}\{1 - \tfrac{1}{2}\theta^2\}\right]$$

$$= \frac{(2n+1)\,h^2 f\theta^2}{2\,(h-f)^2}.$$

The radius of curvature, ρ, at the vertex D, is the limit of $\tfrac{1}{2}TN^2/DN$ as θ approaches zero. Hence

$$\rho = f/(2n+1).$$

Inserting the value of n corresponding to each case, we find that, if ρ_1, ρ_2 be the radii of curvature of the primary and secondary image surfaces at D,

$$\rho_1 = \mu f/(3\mu+1), \qquad \rho_2 = \mu f/(\mu+1). \quad \ldots\ldots\ldots(4)$$

These radii depend only on the lens and not on the distance of the line CH from the lens. They are not changed by diminishing the aperture of the lens. Thus, the radius of curvature of the large primary curve at G (Fig. 181) is the same as that of the small primary curve at F, and similarly for the secondary curves.

It is thus impossible to form upon a flat screen a perfect image of a flat object—such as a lantern slide—or of a distant landscape by a simple lens. When two or more lenses separated by intervals are used, it is possible, by proper design, to make the two image surfaces practically coincide with the plane DK over a considerable area.

295. Forms of image surfaces. We shall now show that the secondary curve, i.e. the curve in which the secondary surface is cut by a plane containing the axis of the lens, is a conic section.

Since $\mu^2\cos^2\theta' = \mu^2 - \mu^2\sin^2\theta' = \mu^2 - \sin^2\theta$, and since $u = h/\cos\theta$, we have, by (4), § 289,

$$\frac{1}{v_2} + \cos\theta\left\{\frac{1}{h} + \frac{1}{(\mu-1)f}\right\} = \frac{(\mu^2 - \sin^2\theta)^{\frac{1}{2}}}{(\mu-1)f}. \quad \ldots\ldots(5)$$

If we multiply by v_2, and then replace $v_2 \cos \theta$ by x and $v_2 \sin \theta$ by y, we have, after squaring,

$$\left[1 + x \left\{ \frac{1}{h} + \frac{1}{(\mu - 1)f} \right\} \right]^2 = \frac{\mu^2 (x^2 + y^2) - y^2}{(\mu - 1)^2 f^2}.$$

Hence

$$x^2 \left\{ 1 - \frac{2f}{(\mu + 1) h} - \frac{(\mu - 1) f^2}{(\mu + 1) h^2} \right\}$$
$$+ y^2 - 2xf \left\{ \frac{1}{\mu + 1} + \frac{(\mu - 1) f}{(\mu + 1) h} \right\} = \frac{\mu - 1}{\mu + 1} f^2. \quad \dots(6)$$

Thus the curve is a conic section. It cuts Oy, which passes through the centre of the lens, in the fixed points $y = \pm f \{(\mu - 1)/(\mu + 1)\}^{\frac{1}{2}}$. When $h > f$, i.e. when the image point at D is *real*, the curve is an ellipse. If h be infinite, the curve is the circle

$$\left(x - \frac{f}{\mu + 1} \right)^2 + y^2 = \left(\frac{\mu - 1}{\mu + 1} + \frac{1}{(\mu + 1)^2} \right) f^2 = \left(\frac{\mu f}{\mu + 1} \right)^2. \quad \dots(7)$$

The centre is on Ox at distance $d = f/(\mu + 1)$ from O towards D and the radius is $r = \mu f/(\mu + 1)$. This circle passes through F in Fig. 181.

If $h = f$, we have the parabola

$$y^2 = \frac{2\mu f}{\mu + 1} \left\{ x + \frac{(\mu - 1) f}{2\mu} \right\}. \quad \dots\dots\dots\dots(8)$$

The vertex is on Ox at distance $\frac{1}{2} (\mu - 1) f/\mu$ from O on the side opposite to D and the concavity is towards D. The distance from focus to vertex is $\frac{1}{2} \mu f/(\mu + 1)$. This parabola is shown in Fig. 181, for $\mu = 1 \cdot 5$.

If we treat (3), § 289, in a similar manner and write $\cos^2 \theta = x^2/(x^2 + y^2)$, we obtain an equation in x and y of the sixth degree, containing no constant term. Hence, when $x = 0$, $y = 0$, and thus the primary curve passes through O as in Fig. 181.

296. Experimental details. Two straight lines CM, CH (Fig. 182) meeting at right angles, are set out on a horizontal drawing board. The cross-wire holder G has a vertical plane face; the vertical and horizontal wires lie in this plane. On this face is drawn a vertical line through P, the intersection of the wires, to allow P to be projected on the board. The centre of the lens is vertically above a point O on CM and its axis is parallel to CM. The locator J has a vertical plane face with a

vertical line ruled on it, to allow the point located to be projected on the board, and J is adjusted so that Q, one of the focal

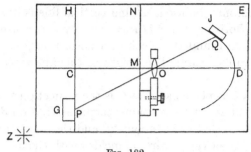

FIG. 182

lines, is in focus on the vertical line ruled on J. For each of
a series of positions of G along CH, the two focal lines are
located by J, and their projections are marked on the board.

When a flame is not available for illuminating the G-wires,
J must have cross-wires which are viewed by the eye at E, since
the illumination will not be sufficient to give satisfactory focal
lines on card or ground glass at Q. In this case, J may be a
vertical metal strip (Fig. 183) with a heavy base. The strip is
pierced by a hole one cm. in diameter, and across this hole on
the face A are fixed fine cross-wires.

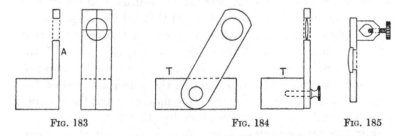

FIG. 183   FIG. 184   FIG. 185

Since the curves will be badly unsymmetrical unless the axis
of the lens be nearly parallel to CM, a suitable lens holder is
used. The lens, of focal length 6 cm., is "edged" by an optician,
and is fitted into a recess, 1·8 cm. in diameter (turned out in
a lathe), in a metal strip with plane faces (Fig. 184). The strip
is fixed to a rectangular block T by a screw which allows the
height of the lens to be adjusted. If MN be drawn perpendicular

to CM and if the face of T be in contact with MN, as in Fig. 182, the axis of the lens will be parallel to CM.

For G, a rectangular block, with a white card fixed to a vertical face, may be used. A fine vertical line is drawn on the card and a horizontal line is furnished by a *fine* wire or black thread passing round the block in a horizontal plane and kept tight by a rubber band. The height of this line is thus capable of adjustment.

The block G is placed so that C is the projection of P, and the horizontal line is set to the same height as that on J. The line MN is ruled so that CO is about $2f$, and the lens is adjusted so that the image of the G-wires coincides with the J-wires, when the vertical line on J intersects CM in D. (A straight-edge, held by weights, guides the base T along MN during this adjustment.) Then G is moved along CH by steps of 1 cm. and at each stage the positions of the horizontal and vertical focal lines are recorded. The coincidence between the horizontal wire of J and the horizontal focal line is tested by parallax and similarly for the vertical line. The projection of the lens centre is the intersection of CD with the projection of PQ_1Q_2.

When a flame is used, it is placed at Z (Fig. 182) and G has an opening across which the wires are stretched. A piece of (finely) ground glass, fixed, with its plane vertical, to a block of wood, is now used for J. A vertical line is drawn on the ground side, which faces O. The lens is set so that (1) the height of the image of the horizontal wire of G above the board equals that of P (the intersection of the wires), and (2) the image of the vertical wire is vertically above D when P is vertically above C.

Smooth curves are drawn through the points marked on the board. The radii of curvature at D are measured and their values are compared with the values given by (4), § 294.

For the curves corresponding to $h = \infty$, a horizontal collimator (Fig. 162) with vertical and horizontal cross-wires, adjusted for "infinity," and backed by a flame, is used. The lens at O, with $f = 15$ to 20 cm., is mounted in a plate (Fig. 185) fitted with a clamp for fixing to a horizontal rod carried by a support which stands on the drawing board. The rod should be long enough to allow J to be brought close to O. The lens projects slightly from the mount so that its centre can be projected on

the board by aid of a set-square. Ground glass is used to locate
the focal lines. The collimator is set at various angles to the
axis of the lens (compare Fig. 186), and the positions of the two
focal lines are recorded for each setting. If the whole of the
horizontal line cannot be sharply focused, the vertical line on
J is set on the part that is sharp. A circle is drawn through the
points on the board corresponding to the horizontal focal line,
and its centre and radius are found. The results are compared
with those given by (7).

297. Practical example. Miss Mary Taylor observed the secondary
(horizontal) focal line, using a collimator. The lens was equi-convex, 0·19 cm.
thick, with $f = 15\cdot33$ cm., $\mu = 1\cdot53$. Radius of circle was $r = 9\cdot15$ cm. and
distance of its centre from centre of lens was $d = 6\cdot15$ cm. Hence, by measurement, $r + d = 15\cdot30$.

By § 295, $r = \mu f/(\mu + 1) = 1\cdot53 \times 15\cdot33/2\cdot53 = 9\cdot27$ cm.

Also, $d = f/(\mu + 1) = 15\cdot33/2\cdot53 = 6\cdot06$ cm.

Hence, by calculation, $r + d = 15\cdot33$ cm.; this distance necessarily equals f.

EXPERIMENT 56. **Curvature of field of lens system and
effect of position of stop.**

298. Method. The apparatus is arranged for the investiga-
tion of the "defects" of astigmatism and curvature of the field,
when, as in landscape photography, the objects are at very
great distances from the lens.

The lens L (Fig. 186) which is to be tested, is fixed, with its
axis horizontal, to a suitable stand. A screen J, of finely ground

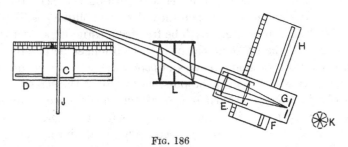

FIG. 186

glass, is mounted on a carriage C, sliding on a short optical
bench. The ground surface faces L, and on it is ruled a horizontal
line divided to half centimetres. The frame supporting the

ground glass is clamped to a vertical rod rising from the carriage (see Fig. 131). To obtain the effect of distant objects, a collimator EG is used. A projection lens E, with its aperture diminished by a stop, is mounted, with its axis horizontal, on a board F. A pair of *fine* horizontal and vertical wires, stretched across an opening in a metal plate, intersect at G (see § 266). The directions of the wires must be accurately adjusted if well-defined focal lines are to be obtained. Since it is convenient to be able to move E parallel to itself, the board is clamped to a cross-rod fixed to the vertical rod of a carriage sliding on the bench H, and the axis of E is adjusted to the same height as that of L. A flame K, placed at a safe distance behind the cross-wires, gives the necessary illumination. An electric lamp may be used if a piece of ground glass be placed between the lamp and the wires. The ground side of the glass should be slightly oiled so that it may be *nearly* transparent.

The benches D, H and the stand supporting L rest on a table. Care must be taken that the axis of L is parallel to the direction of the bench D, and also that the plane of the screen J is perpendicular to that direction. If L be suitably mounted, this setting, which is important if symmetrical curves are to be obtained, may be made by aid of set-squares and straight-edges.

The collimator is first placed with its axis approximately coincident with that of the lens L. The carriage C is then adjusted so that the cross-wires are sharply focused on the screen J, and the bench reading of the carriage is taken and also the reading of the image of the vertical wire on the scale on the screen. The collimator is then moved to an oblique position, as in Fig. 186. Unless the lens L be free from astigmatism for this direction of the incident beam, it will no longer be possible to adjust C so that sharp images of *both* wires are formed on the screen. It will, however, be possible to adjust the screen so as to obtain either a sharp image of the vertical wire, formed by primary focal lines, or a sharp image of *a part* of the horizontal wire, formed by secondary focal lines. The two readings of C will differ from each other, and one, at least, will differ from the reading found for the symmetrical position of the collimator, unless the lens be free from astigmatism and from curvature of field. The observations are repeated for a number of positions of the collimator on either side of the symmetrical position, the obliquity being increased until the mounting of the lens L cut off the light.

Unless the lens L be well corrected, it will be necessary to stop it down considerably to obtain a sharp image of the vertical wire when the obliquity is large. Even so, it will not always be possible to obtain a sharp image of the *whole* of the horizontal wire, though it will be possible to get a sharp image of any particular point on that wire by suitable adjustment of the screen J.

Two curves, V and H, are then plotted. In the primary curve V, the abscissa is the scale reading on the screen J of the sharp image of the vertical wire, and the ordinate is the bench reading of the carriage C. Each, of course, may be taken from some convenient zero. In the secondary curve H, the abscissa is the scale reading on the screen J of the *sharply focused point* of the image of the horizontal wire, and the ordinate is the reading of the carriage.

Fig. 187 was obtained in the test of a lantern projection lens of 21·7 cm. focal length. The two image surfaces corresponding

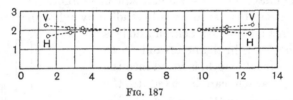

Fig. 187

to V and H are slightly curved but in *opposite* directions. The primary V-surface, on which the image of the vertical wire is focused, is concave towards the lens. With a single thin lens at L, the stop being in contact with the lens, both image surfaces are concave towards the lens, the primary V-surface being more curved than the secondary H-surface.

The diagram shows the improvement that can be effected by suitable design of a lens. The radius of each curve, if treated as a circular arc, is about 55 cm. If a thin lens, with $f = 21·7$ cm. and $\mu = 1·53$, had been used, the radii of the curves V and H would have been 5·9 and 13·1 cm. respectively. (See (4), § 294.)

In this test, the parallel light entered the projection lens at the proper end. If such a lens be turned end for end, so that the parallel light enters at the wrong end, the astigmatic difference and the curvature of the field will be much more marked.

299. Effect of position of stop. The position of the stop has much influence upon the form of the image surfaces. In Fig. 188, parallel light passes first through the stop S and then through the lens. In Fig. 189, it passes first through the lens

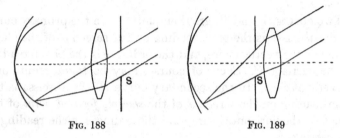

FIG. 188   FIG. 189

and then through S. For a given direction of incidence, the light passes through different parts of the lens in the two cases, and the image surfaces will consequently differ.

The effect of the stop is easily studied by aid of the apparatus shown in Fig. 190. An achromatic lens L, about 5 cm. in diameter and 15 to 20 cm. in focal length, is fitted into the frame F, which can be fixed to the fittings of an optical bench by the clamp C. The stop plate S slides on two rods R, R, and is secured by a screw in the collar K. The lens is used in place of the lens L in Fig. 186, and the observations are made as in § 298. In one set of readings the stop is—say —3 cm. in front of the lens, and in another it is 3 cm. behind the lens. If time allow, a third set may be taken with the stop in contact with the lens.

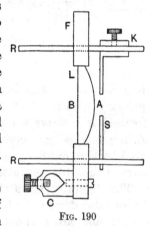

FIG. 190

If the lens be not equi-convex, the light should enter it by the more curved surface, to diminish spherical aberration.

300. Practical example. Mr A. J. Smith used plano-convex achromatic lens L, 5·4 cm. in diam., with $f = 18·1$ cm., fitted as in Fig. 190. Diam. of hole in $S = 1·25$ cm. Light entered L by convex face A. Stop placed (1) 3 cm. in front of A, so that light entered L through S, (2) in contact with A, (3) 3 cm. behind face B. Screen readings of images and bench readings of screen are given in cm. for case (1). Zero of bench D (Fig. 186) was towards L.

Vertical	{Screen	4·96	5·70	7·16	10·30	13·77	15·15	16·04
line	{Bench	16·96	19·17	21·41	23·00	21·33	19·06	15·79
Horizontal	{Screen	3·85	5·12	7·00	10·30	13·90	15·73	17·40
line	{Bench	19·96	21·19	22·28	23·00	22·22	21·26	19·38

In Fig. 191, curves for cases (2), (3) are added; fresh zero used for each case. V and H correspond to vertical and horizontal images. Each curve is concave towards L and V-curve is, in each case, more curved than its fellow H-curve.

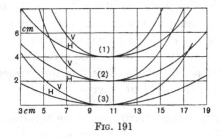

FIG. 191

Radii, ρ_V, ρ_H for curves at vertices are

ρ_V, (1) 4·1, (2) 5·5, (3) 7·0 cm.; ρ_H, (1) 8·0, (2) 10·6, (3) 18·0 cm.

CHAPTER XII

INTERFERENCE AND POLARISATION BY REFLEXION

301. Introduction. The experiments described in the earlier chapters depend on those "laws" of reflexion and refraction which specify the directions of the rays corresponding to reflected and refracted waves, if such exist. These "laws" depend on the facts (1) that light is propagated by waves of some kind, and (2) that the speed with which the reflected and refracted waves sweep along the interface is the same as for the incident wave (§§ 13, 14). They do not depend upon the nature of the waves. The reasoning is merely geometrical and cannot prove the existence of the waves or give their amplitudes and phases.

We will now apply the electromagnetic theory to find the amplitudes and phases of the reflected and refracted waves in terms of those of the incident wave.

302. Boundary conditions. A plane interface S separates two media (K), (K') of inductive capacities K, K'; each has permeability μ. An incident wave advancing in (K) meets S and gives rise to a reflected wave in (K) and a refracted wave in (K').

The electromagnetic principles lead to four boundary conditions which must be satisfied at the interface.

Since the media are non-conducting, if there be no charge on S before the wave meets it, there will be none afterwards. Now take a very short cylinder such that S lies between, and parallel to, the plane ends, and let D_n, D_n' be the normal components of the electric displacement on the ends, each measured in the *same* direction *in space*, so that one is outward and the other inward. Then, by relation I, § 4, we have $D_n - D_n' = 0$. Hence, at S:

I. The normal electric displacement is continuous.

Similarly, by relation II, § 4:

II. The normal magnetic induction is continuous.

Let the plane of Fig. 192 cut the plane S at right angles in TT. Take a rectangle $LMPQ$ in the plane of the figure, with LQ,

MP normal to S; let LM, QP lie on opposite sides of TT and be infinitely close to TT. Let E_t, $E_t{}'$ be the components of the electric force E along LM, QP. Since LQ, MP are infinitesimal, the voltage in the circuit is $(E_t - E_t{}')LM$.

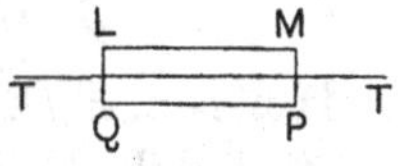

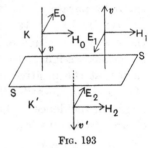

Fig. 192

The flux of induction through the infinitesimal area cannot change at a finite rate, and hence, by relation III, § 4, $E_t - E_t{}' = 0$. Thus:

III. The tangential electric force is continuous.

Similarly, by relation IV, § 4:

IV. The tangential magnetic force is continuous.

303. Normal incidence. The incident wave, with electric and magnetic forces E_0, H_0, and velocity v, advances towards the interface SS (Fig. 193), E_0, H_0 being parallel to S. We take the positive direction of H_1 in the reflected wave and of H_2 in the transmitted wave to be the same as that of H_0. The velocities of the waves in (K), (K') are v, v'. Since, by § 8, right-handed rotation through $\frac{1}{2}\pi$ about the direction of propagation turns E round to H, the positive directions of both

Fig. 193

E_0 and E_2 are away from the reader and the positive direction of E_1 is towards him. Applying III and IV, § 302, we have

$$E_0 - E_1 = E_2, \qquad H_0 + H_1 = H_2. \quad\ldots\ldots\ldots(1)$$

By (4), § 8, $\qquad H_0 = E_0/\mu v, \qquad H_1 = E_1/\mu v, \qquad H_2 = E_2/\mu v'.$

But, if V be the velocity in a vacuum, and n, n' be the refractive indices* relative to a vacuum, $1/v = n/V$, $1/v' = n'/V$. Hence

$$H_0 = nE_0/\mu V, \qquad H_1 = nE_1/\mu V, \qquad H_2 = n'E_2/\mu V. \quad\ldots(2)$$

Then, by (1) and (2),

$$E_1 + E_2 = E_0, \qquad nE_1 - n'E_2 = -nE_0. \quad\ldots\ldots(3)$$

Thus $\quad E_1/E_0 = (n' - n)/(n + n'),\ ,\quad E_2/E_0 = 2n/(n + n'). \quad\ldots(4)$

* In §§ 303, 304, to avoid confusion with permeability, refractive index is denoted by n.

By (2), $H_1/H_0 = E_1/E_0 = (n' - n)/(n + n')$,

$$H_2/H_0 = n'E_2/nE_0 = 2n'/(n + n').   \ldots\ldots\ldots(5)$$

If $n' > n$, E_1/E_0 is positive, and E_1 is in the opposite direction to E_0, while H_1 is in the same direction as H_0. But, if $n' < n$, E_1/E_0 is negative, and E_1 is in the same direction as E_0, while H_1 is in the opposite direction to H_0. Thus, the reflected wave suffers a reversal of phase if $n' > n$, but not if $n' < n$.

In the transmitted wave, E_2, H_2 have the same directions as E_0, H_0; hence there is no reversal of phase, either when $n' > n$ or when $n' < n$.

304. Oblique incidence. The electric force in the incident wave is not, in general, either in or perpendicular to the plane of incidence. The components in these directions may be considered separately.

Case (i). In Fig. 194, E is in the plane of incidence, which is the plane of the figure, and the positive directions assigned to E_0, E_1, E_2 are such that the positive directions of H_0, H_1, H_2 are towards the reader. The angles of incidence and refraction are θ, θ', and, by Snell's law,

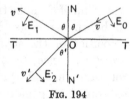

FIG. 194

$$n \sin \theta = n' \sin \theta' \ldots\ldots(5a)$$

By III, § 302, $E_0 \cos \theta - E_1 \cos \theta = E_2 \cos \theta'$, or

$$E_1 \cos \theta + E_2 \cos \theta' = E_0 \cos \theta. \qquad \ldots\ldots\ldots\ldots(6)$$

By IV, § 302, $H_0 + H_1 = H_2$; thus, by (2), $nE_1 - n'E_2 = - nE_0$. Hence, by (5a),

$$E_1 \sin \theta' - E_2 \sin \theta = - E_0 \sin \theta'. \qquad \ldots\ldots\ldots\ldots(7)$$

Solving (6), (7) for E_1, E_2, and using (2), we have

$$\frac{H_1}{H_0} = \frac{E_1}{E_0} = \frac{\sin 2\theta - \sin 2\theta'}{\sin 2\theta + \sin 2\theta'} = \frac{\cos(\theta + \theta') \sin(\theta - \theta')}{\sin(\theta + \theta') \cos(\theta - \theta')}$$

$$= \frac{\tan(\theta - \theta')}{\tan(\theta + \theta')}, \qquad \ldots\ldots\ldots\ldots\ldots(8)$$

$$\frac{nH_2}{n'H_0} = \frac{E_2}{E_0} = \frac{2 \cos \theta \sin \theta'}{\sin(\theta + \theta') \cos(\theta - \theta')}. \qquad \ldots\ldots\ldots\ldots\ldots(9)$$

By (8), $E_1 = 0$ when $\tan(\theta + \theta') = \infty$, or $\theta + \theta' = \frac{1}{2}\pi$; the reflected and refracted rays are then at right angles. These special values of θ, θ' are called the polarising angles. Since $\cos\theta = \sin\theta'$ when $\theta + \theta' = \tfrac{1}{2}\pi$, we have

$$\tan\theta = 1/\tan\theta' = \sin\theta/\sin\theta' = n'/n. \ldots\ldots\ldots(9\,a)$$

Similarly, $\tan\theta' = n/n'$.

Since $-\tfrac{1}{2}\pi < \theta - \theta' < \tfrac{1}{2}\pi$, $\tan(\theta - \theta')$ has the same sign as $\theta - \theta'$ or as $n' - n$. Since $0 < \theta + \theta' < \pi$, it follows that, as θ, θ' increase from zero, $\tan(\theta + \theta')$ becomes infinite only once, viz. when $\theta + \theta' = \tfrac{1}{2}\pi$, and at this value $\tan(\theta + \theta')$ changes sign through infinity. Hence E_1/E_0 changes sign when θ, θ' pass through their polarising values.

The ratio E_2/E_0 is positive for all values of θ.

The normal components of E_0, E_1, E_2 will be found to satisfy condition I, § 302.

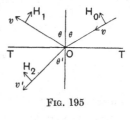

FIG. 195

Case (ii). In Fig. 195, E is perpendicular to the plane of incidence. If the positive directions of E_0, E_1, E_2 be towards the reader, those of H_0, H_1, H_2 are as shown. By III, § 302, $E_0 + E_1 = E_2$,

or

$$E_1 - E_2 = - E_0. \ \ldots\ldots\ldots\ldots\ldots\ldots(10)$$

By IV, § 302, $H_0\cos\theta - H_1\cos\theta = H_2\cos\theta'$; thus, by (2), (5$a$),

$$E_1\cos\theta\sin\theta' + E_2\sin\theta\cos\theta' = E_0\cos\theta\sin\theta'. \ \ldots(11)$$

Solving (10), (11) for E_1, E_2, and using (2), we have

$$\frac{H_1}{H_0} = \frac{E_1}{E_0} = -\frac{\sin(\theta - \theta')}{\sin(\theta + \theta')}, \qquad \frac{nH_2}{n'H_0} = \frac{E_2}{E_0} = \frac{2\cos\theta\sin\theta'}{\sin(\theta + \theta')} \ \ldots(12)$$

The sign of E_1/E_0 is that of $\theta' - \theta$ or of $n - n'$.

The normal components of H_0, H_1, H_2 satisfy condition II, § 302.

305. Retardation due to parallel plate. We now consider the effect of a parallel plate or film of thickness z upon a beam of light. Let the plane of the paper cut the faces of the plate at right angles in AA, BB (Fig. 196). Let μ, μ', μ'' be the refractive indices* of the plate and of the media in contact with it at AA, BB, and let v, v', v'' be the corresponding velocities. Let the ray LO, in the plane of the paper, meet AA in O,

---

* We now return to our usual notation of μ for refractive index.

the angle of incidence being α. Let OS be the reflected and OM, or I, the refracted ray, the angle of refraction being ϕ. The ray I

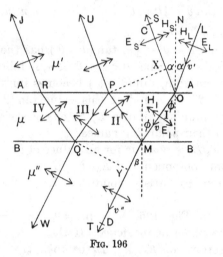

FIG. 196

meets BB in M and gives rise to the reflected ray II and to the refracted ray MT making angle β with the normal, and other rays are formed as in the figure. The rays PU, RJ are parallel to OS, and QW is parallel to MT.

We have $\mu v = \mu' v' = \mu'' v''$, and Snell's law gives

$$\mu \sin \phi = \mu' \sin \alpha = \mu'' \sin \beta.$$

Hence $$(\sin \phi)/v = (\sin \alpha)/v' = (\sin \beta)/v''. \quad \dots\dots\dots\dots(13)$$

Let the time of passage of light from O to P *via* M, or from M to Q *via* P, be t. Then, since $OM = z \sec \phi$,

$$t = 2z (\sec \phi)/v. \quad \dots\dots\dots\dots\dots(14)$$

Draw PX, QY perpendicular to OS, MT; then $XPO = \alpha$, $YQM = \beta$. Since $PO = QM = 2z \tan \phi$, and $\sin \alpha = \sin \phi . v'/v$, we have

$$OX = PO \sin \alpha = 2z \tan \phi . \sin \phi . v'/v = 2z \sec \phi \sin^2 \phi . v'/v \dots.(15)$$

Similarly, $$MY = 2z \sec \phi \sin^2 \phi . v''/v. \quad \dots\dots\dots\dots(16)$$

At the instant when refraction occurs at P, the wave reflected at O has advanced to cut OS in C, where $OC = v't$, and at the same instant PX lies in the front of the wave refracted at P. This line is parallel to the wave front through C. The distance by which X is behind C is $XC = OC - OX$. Since $1 - \sin^2\phi = \cos^2\phi$, we have, by (14), (15),

$$XC = v' \cdot 2z(\sec\phi)/v - 2z\sec\phi\sin^2\phi \cdot v'/v = 2z\cos\phi \cdot v'/v \ldots (17)$$

At the instant when refraction occurs at Q, the wave refracted at M reaches D on MT, where $MD = v''t$, and at the same instant the front of the wave refracted at Q lies behind D by YD, where $YD = MD - MY$. Then

$$YD = 2z\cos\phi \cdot v''/v. \qquad \ldots\ldots\ldots\ldots(18)$$

Each of the distances XC in (μ') and YD in (μ'') is optically equivalent to the distance $2z\cos\phi$ in (μ), i.e. light in (μ) travels over $2z\cos\phi$ in the same time as light in (μ') travels over XC, or light in (μ'') travels over YD.

306. Ratios of electric forces for normal incidence. If, in Fig. 196, the positive direction of H in each wave be normal to the figure and towards the reader, the positive directions of the electric forces $E_L, E_S, \ldots$ in the waves moving along $LO, OS, \ldots$ are as shown by arrows having single barbs. When the incidence is normal, we find, by § 303, $E_S/E_L = (\mu - \mu')/(\mu + \mu')$, and

$$\frac{E_U}{E_{II}} = \frac{2\mu}{\mu + \mu'}, \qquad \frac{E_{II}}{E_I} = \frac{\mu'' - \mu}{\mu + \mu''}, \qquad \frac{E_I}{E_L} = \frac{2\mu'}{\mu + \mu'}.$$

Hence $\qquad a \equiv \dfrac{E_U}{E_S} = -\dfrac{4\mu\mu'(\mu - \mu'')}{(\mu + \mu')(\mu + \mu'')(\mu - \mu')}. \quad \ldots\ldots(19)$

When $\mu'' = \mu'$, $a = -4\mu\mu'/(\mu + \mu')^2$. If $\mu' = 1\cdot5$, $\mu = 1$, as when there is a film of air between two glass plates, $a = -24/25$, and thus the electric force in the U-wave is little less than in the S-wave.

By § 303, $E_{III}/E_{II} = (\mu' - \mu)/(\mu' + \mu)$; since $E_W/E_T = E_{III}/E_I$,

$$b \equiv \frac{E_W}{E_T} = \frac{E_{III}}{E_{II}} \cdot \frac{E_{II}}{E_I} = \frac{(\mu - \mu')(\mu - \mu'')}{(\mu + \mu')(\mu + \mu'')}. \quad \ldots\ldots(20)$$

When $\mu'' = \mu'$, $b = (\mu - \mu')^2/(\mu + \mu')^2$. If $\mu' = 1\cdot5$ and $\mu = 1$, then $b = 1/25$, and thus the electric force in the W-wave is small compared with that in the T-wave.

Since $\qquad E_J/E_U = E_{IV}/E_{II}$, and $E_{IV}/E_{II} = E_{III}/E_I$,

we have $\qquad c \equiv E_J/E_U = E_{III}/E_I = b. \quad \ldots\ldots\ldots\ldots(21)$

307. Ratios of electric forces for oblique incidence.
Case (i). When E is in the plane of incidence and the positive
direction of H in each wave is towards the reader in Fig. 196,
the positive directions of the electric forces are as shown by
the arrows having single barbs, as E_L, E_S.

We have, by (8), (9), $E_S/E_L = \tan(\alpha - \phi)/\tan(\alpha + \phi)$, and

$$\frac{E_U}{E_{II}} = \frac{2\cos\phi\sin\alpha}{\sin(\phi + \alpha)\cos(\phi - \alpha)}, \qquad \frac{E_{II}}{E_I} = \frac{\tan(\phi - \beta)}{\tan(\phi + \beta)},$$

$$\frac{E_I}{E_L} = \frac{2\cos\alpha\sin\phi}{\sin(\alpha + \phi)\cos(\alpha - \phi)}.$$

Hence

$$a_1 \equiv \frac{E_U}{E_S} = -\frac{\sin 2\alpha \sin 2\phi \tan(\phi - \beta)\tan(\phi + \alpha)}{\sin^2(\phi + \alpha)\cos^2(\phi - \alpha)\tan(\phi + \beta)\tan(\phi - \alpha)} \quad ....(22)$$

When ϕ passes through the polarising value given by $\phi + \alpha = \frac{1}{2}\pi$,
or by $\sin\phi = \mu'(\mu^2 + \mu'^2)^{-\frac{1}{2}}$, a_1 becomes infinite and changes
sign, because $\tan(\phi + \alpha)$ changes from $+\infty$ to $-\infty$ when ϕ,
as it increases, passes through this polarising value; for this
polarising value of ϕ, E_U is finite and $E_S = 0$. When ϕ passes
through the polarising value given by $\phi + \beta = \frac{1}{2}\pi$, or by
$\sin\phi = \mu''(\mu^2 + \mu''^2)^{-\frac{1}{2}}$, a_1 again changes sign. When ϕ has
this second polarising value, $E_U = 0$ because $E_{II} = 0$.

By (8), $E_{III}/E_{II} = \tan(\phi - \alpha)/\tan(\phi + \alpha)$, and thus

$$b_1 \equiv \frac{E_W}{E_T} = \frac{E_{III}}{E_I} = \frac{E_{III}}{E_{II}} \cdot \frac{E_{II}}{E_I} = \frac{\tan(\phi - \alpha)\tan(\phi - \beta)}{\tan(\phi + \alpha)\tan(\phi + \beta)} \quad ....(23)$$

It is easily seen that $|b_1| < 1$.

For all values of ϕ, a_1 and b_1 have opposite signs. When ϕ
does not lie between its polarising values, the sign of a_1 is opposite to that of $(\mu'' - \mu)(\mu' - \mu)$. When ϕ lies between these
values, the sign of a_1 is the same as that of $(\mu'' - \mu)(\mu' - \mu)$.

As in § 306, $\qquad\qquad c_1 \equiv E_J/E_U = b_1.$ ....(24)

Case (ii). When E is perpendicular to the plane of incidence
and its positive direction in each wave is towards the reader in
Fig. 196, the positive directions of the magnetic forces H_L,
H_S, ... are as shown by the arrows having double barbs. We
have, by (12), $E_S/E_L = -\sin(\alpha - \phi)/\sin(\alpha + \phi)$, and

$$\frac{E_U}{E_{II}} = \frac{2\cos\phi\sin\alpha}{\sin(\alpha + \phi)}, \quad \frac{E_{II}}{E_I} = -\frac{\sin(\phi - \beta)}{\sin(\phi + \beta)}, \quad \frac{E_I}{E_L} = \frac{2\cos\alpha\sin\phi}{\sin(\alpha + \phi)}.$$

Hence $\quad a_2 \equiv \dfrac{E_U}{E_S} = -\dfrac{\sin 2\alpha \sin 2\phi \sin(\phi - \beta)}{\sin(\phi + \alpha)\sin(\phi + \beta)\sin(\phi - \alpha)}$ ....(25)

By (12), $E_{III}/E_{II} = -\sin(\phi - \alpha)/\sin(\phi + \alpha)$, and thus

$$b_2 \equiv \frac{E_W}{E_T} = \frac{E_{III}}{E_I} = \frac{\sin(\phi - \alpha)\sin(\phi - \beta)}{\sin(\phi + \alpha)\sin(\phi + \beta)}. \quad ......(26)$$

It is easily seen that $|b_2| < 1$.

For all values of ϕ, a_2 and b_2 have opposite signs and the sign of a_2 is opposite to that of $(\mu'' - \mu)(\mu' - \mu)$.

As in § 306, $\qquad c_2 \equiv E_J/E_U = b_2.$ ....(27)

308. Results when $\mu'' = \mu'$. When $\mu'' = \mu'$, the formulae take simpler forms. We now have, in case (i),

$$a_1 = -\frac{\sin 2\alpha \sin 2\phi}{\sin^2(\phi + \alpha)\cos^2(\phi - \alpha)}, \quad b_1 = \frac{\tan^2(\phi - \alpha)}{\tan^2(\phi + \alpha)}....(28)$$

Hence a_1 is negative and b_1 is positive, for *all* values of ϕ.

We also have, in case (ii),

$$a_2 = -\frac{\sin 2\alpha \sin 2\phi}{\sin^2(\phi + \alpha)}, \quad b_2 = \frac{\sin^2(\phi - \alpha)}{\sin^2(\phi + \alpha)}. \quad ......(29)$$

Hence a_2 is negative and b_2 is positive, for all values of ϕ.

309. Interference due to thin plate or film. We consider a transparent plate or film, of thickness z, with parallel faces, which the plane of Fig. 197 cuts at
right angles in AA, BB. Let the index of the plate or film be μ, and the indices of the other media in contact with AA, BB, be μ', μ''. Let monochromatic light from a distant source fall upon the film, $PP_1P_2\ldots$ being a plane wave front, in the medium μ', with its plane perpendicular to that of the figure. Let the wave length *in the medium μ be λ.* An incident ray, such as PQ, gives rise to a reflected ray QR in μ', and to a refracted ray

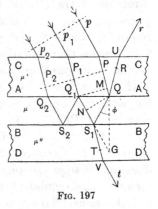

FIG. 197

in the film. The latter ray, on meeting BB, gives rise to a refracted ray in μ'', and to a reflected ray, which on meeting

AA is partly reflected and partly refracted. As the series of reflexions and refractions continues, the amplitudes diminish. Each wave train which returns *through* AA from μ into μ' will travel parallel to QR.

Let $P_1 Q_1$ be a ray (parallel to PQ) which, after refraction at Q_1, reflexion at S_1 on BB and refraction at AA, emerges into μ' along QR. Draw $Q_1 M$, QN perpendicular to PQ, $Q_1 S_1$. Then $Q_1 M$, QN are wave fronts, and the distances MQ, $Q_1 N$ are optically equivalent. Hence, the optical length of the path $P_1 Q_1 S_1 QR$ exceeds that of the path PQR by the distance $NS_1 + S_1 Q$ in μ. If $Q_1 S_1$ produced cut the normal at Q in G, $S_1 G = S_1 Q$, and hence $NS_1 + S_1 Q = NG = 2z \cos \phi$, where ϕ or $Q_1 GQ$ is the angle between the rays *in the film* and the common normal. Similarly, if $P_2 Q_2$, $P_3 Q_3$, ... be incident rays giving rise to a ray along QR, the optical paths from $P_2, P_3, ...$ to R exceed that from P to R by $4z \cos \phi$, $6z \cos \phi$, ... in μ. At the refractions into the film at Q_1, Q_2, ..., at the refraction out of the film at Q, and also at the reflexions at Q_1, Q_2, ..., S_1, S_2, ... definite changes of phase occur in addition to changes of amplitude. But the wave trains which return from μ into μ' will combine into a single train, definitely related in amplitude and phase to the train reflected at Q into μ'. We cannot find the relations between amplitudes and phases in the two trains unless we know, not only the ratios μ'/μ, μ''/μ, but also the direction of the electric force in the incident beam; for some purposes, e.g. EXPERIMENTS 57, 59, this knowledge is not required. We may, however, note that in every instance, as § 307 shows, there is either no change of phase or a change of π, the latter amounting to a reversal.

If z be increased by $\frac{1}{2}\lambda \sec \phi$, where λ is the wave length *in the film*, the optical lengths of the various paths from the wave front $PP_1 P_2 ...$ to R, for rays which enter the film, will exceed the optical length of the path PQR by $2z \cos \phi + \lambda$, $4z \cos \phi + 2\lambda$, $6z \cos \phi + 3\lambda$, ... Thus, the change of thickness does not change the phases of these wave trains, and, consequently, the combined train which *returns* from μ into μ' has the *same* relation as before in amplitude and phase with respect to the train reflected at Q into μ'. If z be such that the resultant train, due to both the combined returning train and the reflected train, has a maximum amplitude, this train will again have a maximum amplitude when z is increased by $\frac{1}{2}\lambda \sec \phi$, and similarly for a minimum. The same result follows if z be increased or diminished by any integral multiple of $\frac{1}{2}\lambda \sec \phi$.

We now consider the light transmitted into μ'' by rays which coincide with the refracted ray S_1T. Draw Q_2M_1 perpendicular to P_1Q_1. Then the optical length of $P_2Q_2S_2Q_1S_1T$ exceeds that of $P_1Q_1S_1T$ by $2S_1Q_1 - M_1Q_1$, or by $2z \cos \phi$, in medium μ, since M_1Q_1, or MQ, in μ' is equivalent to Q_1N in μ. Thus, if z be such that the resultant transmitted train has a maximum amplitude, this train will again have a maximum amplitude if z increase or diminish by an integral multiple of $\frac{1}{2}\lambda \sec \phi$, and similarly for a minimum.

In some cases, the media μ', μ'' are bounded by planes CC, DD parallel to AA, BB, and the whole system is surrounded by air. Let $pp_1p_2 \ldots$ be a wave front in air corresponding to those rays which, by refraction at CC, become PQ, $P_1Q_1, \ldots$, and let Ur be the ray in air due to QRU in μ'. Then the conditions for maximum or minimum amplitude in the ray Ur are the same as for the ray QRU, and the corresponding conditions are the same for the ray Vt as for the ray S_1TV.

If there be *air* between the plates, the angle of incidence of the light on CC is equal to ϕ.

310. Effect of multiple reflexions. When $2z \cos \phi = k\lambda$, where k is a positive integer, the optical paths $[PR]$, $[P_1R]$, $[P_2R], \ldots$ increase by equal steps each equivalent to $k\lambda$ in medium μ. Hence, by § 307, if the amplitude of the wave along QR due to a wave of unit amplitude along PQ be r, the amplitudes of the waves along QR due to P_1Q_1, $P_2Q_2, \ldots$ are ar, arb, $arb^2, \ldots$ where a, b denote a_1, b_1 or a_2, b_2, according to the direction of the electric force. Since at each reflexion or refraction there is either no change or else a reversal of phase, we may treat all the waves along QR as having the *same* phase, for we may regard a reversal of phase as equivalent to a reversal of sign of amplitude. If the resultant amplitude due to all the rays P_1Q_1, $P_2Q_2, \ldots$ be fr, then $fr = ar(1 + b + b^2 + \ldots)$, or $fr = ar/(1 - b)$, since $|b| < 1$. The amplitude due to PQ is r;

if the whole amplitude of the resultant wave train along QR be M, then

$$M = r + ar/(1 - b). \quad \dots\dots\dots\dots\dots(30)$$

If z be increased or diminished by $\frac{1}{4}\lambda \sec \phi$, the amplitudes of the waves along QR due to P_1Q_1, P_2Q_2, ... will be $- ar$, arb, $- arb^2$, ..., if we consider all the waves as having the same phase. If the resultant amplitude due to all these rays be gr, then $gr = - ar (1 - b + b^2 - \dots)$, or $gr = - ar/(1 + b)$, since $| b | < 1$. The amplitude due to PQ is r; if the whole amplitude of the resultant wave train along QR be N, then

$$N = r - ar/(1 + b). \quad \dots\dots\dots\dots\dots(31)$$

We take the intensities X, Y as equal to the squares of the amplitudes. Then

$$X = M^2 = \{r + ar/(1 - b)\}^2, \quad Y = N^2 = \{r - ar/(1 + b)\}^2\dots(32)$$

Since $| b | < 1$, both $1 - b$ and $1 + b$ are positive for all values of ϕ. Hence X is greater or less than Y according as a is positive or negative.

If the light pass through a polariser, such as a Nicol's prism, before falling upon the plates, the direction of the electric force E in the incident beam will be determined by the position of the polariser. By turning the polariser about the axis of the incident beam, we can cause E to be either in the plane of incidence, as in case (i), or perpendicular to that plane, as in case (ii). Thus the phenomena in the two cases can be observed separately.

In case (ii), § 307, in which E is perpendicular to the plane of incidence, the sign of a_2 does not depend upon ϕ, and X is greater or less than Y according as $(\mu'' - \mu) (\mu' - \mu)$ is negative or positive. When there is a film of air or of water (μ) between two glass plates (μ'' and μ'), or when there is a film of soap solution or of glass (μ) in air ($\mu'' = \mu' = 1$), we have ($\mu'' - \mu) (\mu' - \mu)$ positive. In this case, X is less than Y, and there will be relative darkness along QR when $2z \cos \phi = k\lambda$.

In case (i), § 307, E is in the plane of incidence. When ϕ does *not* lie between its polarising values, the sign of a_1 is opposite to that of $(\mu'' - \mu) (\mu' - \mu)$, and hence X is greater or less than Y according as $(\mu'' - \mu) (\mu' - \mu)$ is negative or positive. This agrees with what happens in case (ii) for all values of ϕ, and, consequently, when ϕ does not lie between its polarising values, any thickness z, which gives a maximum of intensity in one case, gives a maximum in the other also.

When, in case (i), ϕ lies *between* its polarising values, the sign of a_1 is the same as that of $(\mu'' - \mu)(\mu' - \mu)$, and now X is less or greater than Y according as $(\mu'' - \mu)(\mu' - \mu)$ is negative or positive. This is in opposition to what happens in case (ii) for these (and all other) values of ϕ and, consequently, any thickness z, which gives a maximum of intensity in one case, gives a minimum in the other.

When, in case (i), $\tan\phi = \cot\alpha = \mu'/\mu$, the ray PQ gives rise to no ray along QR, and S_2Q_1 gives rise to no ray along Q_1S_1. Hence the ray along QR is entirely due to $P_1Q_1S_1$, and is weak, since the angle of incidence at S_1 is near to $\tan^{-1}(\mu''/\mu)$.

When $\tan\phi = \cot\beta = \mu''/\mu$, Q_1S_1 gives rise to no ray along S_1Q; thus the ray QR is due entirely to PQ, and is weak, since the angle of incidence at Q is near to $\tan^{-1}(\mu/\mu')$.

In each instance, since only one ray is concerned, the intensity of the ray along QR will be the same for all values of z.

When ϕ lies between the polarising values, a little light will be reflected at Q and also at S_1 when E is in the plane of incidence, and consequently the intensity of the ray along QR will depend upon z but will be very small for every value of z.

When E is perpendicular to the plane of incidence, there is copious reflexion at both Q and S_1 and consequently strong contrast between brightness and darkness.

When the light is not polarised, each wave may be resolved into two, E being in the plane of incidence in one wave and perpendicular to it in the other; the effects of the two waves will be superposed. When ϕ lies between its polarising values, and when the light is not polarised, the phenomenon is controlled by case (ii). Thus, if the resultant intensities along QR due to both components of E be X_0 and Y_0 when $2z\cos\phi = k\lambda$ and $2z\cos\phi = (k + \frac{1}{2})\lambda$ respectively, X_0 will be greater or less than Y_0 according as $(\mu'' - \mu)(\mu' - \mu)$ is negative or positive.

The reader who pursues the matter may note that $\cot(\phi + \alpha)$, $\cot(\phi + \beta)$ and $\cot(\phi + \alpha)\cot(\phi + \beta)$ are factors in r_1 (E_S/E_L in § 307), in a_1r_1 and in b_1 respectively. Hence, both X and Y

(32) are small when ϕ lies between its polarising values, for
then both $|\cot(\phi + \alpha)|$ and $|\cot(\phi + \beta)|$ are small.

When $\mu'' = \mu'$, $\beta = \alpha$, and the polarising values of ϕ coalesce.
The exceptional conditions no longer exist, and the results of
§ 308 can be used. They show that, in both case (i) and case (ii),
a is negative, and hence there is relative darkness along QR
when $2z \cos \phi = k\lambda$. Using (28) and (29), § 308, we find that,
in both cases

$$f = a/(1 - b) = -1, \quad\ldots\ldots\ldots\ldots\ldots\ldots(33)$$

and thus there is actual darkness along QR when $2z \cos \phi = k\lambda$.
Neither (28) nor (29) gives a simple value for g when ϕ is finite.

EXPERIMENT 57. **Measurement of wave length by
Newton's rings.**

311. Introduction. If a pair of glasses, separated by a
film of air not more than a few wave lengths in thickness, be
examined under suitable illumination by sodium light, a set
of curves alternately bright and dark will be seen. These curves
are known as interference bands. If the thickness of the film,
within a small area surrounding a point Q of the film, though
not constant, do not vary by more than a very small fraction
of a wave length, the illumination which an eye at a distance
perceives at Q will be practically the same as if the whole film
were of uniform thickness equal to that at Q. The value of
$z \cos \phi$ is, by § 309, constant along any curve; here z is the
thickness of the film and ϕ the angle between the rays in the
film and the normal to the plates. If ϕ be small, $\cos \phi$ is very
nearly equal to unity and then the curves become practically
contour lines.

If an accurate surface, e.g. a face of a good prism, be available
as a standard, the character of one surface S of a glass plate
is at once revealed if the plate be laid upon the prism, with
S towards the prism, and be illuminated with sodium light.
Dust will prevent absolute contact, and interference bands
will be seen. If these be straight, parallel and *evenly* spaced,
the surface S is good.

The method of testing the fit between two glass surfaces by
interference bands is fundamental in the work of optical manu-

facturers. Errors of a ten-thousandth of a millimetre are easily detected by it.

When one surface is spherical and the other either plane or spherical and the incident light is, practically, normal to the surfaces, the curves will be concentric circles. If the incident light fall obliquely upon the film, so that the rings are necessarily viewed obliquely, they will, by perspective, appear as ellipses.

312. Newton's rings. When the convex surface of a lens is placed in contact with a plate of glass and the incident light is, practically, normal to the plate, a series of concentric rings may be seen surrounding the point of contact. If monochromatic light be used, many rings, alternately bright and dark, will be seen. With white light, the rings will be coloured, and will be few in number. The scale of the system will be greater, the greater the radius of the surface in contact with the plate. The rings may be observed with either reflected or transmitted light; with transmitted light the contrast between bright and dark is feeble.

From measurements made on such a system, Newton concluded that the distance which, in his emission theory, was the length of a "fit," was 1/89000 inch for yellow light. According to the wave theory, the distance in question is half the wave length. This first measurement of the wave length of "yellow" light gave $\lambda = 5.7 \times 10^{-5}$ cm.

The formation of the rings was first explained on the wave theory by Young, who showed it to be due to the interference of different streams of light.

If the convex face of a lens of radius r *touch* a plane plate of glass at O, and if the thickness of the film at distance ρ from O be z, then $z(2r - z) = \rho^2$. Near O, z is negligible in comparison with $2r$, and we write

$$z = \tfrac{1}{2}\rho^2/r. \quad \dots\dots(1)$$

If, instead of a plane, a *concave* surface of radius s be used, where $s > r$, we have

$$z = \tfrac{1}{2}\rho^2(1/r - 1/s). \quad \dots\dots(2)$$

Dust may prevent actual contact between the glass surfaces, and the surfaces will be distorted if each exert a force upon the other (see § 149). We are, thus, unable to rely upon z being zero at the centre of the ring system, and therefore use a method of differences.

If ρ_1 be the radius of the smallest dark *ring* (not the *centre* itself), and ρ_2, ρ_3, ... be the radii of the larger dark rings in succession, and if z_1, z_2, ... be the corresponding thicknesses,

$$z_n - z_1 = \tfrac{1}{2}\,(\rho_n{}^2 - \rho_1{}^2)\,(1/r - 1/s).$$

But, by § 309, $z_n - z_1 = (n-1) \times \tfrac{1}{2}\lambda \sec\phi$, and hence

$$\lambda = (1/r - 1/s)\cos\phi\,(\rho_n{}^2 - \rho_1{}^2)/(n-1). \quad\ldots\ldots\ldots(3)$$

In the simplest case, one surface is convex with radius r and the other is plane, and the incident rays are normal to the plane surface, so that $\cos\phi = 1$. Then (3) becomes

$$\lambda = (1/r)\,.\,(\rho_n{}^2 - \rho_1{}^2)/(n-1). \quad\ldots\ldots\ldots\ldots(3a)$$

313. Experimental details. The apparatus may be arranged as in Fig. 198. The lens L is firmly supported on a stiff slab S of slate or metal having a black surface and resting on a firm table; a portion only of S is shown. On the lens is placed a glass plate C with *plane* faces. The faces may be tested for flatness by the method of

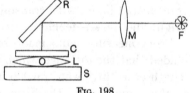

FIG. 198

§ 31 or of § 311. The surfaces in contact should be clean. It is unnecessary (and dirty) to use wax to fix C to L. Light from a *bright* sodium flame F is concentrated by a lens M upon O, the point of contact of C with L, by reflexion at a glass plate R which has plane faces and is inclined at about 45° to the horizontal. The distances FM, MR should each be about twice the focal length of M. If the observer look vertically down upon O, the rings will come into view when R is at the proper angle. A magnifying lens will be useful here.

A travelling microscope is now put into place upon the slab S. If it be focused upon a speck of paper on the top of the plate C, the rings will probably be visible, though not in good focus. They can then be accurately focused. The cross-wire of the microscope, at right angles to the direction of travel, is made to touch the successive dark rings by turning the screw; the darkest part of each ring is used. A dark ring—say about the tenth to the left of the centre—is selected, and the reading for this ring is taken. The cross-wire is then set on the next smaller dark ring, and so on, till the smallest dark ring be reached. The wire is now set on the other side of this ring and then in succession on the other dark rings till the readings on either side of the centre be equal in number. The difference between the two readings for any ring gives its diameter, 2ρ. Denoting the radius of the smallest dark ring by ρ_1, the values of $\rho_1{}^2, \rho_2{}^2, \rho_3{}^2, \ldots$ are plotted against the numbers $1, 2, 3, \ldots$, as in Fig. 199. From the straight line lying most evenly among the plotted points, $(\rho_n{}^2 - \rho_1{}^2)/(n - 1)$ is found.

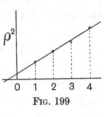

Fig. 199

We may also find $(\rho_2{}^2 - \rho_1{}^2)$, $(\rho_3{}^2 - \rho_1{}^2)/2$, $(\rho_4{}^2 - \rho_1{}^2)/3$, ... and take the mean. If the first two or three values be irregular, they are disregarded, as they are more liable to error than the others.

The radius (r cm.) of the face of the lens in contact with the plate is found by Boys's method (§ 157); for accurate work, § 158 is followed. Since the plate C is plane, $s = 0$.

Since the rays transmitted by the microscope are nearly vertical, ϕ is small, and we may therefore put $\cos \phi = 1$ in (3). We then find λ from equation (3 a).

314. Special apparatus. The lens L is fixed in a block of wood KK (Fig. 200) provided with a clamp H adapted to the fittings of an optical bench. The lens is thus protected against

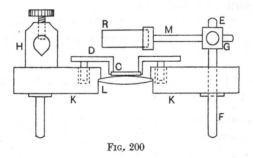

Fig. 200

injury. The plane plate C is cemented into the end of a short tube soldered into a hole in the disk D. This disk has three feet, which drop into three holes in the board with plenty of clearance. The plate C rests on the lens and takes up a definite position, provided the feet do not touch the sides of the holes. The feet are long enough to prevent C from touching a table and so receiving scratches. A small pillar EF passes through KK. By a metal block G pierced by two holes, and fitted with two screws, the glass plate R, attached to the rod M, can be put in any position either above or below the board. The lower part of the pillar forms a foot. Two other feet are provided at the other end of the board.

When transmitted light is to be used, the block G is attached to the lower part of the pillar EF, and light from the flame is sent upwards through the system by R. The measurement of the rings and the calculation of the wave length are carried out exactly as in the case of reflexion.

315. Practical example. Mr J. A. Pattern used lens of thickness $c = \cdot19$ cm., $\mu = 1\cdot52$. The A-face touched plate; by Boys's method (EXPERI-MENT 31), radius of A-face was $r = 85\cdot60$ cm. The table gives radii of dark rings for reflected sodium light, and $(\rho_n{}^2 - \rho_1{}^2)/(n-1)$ is denoted by D^2.

Values * were rejected as irregular.

n	$100\,\rho$ cm.	$1000\,\rho^2$ cm.^{2}	$1000\,D^2$ cm.^{2}	n	$100\,\rho$ cm.	$1000\,\rho^2$ cm.^{2}	$1000\,D^2$ cm.^{2}
1	6·725	4·52		6	17·25	29·76	5·048
2	9·850	9·70	*	7	18·60	34·60	5·013
3	12·35	15·25	*	8	19·90	39·60	5·011
4	14·00	19·60	5·027	9	21·30	45·37	5·106
5	15·65	24·49	4·992	10	22·30	49·73	5·023

Mean $D^2 = 5\cdot031 \times 10^{-3}$ cm.^{2}. Hence
$$\lambda = (\rho_n{}^2 - \rho_1{}^2)/\{(n-1)\,r\} = D^2/r = 5\cdot877 \times 10^{-5} \text{ cm.}$$
When the rings were observed with transmitted light, $D^2 = 5\cdot071 \times 10^{-3}$ cm.^{2}, and $\lambda = 5\cdot924 \times 10^{-5}$ cm.

EXPERIMENT 58. **Measurement of refractive index of a liquid by Newton's rings.**

316. Method. We assume that the incidence of the light is normal, so that $\cos\phi = 1$. The value of $D^2 = (\rho_n{}^2 - \rho_1{}^2)/(n-1)$ is found when the medium between the plate and the lens is air. The plate is removed, a *small* drop of water or other liquid is put on the lens, and the plate is replaced. A ring system will again be seen; it will be on a smaller scale and the contrast between bright and dark will be *much* less than with air. The radii σ_1, σ_2, ... of the dark rings are measured, and the mean value of $E^2 = (\sigma_n{}^2 - \sigma_1{}^2)/(n-1)$ is found.

Then, by (3 a), § 312, if λ and λ' be the wave lengths of sodium light in air and in the liquid,

$$\lambda = (\rho_n{}^2 - \rho_1{}^2)/\{(n-1)r\} = D^2/r, \quad \lambda' = (\sigma_n{}^2 - \sigma_1{}^2)/\{(n-1)r\} = E^2/r.$$

But $\lambda' = \lambda/\mu$, where μ is the index of the liquid relative to air. Thus

$$\mu = \lambda/\lambda' = D^2/E^2. \qquad\qquad (1)$$

317. Practical example. Mr J. A. Pattern used lens of § 315. Radii σ_1, σ_2 ... of dark rings observed with reflected sodium light were measured; $(\sigma_n{}^2 - \sigma_1{}^2)/(n-1) = E^2$. Experiment gave mean $E^2 = 3 \cdot 810 \times 10^{-3}$ cm.². Mean D^2 found in § 315 for reflected light was $5 \cdot 031 \times 10^{-3}$ cm.². By (1),

$$\mu = D^2/E^2 = 5 \cdot 031/3 \cdot 810 = 1 \cdot 32.$$

EXPERIMENT 59. **Measurement of angle between two nearly parallel plates.**

318. Method. Let two plates of glass having plane faces be placed nearly in contact, and let the opposed faces make a very small angle θ radians with each other. If the plates be illuminated with sodium light, as in Fig. 198, a series of parallel bright and dark bands, at equal intervals, will be seen. These bands may be used to measure θ.

If the thickness of the air film at a point on a dark band be y, the thickness at the next dark band on one side will be $y + \frac{1}{2}\lambda$, and the thickness will increase by $\frac{1}{2}\lambda$ for each successive band.

Let one dark band, denoted by O, be chosen as origin. Let the distances of the mth and nth dark bands, M and N, from O, measured at right angles to the bands, be x_m and x_n; both bands are on the same side of O. Then, if y_m, y_n be the thicknesses, $y_n - y_m = \frac{1}{2}(n-m)\lambda$. Hence, in circular measure,

$$\theta = \frac{y_n - y_m}{x_n - x_m} = \frac{(n-m)\lambda}{2(x_n - x_m)}. \qquad\qquad (1)$$

The sodium "line" has two members D_1 and D_2 of wavelengths $\lambda_1 = 5 \cdot 896 \times 10^{-5}$ cm. and $\lambda_2 = 5 \cdot 890 \times 10^{-5}$ cm., and hence λ_1 exceeds λ_2 by about $\frac{1}{10}$ per cent. Thus, approximately,

$500\lambda_1 = (500 \cdot 5)\lambda_2$. Hence, if $y = 500\lambda_1 = \cdot 029$ cm., the bright bands due to one member will coincide with the dark bands due to the other, and the contrast will be weak. Hence the greatest distance between the plates should be small compared with $\cdot 029$ cm.

It may happen that although the bands be parallel they are not evenly spaced. This indicates that the plates are not plane. In this case, the value of θ given by (1) will be very nearly equal to the actual angle at a point midway between M and N, i.e. where $x = \frac{1}{2}(x_m + x_n)$.

If we observe the bands of orders $0, p, 2p, 3p, \ldots$ and find, by (1), the value of θ at the centre of each of the ranges 0 to $2p$, $2p$ to $4p$, $\ldots$, we have

$$\theta_p = p\lambda/x_{2p}, \quad \theta_{3p} = p\lambda/(x_{4p} - x_{2p}), \quad \theta_{5p} = p\lambda/(x_{6p} - x_{4p}) \ldots \ldots (2)$$

Let
$$\theta_{3p} - \theta_p = \phi_{2p}, \quad \theta_{5p} - \theta_{3p} = \phi_{4p}, \ldots, \quad \ldots \ldots (3)$$

and let $R_{2p} = (x_{3p} - x_p)/\phi_{2p}, \quad R_{4p} = (x_{5p} - x_{3p})/\phi_{4p} \ldots \ldots (4)$

By (4), $R_{2p}, R_{4p}, \ldots$ are approximately the radii of relative curvature at the points defined by $x_{2p}, x_{4p}, \ldots$.

The bands are measured by a travelling microscope.

319. Experimental details. The plates may be squares about 3 cm. on the side cut from plate glass not less than $0 \cdot 7$ cm. thick. From a number of such plates, two are selected which give very broad bands when placed as nearly in contact as dust will allow. The bands are observed as in Fig. 198. The plates are then cleaned and are put together with a strip of tinfoil near one edge. The plates are held between the fingers and a little hard wax, such as Everett wax, is deposited along the joint by aid of a hot wire. The edges are then heated, a part at a time, by a very small gas flame, until the wax begin to run between the plates by capillary action. As soon as this occurs, the heating of that part is stopped. If the heating be continued too long, the wax will fill the whole space between the plates. The band of wax may be about $0 \cdot 5$ cm. wide.

When the bands are examined, it may be found that, although straight, they are not in the desired direction. The plates are warmed cautiously and are pressed so as to make the bands parallel to the strip of foil; the pressure is continued

while the plates cool. The angle between plates so prepared
will remain constant, unless, of course, the plates be heated.

The plates used in § 320 were of optical quality and had very
nearly plane surfaces. Their length, measured at right angles
to the bands, was 4·4 cm.; the width was 2·5 cm. and the
thickness 0·35 cm. The relative curvature disclosed by the
measurements is due to the distorting effect of the wax. Small
and thick plates must be used if equidistant bands are to be
obtained.

320. Practical example. G. F. C. Searle obtained the following
results, with $p = 10$; sodium light, mean $\lambda = 5 \cdot 893 \times 10^{-5}$ cm., was used. The
table gives θ_{10}, θ_{30}, ... and ϕ_{20}, ϕ_{40}, ... in radians.

Band	Microscope reading	Change for 20 bands	θ	ϕ
0	·0635 cm.			
10	·3050	·4737 cm.	$1 \cdot 244 \times 10^{-3}$	
20	·5372	·····	· · · · ·	$1 \cdot 38 \times 10^{-4}$
30	·7534	·4265	$1 \cdot 382 \times 10^{-3}$	· · · · ·
40	·9637	·····	· · · · ·	$1 \cdot 07 \times 10^{-4}$
50	1·1651	·3957	$1 \cdot 489 \times 10^{-3}$	· · · · ·
60	1·3594	·····	· · · · ·	$1 \cdot 22 \times 10^{-4}$
70	1·5441	·3659	$1 \cdot 611 \times 10^{-3}$	
80	1·7253			

Thus θ varies from $1 \cdot 244 \times 10^{-3}$ to $1 \cdot 611 \times 10^{-3}$ radians, or from 4′ 17″ to
5′ 32″. By (4),
$$R_{20} = (x_{30} - x_{10})/\phi_{20} = \cdot 4484/(1 \cdot 38 \times 10^{-4}) = 3249 \text{ cm.}$$
Similarly, $R_{40} = 3848$ and $R_{60} = 3107$ cm. Mean $R = 3401$ cm.

EXPERIMENT 60. **Measurement of thickness of air film
by interference method.**

321. Method. In Newton's experiment, the wave length is
found from the radius of the curved surface and from the
squares of the radii of the rings, and the actual shortest distance
between lens and plate does not concern us. In the present
experiment, however, we make a comparison between the wave
length and the smallest distance between two very nearly
parallel plates of glass of indices μ', μ''; the film of air between
the plates is of unit index. In Newton's experiment it was
unnecessary to know what change of phase (if any) occurs on

reflexion or refraction at the various surfaces, but now this knowledge is required.

We suppose, in the first instance, that the glass plates are identical in quality, so that $\mu'' = \mu'$. Then § 308 shows that the ratio a is negative, whether the electric force be in or perpendicular to the plane of incidence. By (32), § 310, the intensity of the resultant train of waves along QR (Fig. 197) will be a minimum whenever NG or $2z \cos \phi$ is an integral multiple of λ, the wave length in the air film. Here z is the distance between the plates. By (33), § 310, each minimum will be zero, since $\mu'' = \mu'$. Since the plates of glass have (practically) parallel faces, the angle of incidence on the outer face of the plate CC (Fig. 197) may be taken as equal to ϕ.

In the general case, μ' and μ'' are not identical. But § 310 shows that, unless the electric force perpendicular to the plane of incidence be destroyed by accurate polarisation of the incident light, there will be relative darkness along QR whenever NG or $2z \cos \phi$ is an integral multiple of λ.

If $2z = p\lambda$, where p is a positive integer, there will be darkness * when the incident beam is normal to CC, and also when NG/λ or $2z \cos \phi/\lambda$ has the integral values $p - 1$, $p - 2$, ... $p - n$, ... In general, however, $2z = (p + q)\lambda$, where q is a positive fraction, and there will not be darkness when $\phi = 0$. To obtain the first darkness, we must increase ϕ from zero to ϕ_1, where $2z \cos \phi_1 = (p + q)\lambda \cos \phi_1 = p\lambda$, or

$$(p + q) \cos \phi_1 = p. \quad \dots\dots\dots\dots\dots(1)$$

The second darkness occurs when $(p + q) \cos \phi_2 = p - 1$, and the nth darkness when

$$(p + q) \cos \phi_n = p - (n - 1). \quad \dots\dots\dots\dots(2)$$

Hence $$p + q = (n - 1)/(\cos \phi_1 - \cos \phi_n). \quad \dots\dots\dots\dots(3)$$

If we take $n = 2, 3, ...$, we obtain a series of values of $p + q$. The calculations are simplified if we find the values of $(\cos \phi_1 - \cos \phi_n)/(n - 1)$ and take the reciprocal of the mean as equal to $p + q$. If $\cos \phi_n$ be plotted against n, the value of $p + q$ may be deduced from the resulting straight line.

---

* The darkness will be absolute when $\mu'' = \mu'$; when the incident light is not polarised and μ'' is not equal to μ', the darkness is relative.

322. Experimental details. The apparatus may be arranged as in Fig. 201. The two plates of glass F enclosing the air film are attached to the shaft A, so that the opposed faces of the plates are parallel to the axis of A, which, ideally, is midway between them. The shaft carries the divided circle S, which is read by an index V, or, better, by two index points at opposite ends of a diameter (§ 57). The observer looks through the spy-hole J and views F by reflexion at the *fixed* plane mirror M.

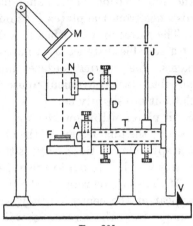

FIG. 201

The function of the plane mirror N, of unsilvered glass, is to provide a beam of incident light of suitable direction in the case in which MF is normal or nearly normal to the plates F. The mirror N is carried by a rod C held in the arm D, which can be turned into any position on the bearing tube T.

Fig. 202 shows two positions N_1, N_2 of the reflector and the corresponding positions F_1, F_2 of the plates. When the incidence

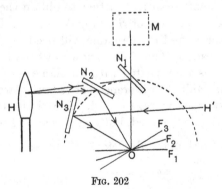

FIG. 202

reaches 20° to 30°, N may be discarded, as the direct light from the *bright* sodium flame H will serve, especially if the burner be raised above the level of F. For larger angles of incidence, the burner may be placed at H' on the other side of the plates, the reflector being at N_3. As MO, the line of sight, does not pass through N_3, silvered glass may now be used to improve the illumination. The flame must be kept at such a safe distance from the plates F as not to heat them appreciably.

The plates may be stuck together with wax, being kept apart at a suitable distance by a narrow annulus of tinfoil near their edges. The internal surfaces must, of course, be as clean as possible. It is convenient to use plates which are not absolutely flat but are slightly convex with a radius of 100 or 200 metres. Such plates will show Newton's rings large enough to be well seen by the unaided eye of the observer who views them through the spy-hole. In this case, it is the shortest distance between the plates which is found in the experiment.

If the shaft be rotated, the rings will either contract and disappear at the centre or they will expand and fresh rings will emerge from the centre, according to the direction of rotation. But there is one position such that, when the shaft is turned from it in either direction, the rings expand; it is supposed here that the air film is thinnest at the centre. This zero position corresponds to normal incidence. The zero reading of S is observed.

The shaft is now turned, until the first dark spot appear in the midst of the ring system, and is adjusted so that the *central point* of the spot is as dark as possible. The reading of S then gives ϕ_1. The shaft is turned further to obtain the 2nd, 3rd, ... dark spots, the readings of S giving ϕ_2, ϕ_3, ... The positions of the reflector N and of the flame will need readjustment for nearly every reading. The circle is now put back to its zero, and the observations are repeated with the flame on the other side of the plates. Half the difference between the two readings for the first dark spot is taken as ϕ_1, and similarly for ϕ_2, ... The zero reading is not used, but it serves as a check.

323. Practical example. Mr T. G. Wilson used sodium light; $\lambda = 5\cdot893 \times 10^{-5}$ cm.

n	ϕ	$\cos \phi$	$\dfrac{\cos \phi_1 - \cos \phi_n}{n-1}$	n	ϕ	$\cos \phi$	$\dfrac{\cos \phi_1 - \cos \phi_n}{n-1}$
1	10° 0′	·9848	—	5	31° 30′	·8526	·03305
2	17 52	·9518	·03300	6	34 52	·8205	·03286
3	22 52	·9214	·03170	7	37 45	·7907	·03235
4	27 22	·8881	·03223	8	40 45	·7576	·03246

Mean value of $1/(p+q) = (\cos \phi_1 - \cos \phi_n)/(n-1) = \cdot03252$. Hence $p + q = 30\cdot75$. Distance between plates $= z = \frac{1}{2}(p+q)\lambda = 9\cdot06 \times 10^{-4}$ cm.

EXPERIMENT 61. **Determination of polarising angle.**

324. Introduction. When a ray moving in any medium meets a new medium, it gives rise, in general, to a reflected as well as to a refracted ray. But (§ 304), if the electric force in the incident ray be in the plane of incidence, there will be no reflected ray if the sum of the angles of incidence and refraction be $\frac{1}{2}\pi$. Let η, η' be the polarising angles, i.e. the angle of incidence and the corresponding angle of refraction for which there is no reflected ray. Then, if in $(9a)$, § 304 we put $n = 1$, $n' = \mu$, we have $\tan\eta = \mu$. Hence, if we can find η, we can deduce μ.

A plate of glass B for which the polarising angle β is known can be employed to determine the polarising angle η for a given plate A. Take rectangular axes Ox, Oy, Oz, and let A be placed so that Oz lies in its reflecting surface. At a distance from O along OX place B so that its reflecting surface cuts Ox in O' and the plane of xy in a line $O'y'$ parallel to Oy, and let the angle between OO' and the normal to B be the polarising angle proper to B. Then, a ray OO' meeting B will give rise to no reflected ray, if the electric force in the ray OO' be parallel to Oz. Hence, if any ray along OO' give rise to a ray reflected from B, the electric force in the ray OO' has a component parallel to Oy. If a ray PO in the plane of xy fall on A, and A be so turned about Oz that the reflected ray passes along OO', the existence of a ray reflected from B shows that PO did not make angle η with the normal to A. If, however, the direction of PO and the position of A be adjusted so that no ray is reflected from B, the angle between OO' and the normal to A is the polarising angle proper to A.

This arrangement is not convenient in practice, and, in place of the second reflecting plate, a Nicol's prism is used. As double refraction is not considered in this book, it must suffice to say that, if a beam pass through a Nicol, the electric force in the emergent beam is parallel to a plane fixed relative to the Nicol.

If a ray PO (Fig. 203) in air fall upon a reflecting surface S at O, and the reflected ray OQR pass through the Nicol N, the intensity of the light emerging from N will vary considerably as N is turned about OQ as axis. The position of N which

gives minimum intensity is easily found. If S be turned about
an axis through O perpendicular to
the plane of incidence, and if N be
turned about the same axis so as
always to receive the reflected ray, a
range of values of θ, the angle of incidence, will be found over which the
light transmitted by the Nicol is very
small. It will be difficult to determine

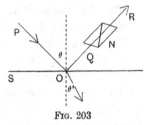

FIG. 203

nearer than one or two degrees the value of θ which gives the
minimum intensity.

When, for a given θ, the Nicol is turned about the reflected
ray as axis until the intensity be a minimum, the component of
the electric force perpendicular to the plane of incidence is sup-
pressed by N, and in the ray emerging from N the electric force
is in the plane of incidence. If E_0, E_1 be the components in this
plane of the electric forces in the rays PO, OQ, then, by (8),
§ 304, $E_1/E_0 = \tan (\theta - \theta')/\tan (\theta + \theta')$, where θ' is the angle of
refraction. If I, I_0 be the intensities, $I/I_0 = (E_1/E_0)^2$. The lower
curve in Fig. 204 shows how I/I_0 depends upon θ when a ray

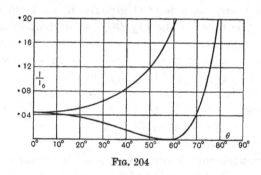

FIG. 204

in air of unit index falls at angle θ upon glass of index $\mu = 1 \cdot 528$,
E being in the plane of incidence. When $\theta = 0$,

$$I/I_0 = (\mu - 1)^2/(\mu + 1)^2 = \cdot 0436.$$

As θ increases from $0°$ to the polarising value $\eta = 56° \ 48'$, I/I_0
diminishes and vanishes when $\theta = \eta$. As θ further increases, I/I_0
increases rapidly and reaches unity when $\theta = 90°$. Near $\theta = \eta$,
I/I_0 is very small, so small, in fact, that any stray light will

make the observation of the minimum difficult. Accurate determination of η is made additionally difficult by the want of symmetry on the two sides of the minimum. We can show that

$$\frac{I}{I_0} = (\theta - \eta)^2 \frac{(\mu^4 - 1)^2}{4\mu^6} + (\theta - \eta)^3 \frac{(\mu^4 - 1)^2 (\mu^4 + \mu^2 + 2)}{4\mu^9} + \ldots.$$

When $\mu = 1{\cdot}528$, $I/I_0 = {\cdot}389\,(\theta - \eta)^2 + 1{\cdot}07\,(\theta - \eta)^3 + \ldots.$ Thus, if $\theta - \eta = {\cdot}1$ radian, $I/I_0 = {\cdot}00496$, but if $\theta - \eta = -{\cdot}1$, $I/I_0 = {\cdot}00282$. For equal distances from the actual zero of illumination, I/I_0 is so much smaller when $\theta < \eta$ than when $\theta > \eta$ that the observer naturally tends to select a value of θ which is too small.

The upper curve in Fig. 204 shows how I/I_0 depends upon θ when the electric force in the incident ray is perpendicular to the plane of incidence. When $\theta = 0$, $I/I_0 = (\mu-1)^2/(\mu+1)^2 = {\cdot}0436$, as before. As θ increases from $0°$ to $90°$, I/I_0 constantly increases and reaches unity when $\theta = 90°$.

325. Method. The measurements may be made on the simple spectrometer shown in Fig. 19, if a special observing telescope be used in place of the tube shown on the right-hand carriage. The telescope T (Fig. 205) has an objective L_1 ($f_1 = 10$ cm.), and the needle

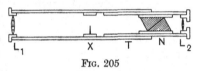

FIG. 205

X, mounted in a short length of tube, is set in the focal plane of L_1. The eye-piece L_2 fits into a draw-tube containing the Nicol N and is of low enough power ($f_2 = 6$ cm.) to allow the observer to focus X by varying the distance L_2X. A simple tube, having a vertical wire at its outer end, is used in place of the tube with fittings shown on the left-hand carriage, and is adjusted so that the wire (W) is in the focal plane of the lens borne by the carriage. This lens, with W, forms a collimator (C). By adjusting the carriages, the image of W may be made to coincide with X.

A face of a prism is used as the reflecting surface. The prism is partially surrounded by a shield, to cut off stray light. The base of the shield is a circular metal plate which rests on the central platform of the spectrometer; it has a pivot fitting a central hole in the platform and can thus turn about the axis of the spectrometer. An equal plate forms the top of the shield.

The plates are connected by a semi-cylinder of sheet metal of about the same radius as the plates. The inside of the shield is blacked. By a screw passing through the top of the shield, the prism can be clamped in the shield so that its reflecting face (S) lies in the plane of the edges of the semi-cylinder. The axis of the spectrometer will then, approximately, lie in the plane of S.

The index μ of the prism is deduced from the angle (i) between two faces and the corresponding minimum deviation (D). The measurements are made on the spectrometer by aid of the collimator C and the telescope T. Then

$$\mu = \sin \tfrac{1}{2}(D + i)/\sin \tfrac{1}{2}i. \quad \dots\dots\dots\dots\dots(1)$$

The face S is cleaned with soap and water and is not touched afterwards. The prism is then put into the shield and is (lightly) clamped in position.

The C carriage is set to, and is kept at, a definite reading. Before the shield is placed on the spectrometer, the T carriage is set so that, T being in line with C, the image of W coincides with X; this gives the zero reading for T.

A sodium flame (F) is placed in line with C; by a screen, the light reaching the prism from F is limited to that which passes through the C tube.

The shield containing the prism is now put in place and is set so that θ, the angle of incidence of the light upon S, is about $60°$. Then T is placed so that the image of W is seen by reflexion. If the Nicol N be turned about the axis of T, the intensity varies greatly, and N is set to give the least illumination for the particular position of T. The shield is now moved to alter θ, and the image of W is followed up with T. Using both hands, the observer can, with a little practice, turn T twice as fast as the prism, and thus keep W in sight. He seeks the position of T for which the intensity is a minimum, while, at the same time, the image of W is at X. The minimum is ill-defined and discrepancies of one or two degrees will occur among the readings. The difference between the mean of these readings for T and the zero reading for T gives a value for $\pi - 2\eta$, where η is the polarising angle. The telescope is turned back to its zero and through a further $60°$, and the prism is turned to reflect the light into it. A fresh set of readings is taken, and

from their mean a second value of $\pi - 2\eta$ is found. The mean
of the two values of η is taken, and the index is found by

$$\mu = \tan \eta. \qquad\qquad\qquad (2)$$

326. Practical example. The prism used by G. F. C. Searle had
$i = 60° 12'$, and, with sodium light, $D = 48° 36'$. Hence, by (1), $\mu = 1\cdot621$,
and $\tan^{-1} 1\cdot621 = 58° 20'$.

Zero reading of T, $179°\cdot3$. Mean of five readings of T for minimum intensity
on one side of zero, $114°\cdot0$. Mean on other side $245°\cdot3$. Values of $\pi - 2\eta$,
$65°\cdot3$, $66°\cdot0$; mean $65° 39'$. Polarising angle $= \eta = \frac{1}{2} (\pi - 65° 39') = 57° 10'$.
Index $= \mu = \tan \eta = 1\cdot55$.

EXPERIMENT 62. **Resolving power of a telescope.**

327. Introduction. The theory of geometrical optics asserts
that, when a distant luminous *point* lies on the axis of a converging lens, a *point* image of the source is formed in the focal
plane, and that there is no illumination at any other point in
that plane. It is here supposed, of course, that the lens has
been designed to be free from spherical aberration for the particular position of the luminous point.

The simple geometrical theory rests on the assumption that
the wave length of light is not merely small but is accurately
zero. When, however, we take account of the finite, though
small, wave length λ, we find, in accordance with observation,
that the disturbance in the focal plane is not confined to a
single point, but that a diffraction pattern, surrounding the
"geometrical" image and proportional in linear magnitude to λ,
is formed in that plane. If the light pass through a hole in a
screen placed in front of the objective of a telescope, the geometrical form of the pattern will depend upon that of the hole,
and its linear dimensions will be inversely proportional to those
of the hole.

If two *independent* sources of illumination, such as two stars,
each lying on or very near to the axis of the lens, be observed
simultaneously, there will be in the focal plane two diffraction
patterns, which will more or less overlap each other. If the
angle subtended by the two sources at the objective of a
telescope be less than some definite angle, the two patterns
will overlap to such an extent that the observer will be unable
to decide that the disturbance is due to two sources rather than

to a single source. When he can decide that he is viewing *two* separate sources, the telescope is said to resolve them.

328. The diffraction pattern. Let a point emitting light of wave length λ be situated at A (Fig. 206) on the axis of a lens

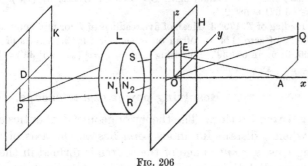

FIG. 206

L. Let the rays from A, when the full aperture is used, come to a focus at D on a plane K which is transverse to the axis of L. If a screen H, having a small opening, be placed between L and A, a diffraction pattern will be formed on K.

Let the axis of L cut the plane of H in the point O, which is taken as origin. Let Ox coincide with the axis of L, the axes Oy, Oz lying in the plane H. Let P be any point on K near D, and let Q be the image of P formed by L. Then Q lies in the transverse plane through A, since D is the image of A. Let E be any point of the opening in H, and let QO, QE meet L in R, S.

On account of diffraction, light from A, after passing O and E, is not confined to the continuations of the straight lines AO, AE, but some will travel along OR, although AOR is not straight, and similarly some will travel along ES. Thus every ray from A which passes through the aperture gives rise to some disturbance at P. The amplitude of the resultant disturbance at P depends upon the form and the size of the hole.

The x coordinate of E is zero; let the y, z coordinates be η, ζ. Let $OA = a$. Then the x coordinate of Q is a; let the other coordinates be y, z. Since P is the geometrical image of Q by L, the y, z coordinates of P are proportional to those of Q and are, in fact, $Y = - y \cdot DN_1/N_2 A$ and $Z = - z \cdot DN_1/N_2 A$, where N_1, N_2 are the nodal points of L.

If $[AESP]$ denote the optical length of the path from A to P by the route indicated, and if the path difference $[AESP]-[AORP]$ be denoted by Δ,

$$\Delta = AE + [ESP] - AO - [ORP]. \qquad \dots\dots\dots(1)$$

Since Q is the image of P, $[QESP] = [QORP]$, and thus

$$QE + [ESP] = QO + [ORP]. \qquad \dots\dots\dots(2)$$

Hence, (1) becomes

$$\Delta = AE - AO - (QE - QO) = \frac{AE^2 - AO^2}{AE + AO} - \frac{QE^2 - QO^2}{QE + QO}$$

$$= \frac{\eta^2 + \zeta^2}{AE + AO} - \frac{(y-\eta)^2 + (z-\zeta)^2 - \{y^2 + z^2\}}{QE + QO}.$$

Since OE is very small compared with OA, we need only retain the first powers of η, ζ, and may replace $AE + AO$ by $2a$. Since the angle QOA is small we may replace $QE + QO$ by $2a$. Then

$$\Delta = (y\eta + z\zeta)/a. \qquad \dots\dots\dots\dots\dots(3)$$

If the disturbance at P, due to light transmitted by an element of area $d\eta\,d\zeta$ at O, be $g \sin (2\pi vt/\lambda).d\eta\,d\zeta$, and if $d\omega$ be the disturbance at P due to light transmitted by an element $d\eta\,d\zeta$ at E, then

$$d\omega = g \sin \left\{\frac{2\pi}{\lambda}\left(vt - \frac{y\eta + z\zeta}{a}\right)\right\}.d\eta\,d\zeta. \qquad \dots\dots\dots(4)$$

If we integrate this expression over the area of the hole, we shall obtain the resultant disturbance at P. For our purpose we consider a rectangular hole, the sides being $2b$, $2c$ parallel to Oy, Oz respectively. Then

$$\omega = g \int_{-b}^{b}\int_{-c}^{c} \sin \left\{\frac{2\pi}{\lambda}\left(vt - \frac{y\eta + z\zeta}{a}\right)\right\}.d\eta\,d\zeta.$$

Integrating with respect to ζ, we have

$$\omega = \frac{ga\lambda}{2\pi z}\int_{-b}^{b}\left[\cos\left\{\frac{2\pi}{\lambda}\left(vt - \frac{y\eta + zc}{a}\right)\right\} - \cos\left\{\frac{2\pi}{\lambda}\left(vt - \frac{y\eta - zc}{a}\right)\right\}\right]d\eta$$

$$= \frac{ga\lambda}{\pi z}\sin\left\{\frac{2\pi zc}{\lambda a}\right\}.\int_{-b}^{b}\sin\left\{\frac{2\pi}{\lambda}\left(vt - \frac{y\eta}{a}\right)\right\}d\eta.$$

When we integrate with respect to η, we obtain, in a similar way,

$$\omega = \frac{ga^2\lambda^2}{\pi^2 yz}\sin\left\{\frac{2\pi yb}{\lambda a}\right\}.\sin\left\{\frac{2\pi zc}{\lambda a}\right\}.\sin\left(\frac{2\pi vt}{\lambda}\right)$$

$$= 4bcg.\frac{\sin(2\pi yb/\lambda a)}{2\pi yb/\lambda a}.\frac{\sin(2\pi zc/\lambda a)}{2\pi zc/\lambda a}.\sin\left(\frac{2\pi vt}{\lambda}\right). \qquad \dots(5)$$

Since the limit of $(\sin \theta)/\theta$ is unity as θ approaches zero, the disturbance at D, corresponding to $y = 0$, $z = 0$, is

$$\omega_0 = 4bcg \sin (2\pi vt/\lambda). \quad\quad\quad\quad\dots\dots\dots\dots\dots(6)$$

We see, by (5) and (6), that the phase of the disturbance at any point P on K is either the same as or opposite to the phase of the disturbance at D.

If I be the intensity at P, i.e. the square of the amplitude, and if I_0 be the intensity at D, we have, by (5) and (6),

$$I = I_0 \cdot \frac{\sin^2\phi}{\phi^2} \cdot \frac{\sin^2\psi}{\psi^2}, \quad\quad\quad\dots\dots\dots\dots\dots(7)$$

where $\quad\quad \phi = 2\pi yb/\lambda a, \quad\quad \psi = 2\pi zc/\lambda a. \dots\dots\dots\dots(8)$

There is darkness when $\phi = \pm \pi, \pm 2\pi, \pm 3\pi, \dots$ and similarly when $\psi = \pm \pi, \pm 2\pi, \dots$, and hence the diffraction pattern is crossed by two sets of lines of darkness, which divide it into rectangles. When ϕ increases by π, y increases by $\frac{1}{2}\lambda a/b$, and when ψ increases by π, z increases by $\frac{1}{2}\lambda a/c$. If the corresponding steps in Y, Z on the plane K be β, γ, then

$$\beta = \lambda a \cdot DN_1/(2b \cdot N_2 A) \quad \text{and} \quad \gamma = \lambda a \cdot DN_1/(2c \cdot N_2 A).$$

Hence $\beta/\gamma = c/b$, and thus the rectangles on K are similar to the rectangular opening in H but are turned through a right angle. It must be remembered that the lines $Y = 0$ and $Z = 0$ in the plane K are exceptional and are not completely lines of darkness. Thus, along $Y = 0$, $I = I_0 (\sin^2 \psi)/\psi^2$, and therefore I is zero only at the points corresponding to $\psi = \pm \pi, \pm 2\pi, \dots$.

If, instead of a luminous point, an illuminated slit parallel to Oz be used, each point of the slit will act as an *independent* source. The line $Y = 0$ in the plane K will be the locus of the centres of an infinity of overlapping interference patterns. For each pattern, $I = 0$ when $\phi = \pm \pi, \pm 2\pi, \dots$, and hence the dark lines parallel to Oz will remain. On account of the overlapping of the patterns, the intensity along any line $Y = \text{constant}$ will be independent of Z.

For each of the overlapping patterns, the intensity along the line $Y = \text{constant}$ contains the factor $(\sin^2 \phi)/\phi^2$, and hence the resultant illumination along this line due to all the overlapping but independent patterns will be proportional to $(\sin^2 \phi)/\phi^2$. Thus the pattern is now represented by

$$I = I_0 (\sin^2 \phi)/\phi^2, \quad \ldots\ldots\ldots\ldots\ldots\ldots(9)$$

where I_0 now denotes the resultant intensity at D.

As ϕ increases from zero, $\sin^2 \phi$ varies over the range 0 to 1, while ϕ^2 continually increases. Hence the curve showing I/I_0 as a function of ϕ consists of a series of waves of rapidly diminishing amplitude, as shown in the table and in Fig. 207.

ϕ	I/I_0	ϕ	I/I_0	ϕ	I/I_0
0	1·0000	$5\pi/4$	$8/25\pi^2 = \cdot0324$	$5\pi/2$	$4/25\pi^2 = \cdot0162$
$\pi/4$	$8/\pi^2 = \cdot8106$	$3\pi/2$	$4/9\pi^2 = \cdot0450$	$11\pi/4$	$8/121\pi^2 = \cdot0067$
$\pi/2$	$4/\pi^2 = \cdot4053$	$7\pi/4$	$8/49\pi^2 = \cdot0165$	3π	0
$3\pi/4$	$8/9\pi^2 = \cdot0901$	2π	0	$13\pi/4$	$8/169\pi^2 = \cdot0048$
π	0	$9\pi/4$	$8/81\pi^2 = \cdot0100$	$7\pi/2$	$4/49\pi^2 = \cdot0083$

When I/I_0 is plotted against ϕ, the central band of illumination in Fig. 207 has width 2π; each of the other bands of illumination has width π.

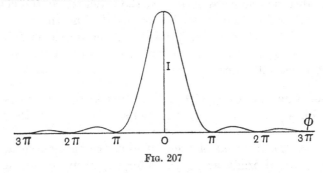

Fɪɢ. 207

329. Case of two slits. If two parallel illuminated slits be observed, each will give rise to a diffraction pattern, and, since the slits act as independent sources, the resultant intensity is the sum of the intensities due to the separate slits. Thus, when each slit is equally illuminated, the resultant illumination can be found either by calculation or by drawing two curves like Fig. 207, and then adding their ordinates. Fig. 208 shows the case in which the chief maximum of one diffraction pattern coincides with one of the first minima of the other. The result is a band of illumination of width 3π with a minimum at its centre and a maximum on either side at distance $\frac{1}{2}\pi$ from the centre. At each of these maxima the intensity is I_0, and at the

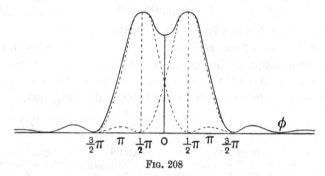

$$\frac{3}{2}\pi \quad \pi \quad \tfrac{1}{2}\pi \quad \mathrm{O} \quad \tfrac{1}{2}\pi \quad \pi \quad \frac{3}{2}\pi$$

Fig. 208

central minimum it is $2\,(4/\pi^2)\,I_0$ or $\cdot 8106 I_0$. On either side of this wide central band is a series of bands of small intensity.

If the distance between the centres of the two patterns be a little less than the critical value π, the central minimum in the resultant pattern will practically disappear. On the other hand, if the distance be greater than π, the intensity at the central minimum will fall markedly below $\cdot 8106 I_0$, and the separation between the centres of the patterns will be unmistakable. We may thus expect that, if the separation fall a little below π, an observer will detect no separation between the patterns.

330. Case of many slits. More precise results can be obtained if a grating of equally spaced slits be used. If the pitch of the grating, i.e. the distance between successive slits, be such that the abscissae of the centres of the patterns measured from a central origin are $\pm\,\tfrac{1}{2}\pi,\ \pm\,\tfrac{3}{2}\pi,\ ...$, when plotted on the scale of ϕ, the intensity at $\pm\,\tfrac{1}{2}\pi$ is I_0 and at the central zero it is, by the table in § 328,

$$2I_0\left(\frac{4}{\pi^2}+\frac{4}{9\pi^2}+\frac{4}{25\pi^2}+\,...\right)=I_0,$$

since $1/1^2 + 1/3^2 + 1/5^2\,... = \pi^2/8$. It is evident, without further calculation, that the intensity must be very nearly constant throughout the pattern. It can, however, be shown that, if the pitch of the pattern be π, or less than π, on the scale of ϕ, the intensity is constant throughout the pattern. If the pitch be increased beyond π, the intensity at the central origin will rapidly diminish relative to the two neighbouring maxima. Thus there will be no resolution at all if the pitch be less than π, but resolution will begin as soon as the pitch is increased beyond π.

When $\phi = \pi$, we have, by (8), $y/a = \lambda/2b$.

If the distance from the central line of one slit to the central line of the next be p cm., the critical case occurs when $y = p$. We then have

$$p/a = \lambda/2b. \quad\quad\quad\quad\quad\quad\quad\quad\quad (10)$$

In this case, the angle subtended by λ at distance $2b$, the width of the slit in the screen H (Fig. 206), equals the angle subtended by the pitch p at distance a. Hence, if we measure a and b and know p, we can find λ.

331. Experimental details. A suitable grating may be made by ruling scratches on the coating on a photographic plate which has been exposed to light and then developed. The pitch p may be ·1 cm.; each scratch should not exceed ·02 cm. in width. If the plate be fixed to the carriage of a travelling microscope, it can be advanced by steps of ·1 cm. A *fixed* straight-edge, set so as nearly to touch the plate, guides the ruling tool. A simple "geometrical" arrangement may be used. A needle A forms the tool and also serves as one of three legs A, B, C of a small metal plate. The angle ABC is approximately $90°$. The needle and the leg B are kept in contact with the fixed straight-edge. The two legs B and C slide on a smooth track formed by a plate of glass or metal fixed to the carriage. Care is needed to rule a grating of satisfactory uniformity.

A serviceable grating may be made by laying a number of straight wires in the threads of two similar screws fixed with their axes parallel and about 2 cm. apart. The wires, of copper, may be straightened by rolling them between two glass plates.

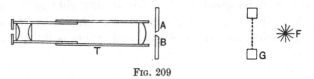

FIG. 209

The apparatus is shown in plan in Fig. 209. The grating G is set up with its rulings vertical and is backed by a sodium flame

F. The telescope T is focused on the grating. A slit AB, adjustable by a screw, is placed in front of the objective of T, the edges of the slit being vertical. If the slit be sufficiently wide, the grating will still be seen distinctly, but with considerable loss of illumination. As the slit AB is narrowed, a stage is reached when the lines begin to be blurred. If the slit be made a little narrower, only uniform illumination will be seen. There may be vertical streaks due to irregularities in the grating, but these are easily distinguished from the evenly spaced blurs which were seen when the slit was a little wider. The observer endeavours to adjust the slit so that this *regular* pattern is just not visible. A little practice is needed to train the judgement, and several trial settings should be made before any readings are recorded.

The width of the slit is measured by a graduated wedge; the width is indicated by the reading on the face of the wedge. The wedge is *gently* inserted into the slit, and the graduation is read. These wedges are supplied by tool-dealers.

The distance a cm. from slit to grating is measured; if a tape measure be used, it should not be attached to telescope or grating lest an accident occur.

The observations are repeated at different distances. For *each* value of a, the telescope is focused on the grating either *before* the slit is put into position, or with the slit wide open.

As a test of the theory, the wave length is found by (10). Thus

$$\lambda = 2bp/a. \dots\dots\dots\dots\dots\dots\dots(11)$$

Great accuracy is not to be expected, but, with care, the errors should be less than five per cent.

The minimum width of the adjustable slit which allows resolution will be greater when the grating slits are of finite width than when their width is infinitesimal. Hence the value of λ, found when the width of the grating slits is finite, will be too large.

332. Practical example. Messrs McKenzie and Dyson used a grating formed by ruling lines on a photographic plate; the pitch was $p = 0.1$ cm. Sodium light, $\lambda = 5.893 \times 10^{-5}$ cm., was used.

Grating to slit = a	Width of slit	Mean width = $2b$	$\lambda = 2bp/a$
385·1 cm.	·220 ·219 cm.	·2195 cm.	$5·700 \times 10^{-5}$ cm.
426·7	·251 ·251	·2510	$5·882 \times 10^{-5}$
455·7	·279 ·274	·2765	$6·068 \times 10^{-5}$
498·6	·305 ·301	·3030	$6·077 \times 10^{-5}$

Mean value of $\lambda = 5·93 \times 10^{-5}$ cm.

EXPERIMENT 63. **Measurement of wave length by interference with two slits.**

333. Introduction. Let B_0B, C_0C (Fig. 210) be two equal and very narrow slits in a plate H at distance b apart. Let B_0C_0 be perpendicular to the slits and let O be its mid-point. Take as axes Ox normal to H, Oy in the plane of H and perpendicular to the slits, and Oz parallel to the slits. Through B_0OC_0 draw a plane $A_0B_0C_0$—the plane of xy—meeting H at right angles Let AA_0 be a straight line parallel to Oz emitting light of wave

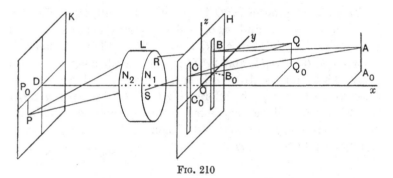

FIG. 210

length λ in air; actually AA_0 is the image formed by an auxiliary lens (Fig. 213) of a narrow slit in a plate, the light coming from a sodium flame behind the plate. Let L be a lens, or system of lenses, with nodal points N_1, N_2, which lie approximately upon Ox. Let Ox meet a plane K normally at D, let P be a point on K near D, and let P_0 be its projection on the plane of xy. We wish to find the illumination at P.

Let Q be the image of P formed by L, and let Q_0 be its projection on Oxy. Then Q lies in the plane conjugate to K with respect to L. Let QB, QC meet L in R, S. On account of diffraction, light from a given point A on AA_0, after passing

B, C, is not confined to the straight lines AB, AC, but spreads out. Thus, light from A will travel along BR, although ABR be not straight, and similarly for CS.

The optical distances $[ABP]$, $[ACP]$ from A to P by the two paths are $AB + [BP]$ and $AC + [CP]$ respectively, where $[BP]$, $[CP]$ involve the indices of the lenses and other media between the planes H and K. But the optical distances $QB + [BP]$ and $QC + [CP]$ are equal, since Q and P are conjugate. Thus

$$[ACP] - [ABP] = AC - AB - (QC - QB). \quad \ldots(1)$$

Let the coordinates of A be α, β, γ, let those of Q be ξ, η, ζ, and let $BB_0 = CC_0 = k$. Then

$$AB^2 = \alpha^2 + (\beta - \tfrac{1}{2}b)^2 + (\gamma - k)^2, \quad AC^2 = \alpha^2 + (\beta + \tfrac{1}{2}b)^2 + (\gamma - k)^2.$$

Hence

$$AC - AB = (AC^2 - AB^2)/(AC + AB) = 2b\beta/(AB + AC).$$

Similarly

$$QC - QB = 2b\eta/(QB + QC).$$

When, as in practice, k/OA_0, γ/OA_0 are small, the difference between $AB + AC$ and $A_0B_0 + A_0C_0$ is very small compared with $A_0B_0 + A_0C_0$. Since b/OA_0 is very small, the difference between $A_0B_0 + A_0C_0$ and $2A_0O$ is very small compared with $2A_0O$. Thus, unless $AC \sim AB$ be *large* compared with λ, the error made in the path difference $AC - AB$ by writing $2A_0O$ for $AB + AC$ will be very small compared with λ. Similarly, we may write $2Q_0O$ for $QB + QC$. Then

$$[ACP] - [ABP] = b\beta/OA_0 - b\eta/OQ_0 = b\,(\sin A_0Ox - \sin Q_0Ox).$$
$$\ldots\ldots\ldots(2)$$

The angles A_0Ox, Q_0Ox are positive when A_0, Q_0 lie within the angle xOy. Equation (2) does not contain ζ or k. Thus, since ζ or QQ_0 is proportional to PP_0, the difference of optical distance is, to our very close approximation, independent of the distance PP_0 and is also independent of k or BB_0. Hence no point on A_0A will cause illumination at any point on PP_0, if, n being a positive or negative integer, Q_0 satisfy the condition

$$b \sin Q_0Ox = b \sin A_0Ox + (n + \tfrac{1}{2})\,\lambda. \quad \ldots\ldots\ldots(3)$$

The points of the illuminated slit A_0A act as *independent* sources, but, if (3) be satisfied, no point on A_0A causes illumination at any point on P_0P, and hence the whole slit A_0A causes no illumination along PP_0. Thus PP_0 is one of a series of dark bands.

If K be a material screen, there will be a bright or a dark band through P according as $b\,(\sin Q_0Ox - \sin A_0Ox)$ is an even or odd multiple of $\frac{1}{2}\lambda$.

If the rays after passing through L pass through other lenses and meet again in P', the image of P by the other lenses, the optical distance from P to P' is the same for all the rays, and hence, if there be darkness at P, there will be darkness at P'. If a photographic plate be placed so that P or P' lies on it, we shall obtain a photograph of the bands. In the present experiment, the bands are observed by aid of a telescope whose objective is L, and P' is that point on the retina where an image of P is formed by the eye-piece and the eye itself.

The *scale* of the system of bands will depend upon the distance between the slits, upon the focal length of the lens L and upon the positions, relative to L, of the screens H and K. But the distinctness of the bands is independent of these quantities. There is no question of focusing the bands, for the observer cannot put them out of focus.

334. Slits of finite width. Since the distance between the centres of the slits is only a few tenths of a millimetre, while each slit must be a few hundredths of a millimetre wide to transmit a satisfactory amount of light, the effect upon the interference pattern of using slits of finite width must be investigated.

We take A_0 at infinity in the plane Oxy; then the wave fronts of the incident light from A_0 are parallel to the slits. Let Fig. 211

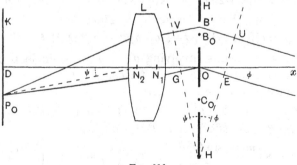

Fig. 211

be a section of the system by the plane Oxy, and let B_0, C_0 be now the centres of the sections of the slits. Let $B_0C_0 = b$, and let the width of each slit be h. The focal plane of L is taken as the plane K. Let EU be an incident front whose normal OE lies in the plane Oxy and makes angle ϕ with Ox. Let GV be a mathematical plane whose normal OG lies in Oxy and makes angle ψ with Ox. Then, if $P_0N_2D = \psi$, all points on GV will be optically equidistant from P_0, and hence, to find the illumination at P_0, we need only consider the phase relations on GV. Let B', in the plane of the figure, be the point in the first slit for which $y = \frac{1}{2}b + r$, and let the disturbance at P_0 which an element of width dr at O would produce, if there were an opening at O, be $g \sin (2\pi vt/\lambda).dr$. Then, if, as in Fig. 211, the optical distance $[UB'VP_0]$ exceed $[EOGP_0]$, the disturbance at P_0 at time t due to dr at B', i.e. due to the wave transmitted from U to P_0, is the same as that at P_0 at $t - \tau$ due to the wave transmitted from E to P_0, where τ is the excess of the time of travel from U to V over that from E to G. Thus

$$\tau = (UB' - EO + B'V - OG)/v = (\tfrac{1}{2}b + r)\,\theta/v,$$

where $$\theta = \sin \phi + \sin \psi.$$

Hence, if the disturbance at P_0 due to dr at B' be $d\omega_B$,

$$d\omega_B = g \sin 2\pi\lambda^{-1}\{vt - (\tfrac{1}{2}b + r)\,\theta\}.dr.$$

The resultant disturbance at P_0 due to the B-slit is

$$\omega_B = g \int_{-\frac{1}{2}h}^{\frac{1}{2}h} \sin 2\pi\lambda^{-1}\{vt - (\tfrac{1}{2}b + r)\,\theta\}.dr$$

$$= (\lambda g/2\pi\theta)\left[\cos 2\pi\lambda^{-1}\{vt - (\tfrac{1}{2}b + r)\,\theta\}\right]_{-\frac{1}{2}h}^{\frac{1}{2}h}$$

$$= (\lambda g/2\pi\theta)[\cos 2\pi\lambda^{-1}\{vt - \tfrac{1}{2}(b+h)\theta\} - \cos 2\pi\lambda^{-1}\{vt - \tfrac{1}{2}(b-h)\theta\}].$$

To find ω_C, we change the sign of b. Thus

$$\omega_C = (\lambda g/2\pi\theta)[\cos 2\pi\lambda^{-1}\{vt + \tfrac{1}{2}(b-h)\theta\} - \cos 2\pi\lambda^{-1}\{vt + \tfrac{1}{2}(b+h)\theta\}].$$

Combining the terms involving $b + h$ into a single term and similarly combining the $b - h$ terms, we find, for the whole disturbance, $\omega = \omega_B + \omega_C$, where

$$\omega = (\lambda g/\pi\theta)\{\sin \pi (b + h)\,\theta\lambda^{-1} - \sin \pi (b - h)\,\theta\lambda^{-1}\}.\sin 2\pi vt\lambda^{-1}$$

$$= (2\lambda g/\pi\theta).\cos \pi b\theta\lambda^{-1}.\sin \pi h\theta\lambda^{-1}.\sin 2\pi vt\lambda^{-1}$$

$$= 2gh.\cos\left(\frac{\pi b\theta}{\lambda}\right).\frac{\sin (\pi h\theta/\lambda)}{\pi h\theta/\lambda}.\sin\left(\frac{2\pi vt}{\lambda}\right) = I^{\frac{1}{2}}.\sin\left(\frac{2\pi vt}{\lambda}\right). \quad (4)$$

In (4), I is the "intensity," i.e. the square of the amplitude. Since, as x approaches zero, the limit of $(\sin x)/x$ is unity, the intensity when $\theta = 0$ is $I_0 = 4g^2h^2$. Hence, for any value of θ,

$$I = I_0 \cos^2\left(\frac{\pi b \theta}{\lambda}\right) \cdot \frac{\sin^2\left(\pi h \theta / \lambda\right)}{(\pi h \theta / \lambda)^2} \cdot \quad \dots\dots\dots\dots(5)$$

Thus $I = 0$ when θ has the b-values $\pm \frac{1}{2}\lambda/b$, $\pm \frac{3}{2}\lambda/b$, ... But I also vanishes when θ has the h-values $\pm \lambda/h$, $\pm 2\lambda/h$, ... If h/b be small, I vanishes many times at the b-values before it vanishes at the first h-value $\pm \lambda/h$. If $|\theta|$ be increased beyond λ/h, there will again be a succession of b-values at which I vanishes, and so on. If we draw the curve showing $\sin^2(\pi h \theta / \lambda) \cdot (\pi h \theta / \lambda)^{-2}$ in terms of θ, and then inscribe between it and the axis of θ

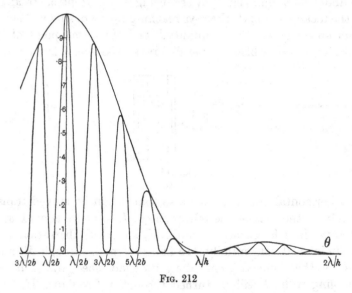

FIG. 212

a series of waves touching that axis at $\pm \frac{1}{2}\lambda/b$, $\pm \frac{3}{2}\lambda/b$, ... we have a representation of the distribution of intensity in the interference pattern. Fig. 212 shows that when $|\theta| > \lambda/h$, I is always small compared with I_0.

335. Experimental details. The twin slits are formed by ruling lines with a needle point on the silvering on a small plate of mirror glass of good quality. The method described in § 331 may be employed. In forming each slit, the plate is advanced by about ·001 cm. after each stroke of the ruling needle so as to remove completely the silver from the opening. The edges of each slit should be cleanly cut; no scraps of silver should remain in the opening. The needle is ground to a chisel edge.

The glass slit-plate is held in a cap R (Fig. 213) which fits on to the telescope T. In the theory of § 333, the distance between O and the lens L may be small or great, but that theory assumes that L is free from spherical aberration. To minimise aberration, it is best to use as far as possible the central part of L, and hence LO should not exceed a few centimetres. To obtain a slit at "infinity," a converging lens J of focal length f is placed between the telescope and the illuminated vertical slit M; f may be about 66·6 cm., the "power" being $+ 1·5$ dioptre. If M be in the focal plane of J, the rays reaching the twin slits will come from an image of M at "infinity," and thus each point of M will give rise to a beam of parallel rays. The lens J is mounted

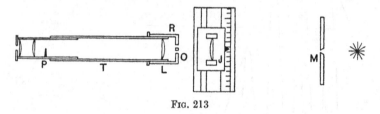

FIG. 213

on a horizontal graduated slide so that it can be moved transversely to the axis of the telescope. If J be plano-convex and its *plane* face be turned towards M, spherical aberration will be diminished. If, initially, the axis of J pass through M, and if J be then moved y cm. along its slide, the parallel beam emerging from J will be turned through y/f radians. If J be moved along its slide, the *dark* bands can be brought, one by one, to coincide with the pointer P. The observer views P through the eye-piece of the telescope; he adjusts the *eye-piece* to suit his own sight.

The *dark* bands are used in preference to the bright bands because the positions of the maxima of brightness depend not only upon b but upon the ratio of h to b.

The slit is illuminated by the flame of a Mekker burner into which sodium chloride is introduced by means of asbestos soaked in brine; if the asbestos enter far into the flame, the temperature is reduced with a loss of brightness. It is necessary to maintain the flame in full brilliancy. The supply of sodium is, therefore, frequently renewed by placing drops of brine upon the asbestos in the flame by aid of a slip of wood.

The cap R and the lens L are removed, and the eye is placed (1) just to the right, (2) just to the left of the telescope. If the slit appear equally bright in both cases, the brightest part of the flame is on the line OM (Fig. 213). With careful attention to the illumination, the bands will be so bright that it will be unnecessary to darken the room.

The telescope T, first focused for "infinity," is placed so that its axis passes through M, and the distance of the lens J from M is adjusted so that, the cap R being absent, the image of M coincides with P, when J is suitably adjusted on its transverse slide. The cap R is now placed on the telescope, the plane of the slits being perpendicular to the axis of T. The twin slits should be as nearly as possible parallel to the slit M.

To identify the centre of the pattern, a patch of silvering, roughly circular in form, not more than $0 \cdot 5$ mm.^{2} in area, may be removed from the band of silver between the slits. The light passing through this opening forms an image of M in the plane of the pointer P. This image will coincide with and reinforce the central bright band in the pattern. This opening in the silvering is, of course, not essential.

The width of the slit M is adjusted to give the most distinct pattern. If M be very narrow, the illumination is so feeble that the bands cannot be satisfactorily located. If M be wide, the pattern will be diffuse and the contrasts will be weak.

In the method used in this experiment, the point P_0 of Fig. 210 remains fixed, as it coincides with the tip of the pointer viewed through the eye-piece. Hence, in Fig. 210, Q_0 is a fixed point, and, therefore, $Q_0 O x$ is a fixed angle, which is zero with perfect alignment.

Since the angle $A_0 O x$ does not exceed one or two degrees, we may replace $\sin A_0 O x$ by ϕ, where $\phi = A_0 O x$. Then, by (3), if there be darkness at P for any value of ϕ, there will again be darkness if ϕ be changed by $n\lambda/b$ in either direction. If a movement of y cm. of the lens J along its transverse slide shift the pattern by n times the distance between adjacent dark bands, the change in ϕ for n bands is y/f. Hence

$$\lambda = by/nf. \quad \dots\dots\dots\dots\dots\dots(6)$$

A value for y/n is found by plotting the readings of the lens index against n, or by finding the change of reading for four or five bands, as in § 336. In the latter case, five or six values of the difference can be obtained; the mean is used.

The slits are measured by a travelling microscope. If this have a single cross-wire, the wire should be perpendicular to the direction of motion of the microscope. The slits, whose edges, taken in order, are D, E and F, G, are set perpendicular to the same direction. The difference between (1) the mean of the readings for D and E, and (2) the mean of the readings for F and G, is taken as the distance between the centres of the slits.

The focal length (f) of the lens J may be found by aid of the telescope, which has been focused for "infinity." The distance MJ is adjusted so that the image of the slit M coincides with P, the cap R being removed. If h be the distance from M to the nearer face of J, $f = h + kc$, where c is the (small) thickness of the lens. If J be equi-convex, $k = \frac{1}{3}$, and if J be plano-convex, the plane surface facing M, $k = \frac{2}{3}$ approximately (see § 198).

If J be plano-convex, f may be found by a pin and a plane mirror (§ 139), the plane face being towards the mirror. Then f equals the distance from the pin to the curved face of J.

In the general case, the telescope will not be *exactly* focused for "infinity," and hence, if the distance MJ be adjusted so that, with the cap R removed, the image of M coincides, without parallax, with the pointer P, M will not be exactly in the focal plane of J. If the distance of M from the appropriate nodal point of J be not f but l, and if the distance of O from the other nodal point be d, it may be shown that

$$A_0 Ox = \frac{y}{f} \cdot \left\{ 1 + \frac{d}{l} \cdot \frac{f-l}{f} \right\}^{-1}.$$

Thus, if both d/l and $(f-l)/f$ be small, y/f is a very close approximation to $A_0 Ox$. Hence, if f be accurately known, we need not measure MJ provided (1) that d/f be small, and (2) that the telescope be nearly focused for "infinity."

336. Practical example. The readings obtained by Mr G. F. H. Harker, when the travelling microscope was set on the edges D, E, F, G of the slits, were

$$\left.\begin{array}{l} D \ \cdot6019 \\ E \ \cdot6073 \end{array}\right\} \text{ Mean } \cdot6046 \text{ cm.} \qquad\qquad \left.\begin{array}{l} F \ \cdot6360 \\ G \ \cdot6415 \end{array}\right\} \text{ Mean } \cdot6388 \text{ cm.}$$

Hence $b = \cdot6388 - \cdot6046 = \cdot0342$ cm.

The focal length of plano-convex lens J, obtained by plane mirror, was $f = 68\cdot64$ cm.

Scale readings of J for successive *dark* bands are given in the table. The bands denoted by $+ 1$ and by $- 1$ are separated only by the central bright band; there is no dark band between them.

Band	Reading	Band	Reading	Difference for six bands
– 6	5·46 cm.	+ 1	6·15 cm.	·69 cm.
– 5	5·56	+ 2	6·28	·72
– 4	5·68	+ 3	6·39	·71
– 3	5·80	+ 4	6·51	·71
– 2	5·90	+ 5	6·61	·71
– 1	6·03	+ 6	6·76	·73

For $n = 6$, mean value of $y = \cdot712$ cm. Hence, by (6),

$$\lambda = by/nf = \cdot0342 \times \cdot712/(6 \times 68\cdot64) = 5\cdot91 \times 10^{-5} \text{ cm.}$$

CHAPTER XIII

DIFFRACTION GRATING AND ZONE PLATE

337. The plane grating. For theoretical purposes, a transmission grating may be supposed to be formed by cutting a large number of equidistant parallel slits, all of the same width, in a very thin opaque lamina, the length of the slits being great compared with their pitch. Let the width of each slit be a cm. and let the shortest distance between one slit and the next be b cm. The distance $a + b$ between the centre lines of successive slits is called the pitch of the grating or the grating interval, and is denoted by d.

Reflexion gratings are formed by ruling, with pitch d, grooves, each of width b cm., upon a polished plane surface of speculum metal, leaving strips of polished surface each a cm. wide. Light is regularly reflected at the polished strips but is merely scattered in all directions by the rough surface of the grooves.

Mr Thorpe has made celluloid casts of metallic reflexion gratings. A thin layer of dissolved celluloid is poured upon the grating. When the film has set, it is stripped off and is then mounted between two glass plates whose surfaces are plane and parallel*. The parts of the film corresponding to the polished strips of the reflexion grating transmit light regularly. The remainder of the film, being rough, scatters light in all directions.

Gratings have been made by ruling grooves with a diamond upon glass. Light is scattered at the rough surface of the grooves and regularly transmitted by or reflected at the polished strips between the grooves.

There will be no concentration of light in any particular direction due to the light scattered at the rough parts of the grating; thus we may ignore this light and may treat the rough parts as perfectly opaque to light in the case of a transmission grating, and as perfectly absorbent in the case of a reflexion grating.

---

* If any solvent for celluloid come in contact with the grating, it will dissolve the film and destroy the grating.

CH. XIII] DIFFRACTION GRATING AND ZONE PLATE   319

338. The axes of a grating. A point O on the centre line of one of the polished strips is taken as origin. The axes are: (1) The normal ON to the plane of the grating. (2) The transverse axis OT, a straight line intersecting the rulings at right angles. (3) The longitudinal axis OL, a straight line parallel to the rulings.

Any plane parallel to the plane ONT is called a principal plane of the grating.

339. Diffraction by transmission grating. We first consider the case in which the width a of the openings is very small compared with the wave length. We shall suppose that the incident rays are parallel to a principal plane and make an angle θ with the normal to the grating. The wave length is denoted by λ.

If we draw a series of consecutive planes of equal phase across the incident beam, these planes will be normal to the rays and will be separated by the common interval λ.

When a wave front falls on an opening in the grating, a disturbance starts out from that opening towards the other side of the grating. When a is very small, the wave front of the transmitted disturbance will be, in our case, half of a circular cylinder. As the beam of incident rays continues to beat upon this opening, a series of cylindrical waves will travel out from it, and the surfaces of equal phase will be coaxal cylinders, with radii differing by steps of λ, and the same is true for every opening.

Let A_1, A_2, ... (Fig. 214) be *narrow* openings in the grating, the distance from centre to centre being d, as before. The plane of the figure is a principal plane of the grating. Let B_1, B_2, ... be the points in which a wave front of the incident beam cuts the rays through A_1, A_2, ... On the other side of the grating draw A_1C_1,

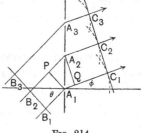

Fig. 214

A_2C_2, ... making an angle ϕ with the normal, the angle ϕ having the same sign as θ when the bent line $B_1A_1C_1$ *crosses the normal* through A_1, as in Fig. 214. Draw C_1C_2, ...

perpendicular to A_1C_1 to cut A_2C_2, A_3C_3, ..., which are
parallel to A_1C_1, in C_2, C_3, ... Then, since a *succession* of
waves of wave length λ falls upon each opening, the disturbances at C_1, C_2, ... will all be in the same phase, if the distances
$B_1A_1C_1$, $B_2A_2C_2$, ... differ by steps of $n\lambda$, where n is an integer.
If circles be drawn through C_1, C_2, ... about A_1, A_2, ... as centres,
they will be the sections by the plane of the diagram of the
cylindrical wave fronts. At a distance of thousands of wave
lengths from the grating, these cylindrical wavelets will merge
into practically a single wave front, and thus A_1C_1 will be the
direction of a diffracted beam.

Draw A_1P, A_2Q perpendicular to B_2A_2, A_1C_1 respectively.
Then the condition that A_1C_1 should be the direction of a
diffracted beam is that $PA_2 - A_1Q = n\lambda$, where n is a positive
or negative integer. But $PA_2 = d \sin \theta$, $A_1Q = d \sin \phi_n$, where
ϕ_n corresponds to n, and thus the condition becomes

$$\sin \theta - \sin \phi_n = n\lambda/d. \quad \ldots\ldots\ldots\ldots(1)$$

Thus, for a given value of θ, there are two diffracted beams for
each value of $|n|$, corresponding to (a) the positive and (b) the
negative value of n, provided that in each case (1) yield a
value of $\sin \phi_n$ lying between $+1$ and -1.

When $\theta = 0$, so that the incident rays are normal to the
grating,

$$\sin \phi_n = n\lambda/d, \quad \ldots\ldots\ldots\ldots\ldots\ldots(2)$$

and then the two diffracted beams of order n are symmetrical
with respect to the normal. The number of pairs of diffracted
beams in this case is the integer next *less* than d/λ.

340. Minimum deviation. The deviation D is $\theta - \phi_n$,
and this has a stationary value, for variations of θ, when
$dD/d\theta = 0$. By (1), $d\phi_n/d\theta = \cos \theta/\cos \phi_n$, and hence

$$dD/d\theta = 1 - \cos \theta/\cos \phi_n.$$

Thus $dD/d\theta = 0$ when $\cos \phi_n = \cos \theta$. The solution $\phi_n = \theta$ corresponds to the beam transmitted without bending, and, by
(1), requires that $n = 0$. The other solution is $\phi_n = -\theta$. Since
we may take θ as *positive*, $\sin \theta = n\lambda/2d$ when the deviation is

stationary for the diffracted beam of (*positive*) order n. If we put $\cos \phi_n = \cos \theta$, $\sin \phi_n = -\sin \theta$ in the expression

$$d^2D/d\theta^2 = (\sin \theta \cos^2 \phi_n - \cos^2\theta \sin \phi_n)/\cos^3 \phi_n,$$

we find that $d^2D/d\theta^2$ has the *positive* value $2 \tan \theta$. Thus, the value of θ for which $\sin \theta = n\lambda/2d$ gives a *minimum* value for D.

341. Effect of finite width of openings. When the openings have a finite width a, we may suppose each to be divided into s narrow strips of width a/s. The disturbances due to the series of corresponding strips of the successive openings will give rise to diffracted beams whose directions are given by (1). Hence the width, a, of the openings will not affect the *direction* of the diffracted beams. The width may, however, affect the *intensity* of the light along any one of the directions.

Let EE' (Fig. 215) be an opening of width a, and F its midpoint. Then, if $SE' - ET = \lambda$, the disturbance from a point on $E'F$ at distance x from E', where $x \not> \frac{1}{2}a$, will be neutralised by that from the point on FE at distance x from F, and the intensity in the direction ET will be zero. The intensity in the direction A_1C_1 (Fig. 214) will be zero if $a = 0$, and will increase as a increases up

FIG. 215

to a certain value when it will fall as a is further increased. In wave length measurements we are not concerned with these effects; we shall not further consider them.

342. Reflexion grating. Suppose, first, that the polished strips are very narrow and let A_1, A_2, ... (Fig. 216) be their sections by a principal plane (§ 338) of the grating. Let B_1B_2 ... be an incident and C_1C_2 ... a diffracted wave front, the rays corresponding to A_1, A_2, ... being $B_1A_1C_1$, $B_2A_2C_2$, ... Let A_1B_1, A_1C_1 make angles θ, ϕ with the normal, the angles having the same sign when C_1 and B_1 lie on the *same* side of the normal through A_1. Draw A_1P, A_1Q perpendicular to A_2B_2, A_2C_2 respectively. Then A_1C_1 will be the direction of a diffracted

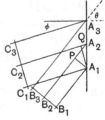

FIG. 216

beam of order n, if $A_2P + A_2Q = n\lambda$. But $A_2P = d \sin \theta$, $A_2Q = d \sin \phi_n$, and so the diffraction equation becomes

$$\sin \theta + \sin \phi_n = n\lambda/d. \qquad \qquad (3)$$

The case when the polished strips are of finite width a may be considered as in § 341.

343. Diffraction of rays incident in any direction. We now consider the case when the incident rays are no longer restricted, as in §§ 339, 342, to be perpendicular to the longitudinal axis OL (§ 338), but have any direction. In Fig. 217, let K be a point on OL. Let QO, AK be parallel incident rays, let OQ'

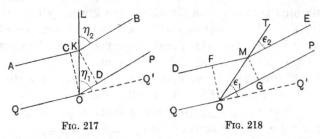

FIG. 217   FIG. 218

be the continuation of QO, and let OP, KB be parallel diffracted rays; QO, OP, OL are not, in general, co-planar. Then planes through O, K, perpendicular to AK, OP and cutting them in C, D, are parallel to the wave fronts. Since K, O are on the *same* opening of the grating, the distances CK, OD from the plane OC to the plane KD are *equal*. Hence, if $\eta_1 = Q'OL = CKO$, and $\eta_2 = BKL = DOK$,

$$\eta_2 = \eta_1. \qquad \qquad (4)$$

In Fig. 218, let OT be the transverse axis and M the centre of the opening next to O, so that $OM = d$. Let QO, DM be parallel incident rays and let OP, ME be parallel diffracted rays. The lines QO, OP, OT are not, in general, co-planar. Draw OF, MG perpendicular to DM, OP. Let QOQ' be à straight line, and let $Q'OT = \epsilon_1$, $POT = EMT = \epsilon_2$. Since OF, MG are parallel to the wave fronts, OG exceeds FM by $i\lambda$, where i is a positive or negative integer. Hence

$$\cos \epsilon_2 = \cos \epsilon_1 + i\lambda/d. \qquad \qquad (5)$$

If, as sometimes, we use n to denote an integer, we have

$$\cos \epsilon_2 = \cos \epsilon_1 + n\lambda/d. \qquad \qquad (6)$$

These results are noteworthy. By (4), $\eta_2 = \eta_1$ for all values of i and λ, and for all angles between the incident rays and OT. By (5), ϵ_2 depends only upon ϵ_1, λ and i, and not upon the angle between the incident rays and OL. The angles η_2 and ϵ_2 determine the directions of the two diffracted beams—one reflected and one transmitted. The transmitted beam can be identified, for the angle between it and the forward normal ON is *less than* $\frac{1}{2}\pi$.

EXPERIMENT 64. **Determination of wave length by diffraction grating.**

344. Introduction. A spectrometer with its axis vertical (§ 58) is used. The telescope and collimator are "set for infinity." The line of collimation, or axis, of the telescope is defined by X, the intersection of its cross-wires (§ 46); this axis should be perpendicular to the axis of the spectrometer (§ 53). A fine wire stretched across the slit of the collimator and intersecting the slit at Y defines the axis of the collimator. The wire is adjusted so that its image can be made to coincide with X by turning the telescope about the axis of the spectrometer When X and the image of Y, as seen through the eye-piece, are at the same level, the axis of the collimator is perpendicular to that of the spectrometer. For these adjustments white light may be used.

345. Adjustment of grating. When collimator and telescope have been adjusted, the grating is placed on the spectrometer table with its plane approximately perpendicular to the line joining A, B, two of the three levelling screws A, B, C of the table. If necessary, it may be secured with a little wax. The slit is illuminated by a sodium flame or other source of light having bright lines in its spectrum.

The longitudinal axis OL (§ 338) of the grating must be parallel to the axis of the spectrometer. Since two angles are required to fix the direction of OL, two angular adjustments will generally be required. The first step is to make the plane of the grating parallel to the axis of the spectrometer. The telescope is set at about 90° to the collimator, the spectrometer table is turned so that the grating reflects light from the slit into the telescope, and the levelling screws A, B are adjusted

so that the image of Y coincides with X. If the plates protecting
the grating be not quite parallel and thus there be more than
one reflected image, a rough adjustment must suffice. But it
can be shown by (5), § 343, that when the transverse axis OT
is horizontal and the incident rays are perpendicular to OT,
though not, all of them, horizontal, the deviation of the curved
diffracted image of the *slit* (not the *point* Y), as measured by
X the intersection of the cross-wires of the telescope, is the
same as when the plane of the grating is vertical.

The grating must now be so set in its own (now vertical) plane
that OL is vertical. The spectrometer table is turned so that
the plane of the grating is approximately perpendicular to the
axis of the collimator. The telescope is then set to view in turn
the diffracted images of Y. By § 343, the directions of the
diffracted beams corresponding to the *point* Y make with OL
the same angle ($\frac{1}{2}\pi$) as does the forward direction of the incident beam. The latter is horizontal, but, unless OL be vertical,
the diffracted beams cannot be horizontal. Thus OL can be set
vertical by turning the screw C until the diffracted images of
Y can be brought to coincide with X by moving the telescope
about the axis of the spectrometer. The grating is now in adjustment relative to the spectrometer table.

346. Method of normal incidence. The first step is to set
the plane of the grating perpendicular to the axis of the collimator. The telescope is turned to view the slit through the
grating, and its cross-wire is brought to the image of the slit;
the telescope vernier is then read. This gives the zero reading
of the telescope. The telescope is now turned through 90°. The
grating is set to bring the image of the slit by reflexion to the
cross-wires and the vernier of the spectrometer table is read.
Lack of parallelism of the plates protecting the grating may
prevent accuracy in this adjustment. The grating is then turned
through 45° in the proper direction; its plane is now perpendicular to the axis of the collimator.

The slit is made as narrow as convenient and is illuminated
with the light of which the wave length is to be found. If a
flame be used, it must not be so near as to damage the slit by
heat or by the spluttering of salt.

The telescope is turned to bring each of the first order images in turn to the cross-wire, the vernier is read and the displacement of the telescope from its zero is calculated. If the images of the second and higher orders be observable, the readings are repeated for each pair. If there be two verniers, each is read.

The diffracted image of the slit formed in the focal plane of the telescope objective (focal length $= f$) is curved. Let ρ be the radius of curvature at X, the intersection of the cross-wires. It can be shown by (5), § 343, that, when all the incident rays are perpendicular to OT, $\rho = f \tan \epsilon_2$. In the present method, $\epsilon_2 = \frac{1}{2}\pi - \phi$, where ϕ is the deviation.

Any slight discrepancy (not due to mere errors of observation) between α and β, the deviations to right and left for the beams of any order, is due to a small angle χ between the normal to the grating (assumed horizontal) and the axis of the collimator. By (1), § 339, $\sin \alpha = n\lambda/d + \sin \chi$, $\sin \beta = n\lambda/d - \sin \chi$. Hence

$$n\lambda/d = \tfrac{1}{2}(\sin \alpha + \sin \beta) = \sin \tfrac{1}{2}(\alpha + \beta)\cos \tfrac{1}{2}(\alpha - \beta).$$

If α and β be nearly equal, $\cos \tfrac{1}{2}(\alpha - \beta)$ is very nearly unity, and we may put

$$n\lambda/d = \sin \tfrac{1}{2}(\alpha + \beta). \quad \dots\dots\dots\dots(1)$$

Hence we may use the sine of the mean deviation instead of the mean of the sines of the deviations.

If time allow, the observations may be repeated after the spectrometer table has been turned through 180°.

347. Practical example. Mr J. A. Pattern used normal incidence; grating had $d = 1 \cdot 75257 \times 10^{-4}$ cm. Zero reading of telescope, 210° 2′ 0″, was used in finding deviations on either side; telescope had only one vernier. The sodium lines D_1, D_2 were resolved; D_1 has greater λ than D_2 and is more deviated. Results for D_1 follow:

	First order		Second order	
Readings	229° 36′ 30″	190° 17′ 45″	252° 11′ 45″	167° 38′ 0″
Deviations	19 34 30	19 44 15	42 9 45	42 24 0
Mean deviations	19 39 22		42 16 52	
Wave length for D_1	$5 \cdot 8952 \times 10^{-5}$ cm.		$5 \cdot 8954 \times 10^{-5}$ cm.	

The mean deviations for line D_2 were 19° 38′ 0″, 42° 13′ 45″ giving $5 \cdot 8886 \times 10^{-5}$, $5 \cdot 8895 \times 10^{-5}$ cm. for wave length for D_2.

EXPERIMENT 65. **Measurement of wave length by auto-collimating spectrometer.**

348. Method. The spectrometer of § 62 is used. The frame of the grating G (Fig. 220) is fixed to the spectrometer table, which is adjusted so that the longitudinal axis OL (§ 338) of G is parallel to the axis of revolution of the table. This is secured by making (1) the plane of G parallel to the axis of the table, and (2) the transverse axis OT of G perpendicular to the axis of the table (§ 349). The axis (line of collimation) of the telescope is assumed perpendicular to the axis of the table. A plane mirror M is attached to the grating; its plane is parallel to that of G and its reflecting surface is towards G.

Let the table be turned so that the normal to the *mirror* is parallel to the axis of the telescope, the reflecting surface being towards the telescope, and let light from the illuminated slit fall as a parallel beam upon G. This beam gives rise to a number of diffracted beams, whose directions are given by

$$\sin \theta - \sin \phi_n = n\lambda/d, \quad \dots\dots\dots\dots\dots(1)$$

as shown in § 339. The diffracted beam for $n = 0$ is parallel to the incident beam, since $\phi_0 = \theta$. This beam falls normally upon M, is reflected back along its path and passes a second time through G, giving rise to a fresh set of diffracted beams. Of this set, the beam for $n = 0$ suffers no deviation and is therefore parallel to the axis of the telescope. It then enters the telescope and comes to a focus in the focal plane, forming an inverted image of the slit which passes through the tip of the pointer. This description does not require G to be parallel to M. When, as in practice, G is approximately parallel to M, the position of the table when the rays fall normally on M may be called the symmetrical position.

If the table be now turned from this position through about 10° in either direction (in the case of a grating of about 14,000 lines to the inch used with sodium light), a fresh image of the slit will appear. An additional displacement of about 10° will bring another image into view and, when the illumination is sufficient, further images may be found.

We will now consider the conditions determining the angular positions of the various images, when the grating is parallel to the mirror. The coincidence of an image with the pointer shows that at least one of the beams which has suffered one reflexion at M and two diffractions at G is parallel to the incident beam.

Let PQ, QR, RS, SU (Fig. 219) be parallel to the successive beams, and let PQ and QR make the angles $PQN = \theta$ and $RQN' = \phi$ with the normal NQN' at Q to the grating G, ϕ having the same sign as θ when the bent line PQR *crosses* the normal NQN', as in Fig. 219. Then, since M is parallel to G, and SU to PQ, the diffraction angles at S are θ and $-\phi$. Hence the diffraction equations for Q and S are, by (1),

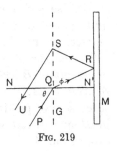

FIG. 219

$$\sin \theta - \sin \phi = p\lambda/d, \qquad \sin \theta + \sin \phi = q\lambda/d, \ldots\ldots(2)$$

where p and q are integers. Hence, if $p + q = n$, we have, by addition,

$$2 \sin \theta = (p + q)\,\lambda/d = n\lambda/d. \quad\ldots\ldots\ldots\ldots(3)$$

Thus, for any given value of n, θ is independent of the particular value of p, and hence, for that value of n, *all* the images given by successive values of p coincide, with the result that only *one* image is seen for that value of n, for *positive* values of θ. There is, of course, a second image of order n for a negative value of θ, and thus there is a single pair of images for each value of n.

When $n = 0$, we have $\theta = 0$ by (3), this case corresponding to the symmetrical position of the grating-mirror system. Making $n = 1, 2, \ldots$, we obtain the values $\theta_1, \theta_2, \ldots$, for the angle between the normal to G and the axis of the telescope, i.e. for the angle through which G is turned from the symmetrical position. We thus obtain the following independent equations:

$$\lambda = 2d \sin \theta_1, \qquad \lambda = \tfrac{1}{2} \times 2d \sin \theta_2, \qquad \lambda = \tfrac{1}{3} \times 2d \sin \theta_3, \ldots.$$

The case in which G is not quite parallel to M may be treated on the lines of § 346.

349. Experimental details. The frame of the grating is secured to the table with a *little* wax. The plane of the grating is set perpendicular to the line joining D, E, two of the three levelling screws D, E, F. The table is then adjusted by D or E

so that when the slit is viewed by reflexion at the glass covering the grating, the lower end of the (inverted) image just touches the horizontal edge of the plate shown in Fig. 39. As there are several surfaces which reflect light, several images will be seen, and these will not quite coincide, since the surfaces are not quite parallel. In this adjustment an opaque screen is placed between G and M.

The mirror M must be of good quality. It is mounted on a brass plate B (Fig. 220) which is adjusted relative to A, the frame of the grating G, by three screws passing through "clearing" holes in B and fitted with springs which press B away from A. The adjustment of M parallel to G is complete when the bright image due to reflexion at the silvered surface of M, with $n = 0$, coincides, as near as may be, with the group of comparatively faint images due to reflexions at the surfaces of the plates pro-

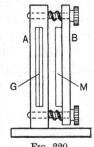

FIG. 220

tecting G. The right-hand edge of the *bright* image is then brought to the pointer and the verniers are read. These readings correspond to the symmetrical position of the table.

The table is then turned so as to bring the other images to the pointer in turn. If the transverse axis OT of the grating be not perpendicular to the axis of the table, the lower ends of the various images of the slit will not coincide with the top of the plate in which the slit is formed. Any necessary adjustment is made by turning the grating in its own plane by the levelling screw F.

The right-hand edge of the image is used throughout; both verniers are read for each setting. With proper adjustments and with a good mirror, the sodium lines will be easily resolved by a grating of 14,000 lines to the inch. The line D_1 is the more deviated and has the greater wave length.

The angle by which the table is displaced from its symmetrical position is calculated for each setting. The two angles found for each value of n should agree closely. The mean of the pair will be independent of any small error in the reading for the symmetrical position and also of any small want of parallelism in either plate covering the grating.

350. Practical example. Grating used by G. F. C. Searle had $d = 1 \cdot 7556 \times 10^{-4}$ cm. The less deviated (D_2) of sodium lines was observed.

Symmetrical position: Vernier I, 160° 8′;   Vernier II, 340° 8′

Diffracted images ($n = 1$):   150° 28′, 169° 46′;   330° 28′, 349° 46′

    „   „   ($n = 2$):   140 31, 179 38;   320 33, 359 38

Values of θ_1 are (I) 9° 40′, 9° 38′; (II) 9° 40′, 9° 38′. Mean 9° 39′.

Values of θ_2 are (I) 19° 37′, 19° 30′; (II) 19° 35′, 19° 30′. Mean 19° 33′.

For $n = 1$, $\lambda = 2d \sin \theta_1 = 5 \cdot 8858 \times 10^{-5}$ cm.

For $n = 2$, $\lambda = d \sin \theta_2 \; = 5 \cdot 8748 \times 10^{-5}$ cm.

Mean $\lambda = 5 \cdot 880 \times 10^{-5}$ cm.

EXPERIMENT 66. **The sloped grating*.**

351. Introduction. Take rectangular axes and let OT, the transverse axis of the grating (§ 338), coincide with OY, as in Fig. 221. Let OL make angle θ with OZ. In the experiment, OZ is vertical and the incident rays are parallel to OX. Then $\eta_1 = \frac{1}{2}\pi - \theta$, $\epsilon_1 = \frac{1}{2}\pi$ (§ 343). Let OP be a diffracted ray, let the plane ZOP cut XOY in OH, and let $HOX = \omega$. Let OP make angle ψ with the plane XOY and therefore $\frac{1}{2}\pi - \psi$ with OZ. Let the direction cosines of OP be l, m, n. Then

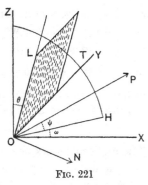

FIG. 221

$$l = \cos \psi \cos \omega, \qquad m = \cos \psi \sin \omega, \qquad n = \sin \psi. \;\ldots(1)$$

The direction cosines of OL are $\sin \theta$, 0, $\cos \theta$, and of OT are 0, 1, 0. Since, by (4), § 343, $\eta_2 = \eta_1$, we have

$$\cos POL = \cos XOL = \sin \theta. \qquad \ldots\ldots\ldots\ldots(2)$$

Hence, using direction cosines to find $\cos POL$†, we obtain

$$l \sin \theta + n \cos \theta = \sin \theta. \;\ldots\ldots\ldots\ldots\ldots\ldots(3)$$

* See "Experiments with a plane diffraction grating," by G. F. C. Searle, *Proc. Camb. Phil. Soc.* Vol. xx, p. 88 (1920). [An error occurs in § 10 of Part ii. The angle should be measured by a circle attached to the tube Q and not by the angular displacement of the *wire* at E.]

† If β be the angle between two straight lines whose direction cosines are f, g, h and f', g', h', then $\cos \beta = ff' + gg' + hh'$.

Now $\epsilon_2 = POT = POY$, and hence $\cos \epsilon_2 = m$. Thus, by (5), § 343, since $\cos \epsilon_1 = \cos \frac{1}{2}\pi = 0$,

$$m = \cos \epsilon_2 = i\lambda/d. \quad \dots\dots\dots\dots\dots(4)$$

Hence m and ϵ_2 are known when i, λ, d are given, and m is independent of θ.

The direction cosines of the normal ON are $\cos \theta$, 0, $- \sin \theta$. Since the beam used in the experiment is transmitted, $\cos NOP$ or $l \cos \theta - n \sin \theta$ is positive; hence $l > n \tan \theta$. Since, by (3), $l = 1 - n \cot \theta$, we have $1 - n \cot \theta > n \tan \theta$, and thus $n < \sin \theta \cos \theta$.

Since $m = \cos \epsilon_2$, we have $l^2 = 1 - m^2 - n^2 = \sin^2 \epsilon_2 - n^2$. By (3), $l \sin \theta = \sin \theta - n \cos \theta$, and thus

$$\sin^2\theta \,(\sin^2\epsilon_2 - n^2) = (\sin \theta - n \cos \theta)^2.$$

Solving for n, and taking the negative sign in the ambiguity, since $n < \sin \theta \cos \theta$, we have

$$\sin \psi = n = \sin \theta \cos \theta - \sin \theta \,(\cos^2\theta - \cos^2\epsilon_2)^{\frac{1}{2}}, \dots(5)$$

or $\qquad n = \frac{1}{2} \sin 2\theta - \sin \theta \,\{(\cos \theta - m)\,(\cos \theta + m)\}^{\frac{1}{2}}; \quad \dots(6)$

equation (6) is used in calculating ψ.

Since $m = \cos \psi \sin \omega$, we have

$$\sin \omega = m/\cos \psi = \cos \epsilon_2/\cos \psi. \quad \dots\dots\dots\dots(7)$$

The angles ψ, ω fix the direction of the diffracted beam and can be calculated when ϵ_2 and θ are known. When $\sin^2\theta > \sin^2\epsilon_2$, $\sin \psi$ is imaginary, and then no diffracted beam of order i will exist. Since ϵ_2 and θ may be restricted to be less than $\frac{1}{2}\pi$, it follows that there is no diffracted beam, if θ exceed the critical value ϵ_2. The critical value of ψ is $\sin^{-1}(\sin \epsilon_2 \cos \epsilon_2)$.

352. Apparatus. The grating G (Fig. 222) is attached to a horizontal shaft A, with its plane parallel to A and its rulings perpendicular to A. A horizontal collimator L has horizontal and vertical cross-wires intersecting at C in its focal plane; these are illuminated by a sodium flame S. The line joining C to the appropriate nodal point of the lens is the line of collimation, or axis, of the collimator. The parallel beam defined by C is parallel to this line. After the light has passed through G, it is received by a goniometer K, and an image of the collimator wires is formed in the focal plane of K. To fix the line of collimation, or axis, of K, cross-wires are placed in the focal plane; they intersect in D. The goniometer is carried on a horizontal shaft B, and its axis is perpendicular to the shaft. One cross-wire is parallel and the other perpendicular to the shaft. The shafts have divided circles E, F, which are read by the pointers U, U', V, V'. A balance weight W is attached to the circle F. The axis of K and the axis of B should each pass approximately through the centre of the grating, and the axis of L should pass through the same point. The angles θ and ψ are measured by the circles E and F.

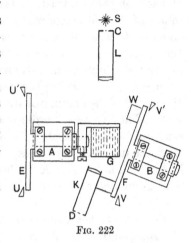

Fig. 222

353. Experimental details. The shaft A is set horizontal by aid of a level. The collimator L is adjusted by an auto-collimating method. The plane of the grating is set horizontal by a level, and the shaft is then turned through 90°, as measured by the circle E, so that the plane of the grating is vertical. Light from a flame is reflected by a plate of glass held at 45° past the cross-wires at C and through the lens on to the grating, and the collimator, previously set for infinity, is adjusted in position so that the image of C, the intersection of its wires, coincides with C itself. If the coincidence be recovered when the grating shaft is turned through 180°, the plane of the grating is parallel to the shaft. The line of collimation of L is then horizontal and is also perpendicular to the grating shaft.

The axis of the goniometer K is set perpendicular to the shaft B by an optical method. An auxiliary collimator, set for "infinity," is placed so that it is approximately perpendicular to B. A plate of plane parallel glass is attached to the circle F near its centre, so that its faces are approximately parallel to the shaft. It is convenient to make the shaft B vertical; the glass plate can then be supported on a small levelling table resting on the circle F. By adjusting both collimator and plate, the faces of the plate are made parallel to the shaft, and the axis of the collimator is made perpendicular to the shaft (see § 53). If each of these adjustments have been correctly made, the image of the collimator wires can, by turning the circle, be made to coincide with those wires, when *either* side of the plate faces the collimator. If the axes of the collimator and of the goniometer meet (approximately) in the centre of the circle F, it will be possible, by turning the goniometer on its shaft, to receive the image of the collimator wires on the focal plane of the goniometer. The "vertical" cross-wire of the goniometer, i.e. the wire perpendicular to the shaft, is now adjusted so that it coincides with the image of the corresponding wire of the collimator. The line of collimation of K is then perpendicular to the shaft B. The goniometer is then put into position and its shaft is levelled.

The axes of the collimator L, of the grating shaft A and of the goniometer shaft B are adjusted to be approximately in the same horizontal plane. The plane of the grating is made vertical, and the goniometer stand is adjusted in azimuth so that one of the diffracted images of the collimator wires can be made to coincide with the goniometer wires by turning the goniometer on its shaft.

When the adjustments already described have been effected, and when the plane of the grating G is vertical, the diffracted beams are parallel to the principal plane ONT (§ 338). If the transverse axis OT be inclined at an angle δ to the grating shaft, the direction of OT will be changed by 2δ if G be turned through 180° about the axis of A from Position 1 to Position 2, when the plane of G is again vertical. The goniometer is turned to receive a diffracted beam when G is in Position 1. If, when G is turned into Position 2, the inclination of the beam to a horizontal plane be found to have changed, the grating must be turned in its own plane until the inclination is the same for both positions.

Since ψ is always small (see § 354), $\cos \psi$ is nearly unity and hence, by (7), $\sin \omega$ has a nearly constant value, for m is independent of θ. Hence, if the image of C, the intersection of the collimator wires, lie on D, the intersection of the goniometer wires, when the plane of G is vertical, the image of C will always

lie near the "vertical" cross-wire, and one setting of the goniometer stand will suffice for all values of θ.

The plane of the grating is made horizontal and the index
reading of E is taken. It is then turned through 90°; it is now
vertical and in its zero position. The goniometer is next adjusted so that the image of C lies on the horizontal wire of K.
The goniometer is then in its zero position and the zero reading
of F can be taken.

The grating is now turned through 10°, 20°, ... on either side
of its zero, the goniometer is turned on its shaft to bring the
image of C on to the horizontal wire of K and the reading of
the circle F is taken in each case. If the grating circle have *two*
indices, the grating is turned through 10°, 20°, ... as indicated
by one index. In reducing the observations, the mean of the
angles furnished by *both* indices is used.

From the values of the interval d and the wave length λ,
$\cos \epsilon_2$ is found by (4), and then the values of ψ corresponding
to the values of θ are calculated by (6). These values are compared with the mean values of ψ given by the goniometer
readings.

When $\theta = \epsilon_2$, we have, by (5), $n = \sin \epsilon_2 \cos \epsilon_2$, and, by (3),
$l = 1 - n \cot \epsilon_2 = \sin^2 \epsilon_2$. Hence, when $\theta = \epsilon_2$,

$$\cos NOP = l \cos \theta - n \sin \theta = 0.$$

Thus, when the critical value of θ is reached, the diffracted
beam is tangential to the plane of the grating.

As θ, and consequently ψ, increases, the angle between the
images of the collimator wires changes. When $\theta = 0$, the images
are at right angles, but the angle diminishes rapidly as θ reaches
its critical value. The theory shows that they are tangential
one to the other when θ has its critical value.

354. Practical example. Dr J. A. Wilcken used a grating having
$d = 1 \cdot 7556 \times 10^{-4}$ cm. and sodium light of mean $\lambda = 5 \cdot 893 \times 10^{-5}$ cm.
Images of *first* order were used and $m = \lambda/d = 0 \cdot 33568 = \sin 19° 36' 49''$.
Thus $\epsilon_2 = \cos^{-1} m = 70° 23' 11''$.

The grating circle was a little eccentric relative to its shaft, and although
the settings were made to integral degrees by one index, the other did not
always read integral degrees. In the table, the *calculated* values of ψ and ω
have been added. The last line gives the *critical* values as calculated.

Mean observed θ	Mean observed ψ	Calculated ψ	Calculated ω
0° 0′ 0″	0° 0′ 0″	0° 0′ 0″	19° 36′ 49″
19 54 0	1 12 45	1 12 33	19 37 7
39 51 0	2 51 45	2 50 18	19 38 19
59 49 0	6 26 15	6 22 31	19 44 27
69 44 0	14 18 0	14 10 3	20 15 19
70 23 11		18 26 0	20 43 18

EXPERIMENT 67. **Measurement of wave length by grating, using convergent light*.**

355. Introduction. Let the plane of the grating be taken as the plane Oyz (Figs. 223, 224) and let the centre line of one of the openings be taken as Oz. Then Oy, Oz are the transverse and longitudinal axes (§ 338) of the grating. Let Ox be the

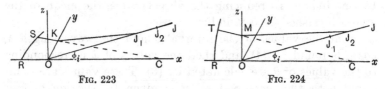

FIG. 223   FIG. 224

direction of the chief ray of a convergent beam moving from left to right, whose rays are directed to C on Ox, where $OC = u$. Let a spherical wave front of the incident beam cut Ox in R to the left of O, and let $RC = r$; then the normals to this front meet in C. The incident beam gives rise to a number of non-spherical diffracted wave fronts. By symmetry, the chief rays of the diffracted beams lie in Oxy. Let OJ be the chief ray of that diffracted beam of order i which makes an acute angle with Oy, and let $JOx = \phi_i$.

There will be two points J_1, J_2 on OJ, in which a neighbouring ray of the diffracted beam will cut OJ, viz. the centres of principal curvature of the diffracted wave front. Let J_1 (Fig. 223) be the point in which rays in the principal section Oxy meet OJ, and let $OJ_1 = v_1$. There will be a focal line through J_1 parallel to Oz. Let J_2 (Fig. 224) be the point in which rays in the principal section JOz meet OJ, and let $OJ_2 = v_2$. There will be a focal line through J_2 in the plane Oxy, but not perpendicular to OJ.

* See "Experiments with a plane diffraction grating," *Proc. Camb. Phil. Soc.* Vol. xx, p. 88 (1920).

On Oy (Fig. 223) take a point K, where $OK = qd$, and let CK meet the spherical front in S; here d is the grating interval and q an integer. Then, the waves reaching J_1 by K and by O will have the same phase at J_1, if $SK + KJ_1 = RO + OJ_1 - qi\lambda$. Since $SK = r - KC$, $RO = r - u$, and $OJ_1 = v_1$, we have

$$KJ_1 - KC = v_1 - u - qi\lambda. \qquad \dots\dots\dots(1)$$

Now $KC = \{u^2 + q^2d^2\}^{\frac{1}{2}}$, and $KJ_1 = \{v_1^2 - 2v_1qd \sin \phi_i + q^2d^2\}^{\frac{1}{2}}$. Expanding the square roots as far as q^2d^2, we have,

$$KC = u + \tfrac{1}{2}q^2d^2/u, \qquad \dots\dots\dots\dots(2)$$

$$
\begin{aligned}
KJ_1 &= v_1 \{1 - \tfrac{1}{2}(2qd \sin \phi_i/v_1 - q^2d^2/v_1^2) \\
&\qquad - \tfrac{1}{8}(2qd \sin \phi_i/v_1 - q^2d^2/v_1^2)^2 - \dots\} \\
&= v_1 - qd \sin \phi_i + \tfrac{1}{2}q^2d^2 \cos^2 \phi_i/v_1. \qquad \dots\dots\dots\dots(3)
\end{aligned}
$$

Since (1) holds for all small values of qd, we have, by (2), (3),

$$d \sin \phi_i = i\lambda, \qquad \dots\dots\dots\dots(4)$$

$$v_1 = u \cos^2 \phi_i. \qquad \dots\dots\dots\dots(5)$$

The direction of OJ is given by (4) and the position of J_1 by (5).

On Oz take M (Fig. 224), where $OM = h$, and let CM meet the spherical front in T. Then, the waves reaching J_2 by M and O will have the same phase at J_2, if $TM + MJ_2 = RO + OJ_2$. Since $TM = r - MC$, $RO = r - u$, and $OJ_2 = v_2$, we have

$$MJ_2 - MC = v_2 - u. \qquad \dots\dots\dots\dots(6)$$

Expanding as far as h^2,

$$MC = \{u^2 + h^2\}^{\frac{1}{2}} = u + \tfrac{1}{2}h^2/u, \qquad MJ_2 = v_2 + \tfrac{1}{2}h^2/v_2.$$

Since (6) holds for all small values of h, we have

$$v_2 = u. \qquad \dots\dots\dots\dots(7)$$

When a finite width of grating is used, the diffracted rays in Oxy do not meet accurately in J_1 but touch a caustic passing through J_1. At J_1 the radius of curvature of the caustic is $3v_1 \tan \phi_i$.

356. Method. On an optical bench slide three carriages D, H, K (Fig. 225); D carries a horizontal glass scale, H carries the grating G (with vertical rulings), whose centre is O, and K carries the converging lens system L, of focal length f. At the end of the bench is a vertical slit illuminated by a sodium flame F; to identify a point E on the slit, a wire may be stretched across

the slit. The divided face of the scale faces the slit, and this face
and the plane of the grating are perpendicular to the bench.

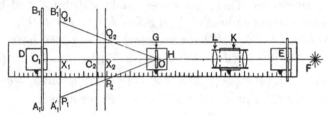

FIG. 225

The line through the nodal points of L is parallel to the bench
and passes through E (see § 210).

The scale is first placed in the position A_1B_1, at a distance
from E exceeding $4f$ by about 30 cm., and the lens is set to form
a real undiffracted image ($i = 0$) on the scale at C_1. The vertical
focal lines of the diffracted beams of order i will, by (5), be at
a distance $OC_1 \cos^2\phi_i$ from O, along lines OP_1, OQ_1, each making
an angle ϕ_i with OC_1. Sharp images of the slit will pass through
P_1, Q_1, and can be focused on the scale if it be moved to $A_1'B_1'$.
With the grating used in § 357, the two sodium lines can easily be
separated. Then $\sin\phi_i = \frac{1}{2}P_1Q_1/OP_1$. Since OP_1 is difficult to
measure, we suppose OX_1 known, where X_1 is the mid-point of
P_1Q_1. If $OX_1 = x_1$, $P_1Q_1 = 2y_1$, $\tan\phi_i = \frac{1}{2}P_1Q_1/OX_1 = y_1/x_1$.

The glass plate protecting the grating prevents an accurate
measurement of OX_1. We therefore move the lens carriage
along the bench so that the undiffracted image is focused on
the scale at C_2. If the scale be moved further towards O,
the diffracted images can be focused at P_2, Q_2. If $Ox_2 = x_2$,
$P_2Q_2 = 2y_2$, $\tan\phi_i = y_2/x_2$. Hence

$$\tan\phi_i = (y_1 - y_2)/(x_1 - x_2). \quad\quad\quad\text{......(8)}$$

Then
$$\lambda = d\sin\phi_i/i, \quad\quad\quad\text{......(9)}$$

where i is the order of the image and d is the grating interval.

Since it is an *angle* we measure, *small* errors of focusing will
be of little account, for, in spite of them, the point in which the
chief ray cuts the scale in each case will be correctly estimated.
To allow a close test of parallax, a few grains of lycopodium
may be placed on AB when the image of the slit does not fall
on a division line.

357. Practical example. G. F. C. Searle used sodium light and a grating with $d = 1\cdot7526 \times 10^{-4}$ cm. Image of first order was used; thus $i = 1$.

Bench reading 118·50 cm., glass scale readings 97·52, 73·84 cm.

„   „   141·21 „   „   „   89·64, 82·18 „

$y_1 = \frac{1}{2}(97\cdot52 - 73\cdot84) = 11\cdot84$ cm., $y_2 = \frac{1}{2}(89\cdot64 - 82\cdot18) = 3\cdot73$ cm.

Also $x_1 - x_2 = 141\cdot21 - 118\cdot50 = 22\cdot71$ cm.

Hence $\tan \phi_1 = (y_1 - y_2)/(x_1 - x_2) = 8\cdot11/22\cdot71 = \tan 19° 39' 8''$.

Then $\lambda = d \sin \phi_1 / 1 = 5\cdot894 \times 10^{-5}$ cm.

EXPERIMENT 68. Measurement of wave length by zone plate.

358. Theory of zone plate. Let $OG_1G_2 \dots$ (Fig. 226) be the section by the plane of the figure of an infinitely thin plane opaque screen, and let $X'OX$ be the normal to the screen at O. Let the spaces to the left and right of the screen be called the object and image spaces respectively. On the screen take G_1 near O, and let $OG_1 = \rho_1$. Let G_2, G_3, ... be such that

$$OG_n{}^2 = \rho_n{}^2 = \rho_1{}^2 + (n - 1)\,k^2.$$

Let narrow circular slits be cut in the screen with O as centre and passing through G_1, G_2, ... This system of screen and slits forms a theoretical zone plate.

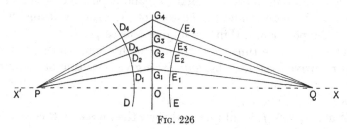

FIG. 226

On $X'OX$ take P, Q at distances u, v from O, and let u and v be positive when P and Q are in the object and image spaces respectively. Then, when ρ_n/u, ρ_n/v are small,

$$PG_n + QG_n = (u^2 + \rho_n{}^2)^{\frac{1}{2}} + (v^2 + \rho_n{}^2)^{\frac{1}{2}}$$
$$= u + v + \tfrac{1}{2}\rho_n{}^2 (1/u + 1/v) \quad \dots\dots\dots\dots\dots(1)$$
$$= u + v + \tfrac{1}{2}\{\rho_1{}^2 + (n - 1)\,k^2\}\{1/u + 1/v\}. \quad \dots(2)$$

By (2), for all values of the positive integer n greater than 1,

$$PG_n + QG_n = PG_{n-1} + QG_{n-1} + \tfrac{1}{2}k^2 (1/u + 1/v).$$

Thus, the paths PG_1Q, PG_2Q, ... increase by equal steps of h, where

$$h = \tfrac{1}{2}k^2 \left(1/u + 1/v\right).$$

Let a train of spherical waves with centre P and wave length λ fall upon the zone plate from the object space, and let D be one of the wave fronts. Let PG_1, PG_2, ... meet D in D_1, D_2, ... Waves will travel out from the slits into the image space. Let E be a sphere about Q as centre, and let G_1Q, G_2Q, ... meet E in E_1, E_2, ... Then the disturbances at E_1, E_2, ... will have the same phase, if the distances $D_1G_1E_1$, $D_2G_2E_2$, ... increase by steps of $p\lambda$, where p is a positive or negative integer. When the distance of E from O is some thousands of wave lengths, the separate wavelets due to the ring slits will merge into a single wave indistinguishable from E. We may thus speak of E as the emergent wave front. The wave D will thus give rise to the wave E, if $PG_n + QG_n$ exceed $PG_{n-1} + QG_{n-1}$ by $p\lambda$. An image of P will then be formed at Q.

Hence Q will be an image of P if $h = p\lambda$. If f_p be the corresponding focal length, and F_p the "power," we have

$$1/u + 1/v = F_p = 1/f_p = 2h/k^2 = 2p\lambda/k^2. \ldots\ldots(3)$$

Thus, the power is proportional to p and is positive or negative with p. We here follow the custom of practical opticians, who treat the power of a thin converging lens as *positive*.

The zone plate thus acts as a lens with a number of positive and negative focal lengths, and, for a given position of a real or virtual luminous point P on $X'OX$, there will be a number of images, some real and some virtual.

When $p_0 = 0$, f_0 is infinite and then the image of P coincides with P itself.

If k^2 be found by measuring the rings with a travelling microscope, λ can be found when f_p and p are known.

The zone plate as just described corresponds to the diffraction grating discussed in § 339. It would produce very feeble images, since the slits are very narrow.

359. Openings of finite width. To obtain strong images, we widen the narrow slits through G_1, G_2, ..., so that their widths are S_1T_1, S_2T_2, ..., the radii of the edges being σ_1,

τ_1, σ_2, τ_2, ..., where $\sigma_n{}^2 = \rho_n{}^2 - z^2$, $\tau_n{}^2 = \rho_n{}^2 + z^2$. By (1) and (3),

$$PG_n + QG_n = u + v + \tfrac{1}{2}\rho_n{}^2/f_p.$$

Similarly, $\qquad PS_n + QS_n = u + v + \tfrac{1}{2}\sigma_n{}^2/f_p.$

Hence

$$PS_n + QS_n = PG_n + QG_n - \tfrac{1}{2}(\rho_n{}^2 - \sigma_n{}^2)/f_p = PG_n + QG_n - \tfrac{1}{2}z^2/f_p.$$

Similarly

$$PT_n + QT_n = PG_n + QG_n + \tfrac{1}{2}(\tau_n{}^2 - \rho_n{}^2)/f_p = PG_n + QG_n + \tfrac{1}{2}z^2/f_p.$$

The effect at Q due to each opening increases as z^2 increases, until $PT_n + QT_n$ differ from $PS_n + QS_n$ by $\tfrac{1}{2}\lambda$. A further increase in z^2 diminishes the effect at Q. For any given focal length, the image will be strongest when $z^2/f_p = \tfrac{1}{2}\lambda$, or $z^2 = \tfrac{1}{2}\lambda f_p$.

Zone plates are generally made so that $z^2 = \tfrac{1}{2}\lambda f_1$. But, by (3), $k^2 = 2\lambda f_1$, and hence $z^2 = \tfrac{1}{4}k^2$. The radii are, therefore, given by

$$\sigma_n{}^2 = \rho_1{}^2 + (n-1)k^2 - \tfrac{1}{4}k^2, \quad \tau_n{}^2 = \rho_1{}^2 + (n-1)k^2 + \tfrac{1}{4}k^2,$$

and thus $\sigma_1{}^2$, $\tau_1{}^2$, $\sigma_2{}^2$, $\tau_2{}^2$, ... increase by steps of $\tfrac{1}{2}k^2$.

It is usual to make $\rho_1{}^2$ equal to either (1) $\tfrac{1}{4}k^2$, or (2) $\tfrac{3}{4}k^2$. In the first case, $\sigma_1{}^2 = 0$, $\tau_1{}^2 = \tfrac{1}{2}k^2$, and then the central opening is circular and not annular. In the second case, $\sigma_1{}^2 = \tfrac{1}{2}k^2$, $\tau_1{}^2 = k^2$, and then the first opening is annular, the lesser radius being equal to that of the circular opening in the first case. The second zone plate can, in fact, be constructed out of the pieces cut out of the screen in forming the first zone plate. In either case, the edges of the openings have radii r_1, r_2, r_3, ..., where

$$r_1{}^2 = k^2/2 = f_1\lambda, \qquad r_2{}^2 = 2k^2/2, \dots \qquad r_n{}^2 = nk^2/2. \quad \dots\dots(4)$$

Each opening will produce a *null* effect at Q, if $PT_n + QT_n$ differ from $PS_n + QS_n$ by $m\lambda$, where m is an integer. Then $m\lambda = z^2/f_p = pz^2/f_1$. But $z^2 = \tfrac{1}{2}\lambda f_1$ in the ordinary zone plate, and thus $m = \tfrac{1}{2}p$. Since m is an integer, the solutions of this equation are $p = 0$, $p = \pm 2$, $p = \pm 4$, ..., and hence, with an ordinary zone plate, made strictly to the above specification, the images, both real and virtual, of even orders will be missing.

360. Construction of a zone plate. Concentric circles with radii proportional to $1^{\frac{1}{2}}$, $2^{\frac{1}{2}}$, $3^{\frac{1}{2}}$, ... are drawn on a large scale on white paper, and alternate zones are blackened. A greatly reduced copy of the drawing is made with a camera on a photographic plate, whose surfaces should be plane. Alternate zones on the photograph are opaque, and thus the conditions contemplated in § 359 are satisfied.

R. W. Wood has constructed phase-reversal zone plates by photographing the drawing on a plate having a film of gelatine impregnated with bichromate of potash. The parts of the film covered by the images of the white zones are, by the action of light, rendered insoluble, and remain when the plate is placed in water, the other parts being dissolved. The endeavour is made to obtain a film of such thickness that sodium light is retarded by an odd number of half wave lengths in passing through it. If this be secured, the rays passing through the zone plate at distances from O whose squares are $k^2/4$, $5k^2/4$, $9k^2/4$, ... will all have the same phase at Q, and this phase will agree with the common phase at Q of the rays passing through at distances whose squares are $3k^2/4$, $7k^2/4$, ... The amplitude of the disturbance at Q will thus be doubled, and the intensity will be four times as great as before.

The images of the edges of the zones will not be quite sharp, and the section of the zone plate will be somewhat as shown in Fig. 227. The rounded parts may be treated as opaque, for

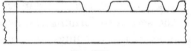

FIG. 227

their action is irregular, and we are then left with a zone plate in which z^2 is less than $\frac{1}{4}k^2$. We shall consequently obtain the images of *even* as well as of odd order.

361. Measurement of zones. The zones are measured by a travelling microscope. The measurements are facilitated by suitable illumination. The light from a *small* flame at a distance of 50 cm. is sent upwards through the zone plate by a plane mirror. The reading of the micrometer screw is taken whenever the cross-wire, which is perpendicular to the direction of motion, is tangential to the image of a circle on the zone plate.

Thus a reading is taken at *each* of the points $R_1, R_2, \ldots R_1', R_2', \ldots$ in Fig. 228. From the readings, the diameters and then the radii $r_1, r_2, \ldots$ of the circles are found. The squares $r_1{}^2, r_2{}^2, \ldots$ are plotted against the integers 1, 2, 3, ... and a value of $r_n{}^2/n$ is deduced from the straight line passing most evenly among the points. It may be more convenient to find $r_n{}^2/n$ for $n = 1, 2, \ldots,$ as

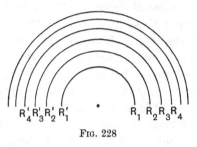

FIG. 228

in § 364, and to take the mean. By (4), we have

$$r_n{}^2/n = k^2/2. \quad \ldots\ldots\ldots\ldots\ldots\ldots(5)$$

362. Measurement of focal lengths. The zone plate G (Fig. 229), mounted in a block with a protecting plate of glass, is attached to a carriage on a long optical bench, with its axis parallel to the bench. A piece of perforated zinc P is used as the object and is illuminated by the sodium flame Z. The images

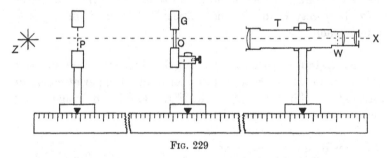

FIG. 229

are observed by aid of a telescope T, adjusted to view objects at a distance of about 1 metre from its objective. A cross-wire at W will be of assistance in accurate focusing. The distance between the objective and the eye-piece of T must then be kept *constant*. The centre P of the object should lie on the line OX, where O is the centre of the zone plate and OX is parallel to the length of the bench. A string stretched parallel to the length of the bench may be used in setting P on to the line OX and in making the axis of the telescope coincide with OX. The telescope carriage is first adjusted so that G is in focus, and the bench reading is taken; this is the zero reading. If T be then moved to the end of the bench and be *gradually* brought up towards G, a series of images of P will be seen, those near G being smaller than those at a distance from G. The bench reading of T is taken for each image when in focus. From several settings for each image the mean reading for that image is found. The difference between the reading for any image and the zero reading gives v, the distance of the image from G. The distance u of P from G is found either from the bench readings of the P carriage by aid of a rod of known length or from the bench reading of T when P is in focus; the zone plate is removed if necessary.

If T be focused on G and then be moved nearer to P, a number of virtual images will be found. For these v is negative.

The "power" F is found in each case by the formula

$$F = 1/f = 1/u + 1/v.$$

When the length of the optical bench is not more than two metres, a collimator with the perforated zinc in its focal plane (Fig. 162) may be used. The "object" is then at infinity and thus in each case $v = f$.

363. Calculation of wave length. The successive positive values of the power F are set out on squared paper at constant intervals as AB, CD, EF, ... in Fig. 230. If no images have been *missed*, the points B, D, F, ... will be on a straight line, which will show the orders of the images. Thus, in Fig. 230, the image for which the power is AB is of the second order. When the *orders* have been settled, F_p/p is found for each

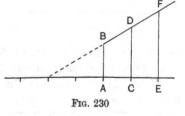

Fig. 230

image, both real and virtual. The mean is calculated, or the best value is found by plotting F_p against p.

We have now found the mean value of F_p/p and the mean value of $\frac{1}{2}k^2 = r_n{}^2/n$. Since, by (3), $\lambda = \frac{1}{2}k^2 F_p/p$, we have

$$\lambda = (r_n{}^2/n)\,(F_p/p). \quad\quad\ldots\ldots\ldots\ldots\ldots(6)$$

364. Practical example. Mr J. A. Pattern measured with a microscope 12 circles on zone plate; results for first three are given.

n	Micro. readings		Rad. $= r_n$	r_n^2	r_n^2/n
1	7·620	6·370 mm.	·6250 mm.	·3906 mm.^{2}	·3906 mm.^{2}
2	7·871	6·115	·8780	·7709	·3854
3	8·065	5·908	1·0785	1·1632	·3877

Mean value of r_n^2/n for the *twelve* circles was ·3866 mm.^{2} or $3·866 \times 10^{-3}$ cm.^{2}.

Distance of perforated zinc from zone plate $= u = 120$ cm. Distances (v) from zone plate of real images $(p = 1$ to $4)$ and virtual images $(p = -1$ to $-4)$ were found for sodium light. Power $= F = 1/u + 1/v$.

p	v	$100\,F_p$	$100\,F_p/p$	p	v	$100\,F_p$	$100\,F_p/p$
	cm.	cm.$^{-1}$	cm.$^{-1}$		cm.	cm.$^{-1}$	cm.$^{-1}$
1	145·0	1·523	1·523	-1	$-42·5$	$-1·520$	1·520
2	45·0	3·056	1·528	-2	$-25·5$	$-3·088$	1·544
3	26·0	4·679	1·560	-3	$-18·5$	$-4·572$	1·524
4	19·0	6·096	1·524	-4	$-14·5$	$-6·063$	1·516

Mean values of F_p/p are $1·534 \times 10^{-2}$ $(p = 1$ to $4)$, $1·526 \times 10^{-2}$ $(p = -1$ to $-4)$. General mean $1·530 \times 10^{-2}$ cm.$^{-1}$.

By (6), $\lambda = 3·866 \times 10^{-3} \times 1·530 \times 10^{-2} = 5·915 \times 10^{-5}$ cm.

EXPERIMENT 69. **Determination of thickness of film by banded spectrum.**

365. Introduction. Each of two glass plates has one surface lightly silvered so as to reflect about as much light as it transmits. The silvered surfaces, placed facing each other, are pressed into contact with a plate of mica of uniform thickness. The central part of the plate of mica is cut out, and thus the central parts of the silvered surfaces are separated by a film of air of definite thickness. We suppose, in the first instance, that each glass plate has plane and parallel surfaces and that the film of air has the constant thickness z.

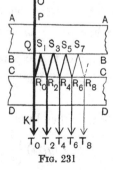

Fig. 231

Let the plane of Fig. 231 cut the faces of the plates at right angles in AA, BB, CC, DD; the faces BB, CC are silvered. Let a train of plane waves, of wave length λ and velocity v in air, fall upon AA, the wave fronts being parallel to the faces of the plates. Let OP be an incident ray, giving rise to an emergent ray R_0KT_0 normal to DD. Some light is reflected at R_0 on CC and some of this reflected light is reflected at BB and strikes CC again, giving rise to some reflected light and some transmitted light. The reflected and transmitted rays all coincide with $OPQR_0T_0$, but, *for the purpose of description only*, they are shown separate from OT_0 in Fig. 231. The process of partial reflexion and partial transmission continues indefinitely, the wave trains becoming progressively weaker.

Let f be the amplitude of the electric force in the ray QR_0. The phase of the ray R_0S_1 will be reversed relative to the ray QR_0; the amplitude will be cf, where c is positive and less than unity. The phase of S_1R_2 will be reversed relative to R_0S_1; the amplitude will be bcf, where b is positive and less than unity.

Let the electric force in the ray T_0 at a point K in air beyond DD be represented by

$$e_0 = g \sin \{2\pi vt/\lambda\}.$$

The amplitudes in T_2 and T_0 are in the same ratio as those in S_1R_2 and QR_0. Hence, the electric force at K in the ray T_2 is

$$e_2 = bcg \sin \{2\pi (vt - 2z)/\lambda\},$$

for, to reach K at the same time as the wave corresponding to T_0, the wave corresponding to T_2 must have left the source at a time earlier by $2z/v$. The double reversal of phase has no resultant effect upon the phase at K.

If the resultant electric force at K due to the whole system of waves be E, then

$$\begin{aligned} E = g\,[&\sin \{2\pi vt/\lambda\} + bc \sin \{2\pi (vt - 2z)/\lambda\} \\ &\quad + b^2c^2 \sin \{2\pi (vt - 4z)/\lambda\} \dots] \\ = g\,[&\sin \alpha + h \sin (\alpha - \beta) + h^2 \sin (\alpha - 2\beta) + \dots], \end{aligned}$$

where $\quad h = bc, \quad \alpha = 2\pi vt/\lambda, \quad \beta = 2\pi.2z/\lambda.$

Using the exponential values of the sines, we have

$$E = (g/2i)\,[e^{\alpha i} \{1 + he^{-\beta i} + h^2e^{-2\beta i} + \dots\} \\ - e^{-\alpha i} \{1 + he^{\beta i} + h^2e^{2\beta i} + \dots\}].$$

Since h lies between zero and unity, the series converge; thus

$$E = (g/2i) \left\{ \frac{e^{\alpha i}}{1 - he^{-\beta i}} - \frac{e^{-\alpha i}}{1 - he^{\beta i}} \right\} = g\,\frac{\sin \alpha - h \sin (\alpha + \beta)}{1 - 2h \cos \beta + h^2}$$

$$= g\,\frac{(1 - h \cos \beta) \sin \alpha - h \sin \beta \cos \alpha}{1 - 2h \cos \beta + h^2}.$$

If we write $E = M \sin \alpha + N \cos \alpha$, and if I be the intensity, i.e. the square of the amplitude, then $I = M^2 + N^2$, and thus

$$I = g^2 \frac{(1 - h \cos \beta)^2 + h^2 \sin^2 \beta}{(1 - 2h \cos \beta + h^2)^2} = \frac{g^2}{1 - 2h \cos \beta + h^2}. \quad ....(1)$$

Since h is positive, it follows that, if g be constant, I has its greatest value $g^2/(1 - h)^2$ when $\cos \beta = 1$ or $\beta = 2n\pi$, where n is an integer, or when $2z = n\lambda$. Further, I has its least value $g^2/(1 + h)^2$ when $\cos \beta = -1$, or $\beta = (2n + 1)\pi$, or when $2z = (n + \frac{1}{2})\lambda$.

366. Banded spectrum. Let the distant source be "white," so that it would produce a continuous spectrum, and let the light, after passing through the plates, fall normally upon a diffraction grating and then be received by a telescope as in the ordinary use of a grating (§ 346). If d be the grating interval and θ be the deviation corresponding to wave length λ in the first order spectrum,

$$\lambda = d \sin \theta.$$

If g be independent of λ and so also of θ,

$$\frac{dI}{d\theta} = \frac{dI}{d\lambda} \cdot \frac{d\lambda}{d\theta}.$$

Since $d\lambda/d\theta$ is not zero unless $\theta = \frac{1}{2}\pi$, a condition unattainable in practice, it follows that $dI/d\theta$ vanishes only when $dI/d\lambda = 0$. Hence, for each value of λ which satisfies $\lambda = 2z/n$, the intensity will be a maximum, and when $\lambda = 2z/(n + \frac{1}{2})$ the intensity will be a minimum. Hence the spectrum will appear "banded." We cannot, of course, *ensure* that g does not depend upon λ, but, if in the non-banded spectrum there be only gradual variations of intensity, as judged by eye, and if the banded spectrum have many bands in its length, we shall commit little error if we assume that the values of λ corresponding to maxima and minima of intensity are independent of the exact distribution of intensity in the non-banded spectrum. If, before incidence on the plates, the light pass through a tank containing a liquid giving rise to absorption bands or through an auxiliary pair of semi-silvered plates, the intensity of the incident light will vary rapidly with λ, and consequently g^2 will vary rapidly with β for certain values of β, and thus we cannot find the maxima and minima of I by considering merely the variations of $1 - 2h \cos \beta + h^2$.

367. Experimental details. In practice, the simple arrangement described in § 366 is modified. A brilliant source of light O (Fig. 232), such as an arc, or a "half watt" lamp is used. During the preliminary adjustments, the pair of plates P is removed from between the lenses L_1, L_2. The source O is set approximately in the focal plane of the lens L_1 by aid of a

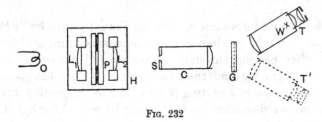

FIG. 232

small plane mirror, suitably placed between L_1 and L_2, which throws the light on to a distant screen—the ceiling will serve. The distance OL_1 is adjusted so that a sharp image of O is formed on the screen.

The lenses L_1, L_2 may be mounted coaxally on a board H, and guides may be provided to secure that the frame, in which the plates are clamped, occupies a definite position between L_1 and L_2. The lenses are preferably achromatic; to diminish aberration, their *curved* faces should be towards P (compare Fig. 76). The focal length of each may be about 10 cm.

The slit S of a spectrometer is placed approximately in the focal plane of L_2 and the distance L_2S is adjusted so that an image of O is formed upon the slit plate. The grating G is set normal to the axis of the collimator C and is adjusted as in § 345.

For the production of a vigorous banded spectrum, good adjustment of the image of O upon the slit S is essential. This setting is easily made if the lamp O and the board H be mounted on a common base. The adjustment of the distance L_2S is satisfactory when, on looking through the lens of the collimator C towards S, the observer sees the slit *very brilliantly* illuminated. To test the alignment of H and C, the eye is moved horizontally so that the line of sight sweeps across the lens of the collimator; if the slit do not appear *brilliant* over the whole range, the direction of the axis of H needs adjustment.

The pair of plates is now put into position at P between the lenses L_1 and L_2. When the telescope T is properly directed, the observer will see the banded spectrum.

Unless the plates have accurately plane faces, the bands will not be straight. This defect does not add any difficulty to the measurements, for W, the *point* of intersection of the cross-wires, is set on the successive bright bands, which is precisely what would be done if the bands were straight.

A totally reflecting prism is placed so as to cover about half the slit and to reflect light from a sodium flame into the collimator. To obtain a bright banded spectrum, the slit may have to be so wide that the two sodium lines are not resolved.

Generally the sodium "line" will lie between two consecutive bright bands; one of these—say the more deviated—is taken as the "zero band," and the telescope verniers are read when W is set on this "zero band." The cross-wire is then set in succession on the 5th, 10th, ... bright bands counting from the "zero band" towards the blue end of the spectrum and the verniers are read. The telescope is then turned back to the "zero band," and W is set in succession on the 5th, 10th, ... bright bands towards the red. More bands can be observed on the blue side of the "zero band" than on the red side.

The telescope is then turned into the position T' (Fig. 232) to receive the other spectrum of the first order, and the observations are repeated.

From the reading of the telescope corresponding to rays which are not deviated by the grating the two values of the deviation for each of the selected bands are found, and the mean of the two is taken.

If the grating interval, d, be not known, it is found optically. The slit is narrowed and the readings for D_1, the more deviated sodium line $(\lambda = 5\cdot8959 \times 10^{-5}$ cm.), are taken, the mean deviation being ϕ. Then $d = 5\cdot8959 \times 10^{-5} \times \operatorname{cosec} \phi$ cm. In § 370, the *less* deviated line D_2 $(\lambda = 5\cdot8900 \times 10^{-5}$ cm.) was used.

Let θ_0 be the deviation for the "zero" bright band and λ_0 the corresponding wave length. Then $z = \frac{1}{2} n_0 \lambda_0$, where n_0 is an integral number. But $\lambda_0 = d \sin \theta_0$ and thus $z = \frac{1}{2} n_0 d \sin \theta_0$. Similarly $z = \frac{1}{2} n_5 \lambda_5 = \frac{1}{2} n_5 d \sin \theta_5$, $z = \frac{1}{2} n_{10} \lambda_{10} = \frac{1}{2} n_{10} d \sin \theta_{10}$, ..., where θ_5, θ_{10}, ... correspond to the 5th, 10th, ... bright bands towards the blue. The wave lengths diminish towards the blue, and thus $n_5 = n_0 + 5$, $n_{10} = n_0 + 10$, ... Hence, taking a step of—say—30 bands, we have

$$\operatorname{cosec} \theta_0 = \tfrac{1}{2} n_0 d/z, \qquad \operatorname{cosec} \theta_{30} = \tfrac{1}{2}(n_0 + 30) \, d/z,$$

and thus $\qquad \frac{1}{2} d/z = \frac{1}{30} (\operatorname{cosec} \theta_{30} - \operatorname{cosec} \theta_0).$ ....(2)

Similarly $\qquad \frac{1}{2} d/z = \frac{1}{30} (\operatorname{cosec} \theta_{35} - \operatorname{cosec} \theta_5),$

and so on. Going in the opposite direction, i.e. towards the red

$$\tfrac{1}{2} d/z = \tfrac{1}{30} (\operatorname{cosec} \theta_{25} - \operatorname{cosec} \theta_{-5}),$$

and so on. With careful work the difference of cosecants will be nearly constant. If the mean difference be Δ, we have $\frac{1}{2} d/z = \frac{1}{30} \Delta$, or $z = 15 d/\Delta$.

If we use the second order spectrum and if ψ_0, ψ_5, ... be the deviations for the bright bands numbered 0, 5, ..., we have, in place of (2),

$$\tfrac{1}{4} d/z = \tfrac{1}{30} (\operatorname{cosec} \psi_{30} - \operatorname{cosec} \psi_0). \quad ....(3)$$

368. The plates. Plates of optical glass with worked faces which are optically plane are, of course, desirable, but satisfactory results may be obtained with selected pieces of plate glass. A number of pieces about 5 cm. square and not less than ·75 cm. thick are obtained from a glazier or shop-fitter; they may be cut from *new* scrap, care being taken to avoid pieces cut from the *edges* of large sheets. The sharp edges and corners may be removed by emery cloth. Each plate is cleaned and both its faces are tested against a face of a good prism by Newton's "rings," using sodium light; the illumination may be effected as in Fig. 198. If the bands be straight, parallel, and evenly and widely spaced the surface is reasonably good. For a careful test, dust must be excluded, since a few specks of dust may render it impossible to place two optically perfect surfaces parallel to each other. If the prism be held in the hand, it will be distorted through inequalities of temperature. When the two best plates have been selected, the better face of each is semi-silvered.

The plates may be mounted in a strong metal frame, as in

Fig. 233. They are separated by a sheet of mica, such as is used by electrical instrument makers in the construction of con-

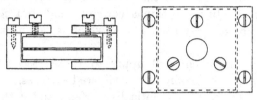

FIG. 233

densers. Such mica is very nearly uniform in thickness. A hole about 1·5 cm. in diameter is cut in the mica and the edges of the hole are carefully freed from any loose pieces of mica. Dust is removed from the silvered plates and from the mica by a soft brush. The plates rest on three projections on one side of the frame and are clamped by three screws opposite to those projections. Since the pressure is taken directly by the mica, there is little tendency to distort the plates. If the plates be not parallel, fringes will be seen when the observer holds the system *at arm's length* and looks through it at a sodium flame. The plates are unsuitable for the experiment if more than one fringe be seen under these conditions.

369. Further discussion. We now consider the general case in which the plates are not strictly plane and parallel and the wave fronts of the incident light are not strictly plane. We suppose that the light after passing through the plates falls on the slit SS' of a spectrometer, as in Fig. 232. Let J (Fig. 234)

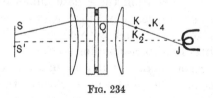

FIG. 234

be a point of the source of light, not necessarily in the plane of the diagram, and S a point of the slit. Light travels out from J in all directions, and one of the rays which has not suffered reflexion will pass through S. Hence the optical distance from J to S by all paths infinitesimally near this ray is the same. If an *image* of J be formed at S, the optical distance from J to S is the same for all rays whose initial paths from J cluster within a cone of small but *finite* angle described about a ray such as JK which is taken as the chief ray.

A similar argument will apply to those rays which have suffered 2, 4, ... reflexions at the silvered surfaces.

If the plates be nearly optically perfect, and if the distance between the silvered surfaces be very small, rays clustered about JK_2, JK_4, ..., where JK_2, JK_4, ... are very close to JK, will form images of J at S. For all these rays, which are nearly normal to the plates, the optical distance from J to S will exceed that for the rays clustered about JK by approximately $2z$, $4z$, ..., where, now, z is the "distance" between the plates in the neighbourhood of Q; we may suppose z measured from one plate to the other along the normal to the first plate.

There will be a resultant disturbance at S due to the rays clustered round JK, JK_2, JK_4, ..., and this resultant will be a maximum for wave lengths such that $2z = n\lambda$, and a minimum for wave lengths such that $2z = (n + \frac{1}{2})\lambda$. If the wave length be continuously varied, the point corresponding to S in the focal plane of the telescope T will describe a horizontal line, which will have points of maximum and minimum intensity.

If, instead of S, a different point S' of the slit be selected, the rays from the corresponding point of the source will traverse a different part of the plates—a part where the value of z is different. The wave lengths corresponding to the maxima and minima will now differ from those determined by S, and, consequently, the angular positions of the maxima and minima on the horizontal line corresponding to S' will not be the same as on the horizontal line corresponding to S.

Thus, the lines of maximum and minimum intensity will not be straight vertical lines, but will be curved, the forms of the curves depending upon the optical imperfections of the plates.

370. Practical example. The plates used by Mr C. Underwood were cut from plate glass and were separated by mica. The readings when telescope was in position T (Fig. 232) are given. The difference of cosecants for 25 bright bands is denoted by Δ.

Band	Deviation	Cosec θ	Band	Deviation	Cosec θ	Δ
− 10	21° 24′	2·74065	15	17° 33′·8	3·31389	·57324
− 5	20 31	2·85323	20	16 58 ·2	3·42617	·57294
0	19 41	2·96893	25	16 25 ·0	3·53831	·56938
5	18 55	3·08459	30	15 53 ·5	3·65204	·56745
10	18 13	3·19886	35	15 23 ·8	3·76648	·56762

Mean Δ = ·57013. Similar readings when telescope was in position T' gave Δ = ·56996. General mean Δ = ·57004.

Mean deviation for D_2 line (the *less* deviated) was $\phi = 19° 38'$; for this line $\lambda = 5\cdot8900 \times 10^{-5}$ cm. Hence

$$d = 5\cdot8900 \times 10^{-5}/\sin 19° 38' = 1\cdot7530 \times 10^{-4} \text{ cm.}$$

Then $\qquad z = \frac{1}{2}.25d/\Delta = 3\cdot844 \times 10^{-3} \text{ cm.} = 65\cdot26\lambda,$

where λ is the wave length of the D_2 line.

EXPERIMENT 70. **Measurement of thickness of air film by circular fringes.**

371. Introduction. A film of air of uniform thickness z is bounded by two parallel plates of glass; the bounding surfaces AA, BB (Fig. 235) are semi-silvered. The other surfaces of the plates are CC, DD. The planes AA, BB, ... are perpendicular to the plane of the figure.

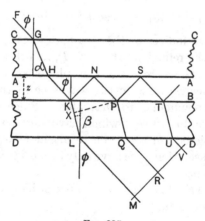

FIG. 235

A ray FG, in the plane of the figure, falls on CC at angle ϕ, and is refracted along GH. At H it meets AA and gives rise to a refracted ray HK, which is parallel to FG and makes an angle ϕ with the normal, since CC is parallel to AA. The ray HK gives rise to a refracted ray and to a reflected ray. The refracted ray KL, which in the plate BD makes an angle β with the normal, gives rise to the emergent ray LM, which makes an angle ϕ with the normal. The reflected ray KN gives rise to a reflected ray NP which, in its turn, gives rise to the refracted ray PQ and the reflected ray PS. The ray PQ gives rise to the emergent ray QR parallel to LM. Each reflexion of the bent ray $HKNPST$ ... at BB gives rise to an emergent ray parallel to LM, but the amplitudes in the rays $LM, QR, UV, ...$ decrease in geometrical progression. Let a plane to which these rays are normal cut them in $M, R, V, ...$

Let $[KNP]$ denote the optical length of the path KNP. Then we have

$$[KNP] = 2KN = 2z \sec \phi.$$

Draw PX perpendicular to KL; then

$$KX = KP \sin \beta = (KP \sin \phi)/\mu,$$

where μ is the index of the plate BD. Since $KP = 2KN \sin \phi$, the optical length of the path KX is given by

$$[KX] = \mu KX = KP \sin \phi = 2KN \sin^2 \phi = 2z \sec \phi \sin^2 \phi.$$

Hence $[KNP] - [KX] = 2z \sec \phi (1 - \sin^2 \phi) = 2z \cos \phi.$

The refractive indices of the two plates may, of course, not be equal, but the reflexions at $K, N, P, ...$ will be governed by the silvering, which reverses the phase at each reflexion, whether the electric force in the waves be in the plane of incidence or perpendicular to that plane. Since the rays LM, $QR, UV, ...$ are due to rays which have suffered 0, 2, 4, ... reflexions at silvered surfaces, the resultant change of phase is zero in every case. Hence we have only to consider the differences of optical distance.

If $[KNP] - [KX]$ be an *even* number of half wave lengths of light *in the film*, so that

$$2z \cos \phi = 2n \cdot \lambda/2, \qquad ....(1)$$

the phases at R, V in the rays QR, UV will be the same as at M in the ray LM, and similarly for all the emergent rays arising from the single incident ray FG. If these (parallel) rays enter a telescope, they will be brought to a single point in the focal plane of the objective, and thus will reinforce each other. If the amplitude at the focus due to the set of rays LM, UV, ... be h and the amplitude due to the alternate set QR, ... be k, then both h and k are positive, and $k < h$. The resultant amplitude due to *both* sets is $h + k$, and the intensity is $(h + k)^2$. Hence, when $2z \cos \phi = 2n \cdot \lambda/2$, the resultant intensity will be strong.

If $[KNP] - [KX]$ be an *odd* number of half wave lengths, so that

$$2z \cos \phi = (2n + 1) \cdot \lambda/2, \qquad \ldots\ldots\ldots\ldots(2)$$

the phase at R in the ray QR will be opposite to that at M in the ray LM and the rays QR, LM will "interfere" when they meet in the focal plane. The phase at V in UV will be opposite to that at R in QR and therefore the same as at M in LM and so on. The amplitudes due to the two sets are now h and $-k$, the amplitude due to both sets is $h - k$ and the intensity is $(h - k)^2$. Hence there will be relative darkness at the meeting point of the rays in the focal plane if $2z \cos \phi = (2n + 1) \cdot \lambda/2$.

If the thickness of the film be exactly an integral number of half wave lengths, so that $2z = p\lambda = 2p \cdot \lambda/2$, where p is an integer, equation (1) will be satisfied if $\phi = 0$ or $\cos \phi = 1$, and hence there will be a maximum of intensity when $\phi = 0$.

When, as will generally happen,

$$2z = (p + q) \lambda, \qquad \ldots\ldots\ldots\ldots\ldots(3)$$

where p is an integer and q is a positive fraction, (1) will not be satisfied by $\phi = 0$. To obtain the first maximum of intensity, we must increase ϕ from zero to ϕ_1 where $(p + q) \lambda \cos \phi_1 = p\lambda$, or

$$(p + q) \cos \phi_1 = p. \qquad \ldots\ldots\ldots\ldots\ldots(4)$$

The second brightness occurs when $(p + q) \cos \phi_2 = p - 1$, and the nth when

$$(p + q) \cos \phi_n = p - (n - 1). \qquad \ldots\ldots\ldots\ldots(5)$$

By (4) and (5),

$$p + q = (n - 1)/(\cos \phi_1 - \cos \phi_n). \qquad \ldots\ldots\ldots(6)$$

More generally,

$$p + q = (n - m)/(\cos \phi_m - \cos \phi_n). \qquad \ldots\ldots\ldots\ldots(7)$$

372. Apparatus. Plates such as those employed in EXPERI-
MENT 69 may be used. The frame holding the plates P (Fig. 236)
is mounted on a carriage A which can turn about a vertical
axis passing through O and has an index moving over a divided

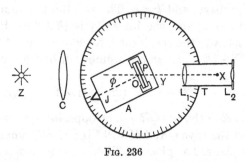

FIG. 236

circle. The plates are mounted centrally with respect to the
circle, and their normal OJ is horizontal. A telescope T, of
low magnifying power, is fixed relative to the circle. A needle
point (or cross-wires) X is adjusted in the telescope so as to be
in the focal plane of the objective L_1; it is viewed through the
eye-piece L_2. The telescope is fixed so that its line of collimation
XY intersects (as nearly as may be) the vertical axis through O.

The table of a spectrometer may serve as the carriage, but
the spectrometer telescope is unsuitable on account of its high
magnifying power.

The circular base shown in Figs. 16, 19 allows a convenient
arrangement. The carriage may be a board provided with three
spherical feet; two feet slide in the circular groove and the third
slides on a flat part of the base near the centre. The carriage
overhangs the central platform. The telescope is mounted on
the circular base. The focal lengths of L_1 and L_2 may be 6 cm.
and 3 cm. respectively. The magnifying power is thus 6/3 or 2.

The apparatus of § 36, modified as described in § 88 (foot-
note), may be used; the two lenses and the needle X are mounted
on the horizontal rod shown in Fig. 21.

Light from the sodium flame Z is concentrated upon the
opening in the plate holder by the condensing lens C.

373. Circular fringes. All the rays which on emerging from
the lens C are parallel to the line of collimation XY will be

parallel to XY after passing through the plates, whether they have or have not suffered reflexion in their passage through the plates, and thus will come to a focus at X. These rays have not come from a *single* source and are therefore not related in phase among themselves. But if a ray parallel to XY fall on the plates when ϕ, the angle between XY and the normal OJ, satisfies (1), each of the rays, such as QR, UV (Fig. 235), due to reflexions at the silvered surfaces, will re-inforce the ray, such as LM, which is transmitted without reflexion. If, on the other hand, ϕ satisfy (2), the phase of the *resultant* disturbance at X due to the reflected rays will be opposite to the phase of the disturbance due to the ray transmitted without reflexion. These statements are true for *every* ray parallel to XY, and thus the resultant illumination at X will be a maximum or a minimum according as ϕ satisfies (1) or (2), in spite of the fact that the incident rays are unrelated in phase.

We now consider all the rays which make an angle ϕ with the normal OJ but are not restricted to be parallel to the plane of Fig. 236. Only those rays which are parallel to XY come to a focus at X. The other rays, or such of them as enter the telescope, come to foci in the focal plane of L_1 at points lying on a curve passing through X. When ϕ is small, the curve will be a circle of radius $f_1\phi$, where f_1 is the focal length of L_1. There will be brightness or darkness at all points of this curve according as there is brightness or darkness at X. When ϕ is large, only portions of rings will be seen.

If several rings are to be visible simultaneously, the condensing lens C must be of large diameter (7·5 cm.) and of comparatively small focal length, and the objective L_1 must subtend a considerable angle at X.

If the surfaces AA, BB (Fig. 235) be not plane and parallel, the smaller dark rings may be rather ill-defined, but their identification will be easy. If a stop (diameter of hole = ·5 cm.) be fitted into the opening in the plate holder, the definition will be improved. The stop plate should be in contact with the outer surface of one of the glass plates so that approximately the same part of the silvered surfaces may be utilised whatever the value of ϕ.

It will not, of course, be possible to obtain a ring system

showing *abrupt* changes from bright to dark, for the intensity must vary gradually as we pass from ring to ring. The concentration of light in the brighter parts of the system will, indeed, be greater for thick than for thin silvering, but the effect will be accompanied by a diminution of intensity.

374. Practical details. The measurements are simple. The carriage A is turned so that the points R_1, R_2, R_3, ... (Fig. 237) on the bright rings are brought in succession to the pointer X, and the readings of the carriage on its circle are recorded. The carriage is then turned to bring the points R_1', R_2', ... to X, and the readings are taken. The difference between the readings for R_1 and R_1' gives $2\phi_1$, and the difference between the readings for R_n and R_n' gives $2\phi_n$. The points R_1 and R_1' need not lie on the *smallest* bright ring, but they must lie on the *same* ring.

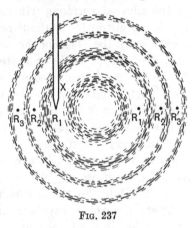

Fig. 237

By (6), § 371, $(\cos \phi_1 - \cos \phi_n)/(n - 1) = 1/(p + q)$, where $\frac{1}{2}(p + q) = z/\lambda$ and z is the thickness of the film. The mean value of $1/(p + q)$ is found for a number of rings. If there be fair agreement between the values, the reciprocal of their mean may be taken as the mean value of $p + q$.

If the smallest bright ring employed be not the first but the mth, the value of $p + q$ is given by (7). In this case, we find the mean value of $(\cos \phi_m - \cos \phi_n)/(n - m)$ for a number of values of n.

The measurements may be repeated with lithium instead of sodium in the flame, the rings now being red. If both sodium and lithium be used, two sets of rings will be seen. From the coincidences between the red and the yellow rings, the ratio of the wave length for lithium to that for sodium can be estimated.

If the distance between the plates be approximately ·0145 cm., and sodium light be used, *two* sets of yellow rings will be visible, one set being due to the D_1 line ($\lambda_1 = 5{\cdot}896 \times 10^{-5}$ cm.) and the other to the D_2 line ($\lambda_2 = 5{\cdot}890 \times 10^{-5}$ cm.). The best effect, in the central part of the ring system, will be obtained when $z = \frac{1}{2}n\lambda_1$, where n is given by $n\lambda_1 = (n + \frac{1}{2})\lambda_2$. Thus

$$z = \tfrac{1}{4}\lambda_1\lambda_2/(\lambda_1 - \lambda_2).$$

The values quoted for λ_1 and λ_2 give $z = {\cdot}0145$ cm., as stated above.

It will be apparent that *two* sets of bright rings are in view, for one set is much brighter than the other. The successive bright rings are due to D_1 and to D_2 alternately. The rings due to the D_2 line are more intense than those due to the D_1 line.

375. Practical example. Mr G. A. Griffiths and Mr C. G. Lyons obtained the following readings, using the plates employed in § 370. Sodium light of mean wave length $5{\cdot}893 \times 10^{-5}$ cm. was used. The difference between the readings for R and R' gives 2ϕ.

n	R	R'	ϕ		$\cos\phi$	$\cos\phi_1 - \cos\phi_n$	$\dfrac{\cos\phi_1 - \cos\phi_n}{n-1}$
1	158°·5	148°·5	5°	0′	·99619		
6	170 ·0	136 ·7	16	39	·95807	·03812	·007624
11	176 ·5	130 ·2	23	9	·91948	·07672	·007672
16	181 ·6	125 ·1	28	15	·88089	·11530	·007687
21	186 ·0	120 ·9	32	33	·84292	·15327	·007663
26	189 ·9	116 ·9	36	30	·80386	·19233	·007693

Mean value of $\dfrac{\cos\phi_1 - \cos\phi_n}{n-1} = {\cdot}007668 = \dfrac{1}{130{\cdot}41}$.

Hence $p + q = 130{\cdot}41$. Then, by (3),

$$z = \tfrac{1}{2}(p+q)\lambda = 65{\cdot}205 \times 5{\cdot}893 \times 10^{-5} = 3{\cdot}843 \times 10^{-3} \text{ cm.}$$

Printed in the United States
By Bookmasters